Sociology in Our Times

SEVENTH EDITION

Diana Kendall

Baylor University

Prepared by

Kathryn Sinast Mueller

Baylor University

THOMSON

WADSWORTH

Australia • Brazil • Canada • Mexico • Singapore • Spain • United Kingdom • United States

ISBN-13: 978-0-495-50665-2
ISBN-10: 0-495-50665-6

Thomson Higher Education
10 Davis Drive
Belmont, CA 94002-3098
USA

For more information about our products,
contact us at:
Thomson Learning Academic Resource Center
1-800-423-0563

For permission to use material from this text or
product, submit a request online at
http://www.thomsonrights.com.
Any additional questions about permissions can be
submitted by email to **thomsonrights@thomson.com.**

From Students to Students about the Study Guide

Dear students,

This study guide is a comprehensive and detailed supplemental source to Kendall's *Sociology in Our Times, Seventh Edition.* Professor Mueller has done a superb job in outlining both the broad and detailed sociological concepts that are found in this textbook. Each individual chapter in this particular study guide carefully examines the overall content of the main text. There is, in fact, a direct correlation between the study guide chapters and the textbook chapters. I recommend this guide as a supplement for any student's Introduction to Sociology course. This guide is a wonderful tool when one is reviewing class lectures, presentations, semester projects, Internet exercises, "understanding the boxes," or any other information that might be pertinent to this class. It is especially helpful during "crunch time," the time right before a big examination. This source has helped me and I know that it will do the same for you!

> Abby Suelflow
> Baylor University, Class of 2008

Dear students,

I recommend reading and studying from this supplemental study guide. This source is complementary to the text. The Internet activities, which are found at the end of each chapter, are both fun and instructive. Try the projects, if you have that option in your class, or if you need to construct a project in another class. Professor Mueller grades the student-submitted projects, and they were awesome! This is a wonderful way to learn the required material.

> Sabrina Stone
> Sociology Major
> Baylor University, Class of 2007

Dear students,

I have known Professor Mueller for years now, and she is my mentor, counselor, sounding board, and friend. Being her teaching assistant has been one of the most rewarding jobs of my life. She has taught me more about sociology than any book has, and Professor Mueller has channeled her plethora of knowledge into this study guide. This study guide will steer you into the path of the sociological imagination. Pay attention because the information is applicable to almost every aspect of your life.

> Rachel E. Lovell,
> Sociology, M.A.
> Baylor University, Class of 2001
> Sociology, Ohio State, PhD. 2007

DEDICATED TO

Kenneth Ray Mueller, "the wind beneath my wings"

Wonderful children: Kristy Mueller Suelflow & Jerry Suelflow,
Kirby Ray Mueller, Keary Kal & Amy Hvasta Mueller

Wonderful grandchildren: Jeremy Allen, Joshua Ray,
Kristyn MaryKathryn Suelflow
Caitlin Virginia, Delaney Rae, and Timothy Patrick Mueller

With Heartfelt Thanks To

Sharon Salon
Abby Suelflow

YOU ARE ALL THE BEST!
AND YOU ARE LOVED!

Table of Contents

PREFACE

Welcome to the study of *Sociology in Our Times*, **7th Edition.** The text, written by my colleague Diana Kendall, is innovative, issue-oriented, interesting, globally conscious, and informative. It's good! You won't want to miss it! This Study Guide is written to assist you mastering the material in your text and enhancing your grade on your instructor's tests. Do not use it as a replacement for the text; it does not contain all of the passages from the text, nor does it depict all of the major topics illustrating the sociological concepts – your text does that! Each chapter in the Study Guide corresponds with the same chapter in *Sociology in Our Times* and has the following components:

Brief Chapter Outline

Chapter Summary

Learning Objectives

Key Terms

Key People

Chapter Outline

Practice Tests

Short Answer and Essay Questions

Student Class Projects and Activities

Internet Activities

InfoTrac Exercises

How To Use This Study Guide

1. <u>Before reading</u> each chapter, look over the Brief Chapter Outline and Chapter Summary to get an overview of the topics in each chapter.

2. <u>As you read</u>, use the Learning Objectives, Key Terms, and Key People as a guide for taking notes. (You may wish to start a sociology notebook with concise responses for study purposes.)

3. <u>Look for recurring</u> themes in each chapter and between chapters. For example, how do functionalist, conflict, and symbolic interactionist perspectives view each of the topics being examined? Outline the key components of each theory in your notebook.

4.   <u>Look at the topics</u> and issues presented through the lenses of race, class, gender, age, and (where applicable) ability/disability.

5.   <u>After you have read</u> the chapter, take the practice test under actual testing situations (i.e., "closed book," sitting at a table, and timing yourself). Grade your practice test using the answers listed in the questions. Use the page numbers indicated on the practice test and in the answers to revisit the text for your information on each item you missed. (You may wish to save the practice tests to use as a self-test before you take an in-class examination.)

6.   <u>Study</u> the short answer and essay questions as a way to develop critical thinking skills and to synthesize information in a logical manner. These questions could also be used to assist your participation in class discussion.

7.   The student class projects, activities, and Internet and InfoTrac exercises are designed to involve you actively in the learning process and to help you recognize the relevance of sociology to your life and to our global society. Some of these projects may be assigned as a part of the course requirements.

Best wishes as you explore the text Sociology *in Our Times*, 7th Edition, and as you use this study guide. May you experience and enjoy the sociological imagination! I invite your questions, comments or suggestions. You may contact me at this address:

Kathryn Sinast Mueller
Baylor University
P.O. Box 97326
Waco, Texas 76798-7326
E-mail address: Kathryn _Mueller@Baylor.edu

1

THE SOCIOLOGICAL PERSPECTIVE

BRIEF CHAPTER OUTLINE

PUTTING SOCIAL LIFE INTO PERSPECTIVE
Why Study Sociology?
The Sociological Imagination
THE IMPORTANCE OF A GLOBAL SOCIOLOGICAL IMAGINATION
THE ORIGINS OF SOCIOLOGICAL THINKING
Sociology and the Age of Enlightenment
Sociology and the Age of Revolution, Industrialization, and Urbanization
THE DEVELOPMENT OF MODERN SOCIOLOGY
Early Thinkers: A Concern with Social Order and Stability
Differing Views on the Status Quo: Stability versus Change
The Beginnings of Sociology in the United States
CONTEMPORARY THEORETICAL PERSPECTIVES
Functionalist Perspectives
Conflict Perspectives
Symbolic Interactionist Perspectives
Postmodern Perspectives
COMPARING SOCIOLOGY WITH OTHER SOCIAL SCIENCES
Anthropology
Psychology
Economics
Political Science
In Sum

CHAPTER SUMMARY

Sociology is the systematic study of human society and social interaction. Sociology enables us to see how individual behavior is largely shaped by the groups to which we belong and the **society** in which we live. The **sociological imagination** helps us to understand how seemingly personal troubles actually are related to larger social forces including those that are related to *global interdependence*, a relationship in which the lives of all people are intertwined closely and any one nation's problems are part of a larger global problem. The first systematic analysis of society is found in the philosophies of early Greek thinkers, such as Plato and Aristotle. With the onset of scientific revolution, social thinkers sought to develop a scientific understanding of social life. The origins of contemporary social thinking can be traced to the scientific revolution of the late seventeenth and mid-eighteenth centuries and to the Age of Enlightenment, where emphasis was placed on the individuals possession of critical reasoning and experience, coupled with a widespread skepticism of the primacy of religion as a source of knowledge, and a strong opposition to traditional authority. Sociology emerged out of the social upheaval produced by **urbanization** and **industrialization** during the nineteenth and early twentieth centuries. Some early social thinkers -- including **Auguste Comte**, **Harriet Martineau**, **Herbert Spencer**, and **Emile Durkheim** -- emphasized social order and stability; others -- including **Karl Marx**, **Max Weber**, and **Georg Simmel** -- focused on conflict and social change. From its origins in Europe, sociology spread to the United States in the 1890s when departments of sociology were established at the University of Chicago and Atlanta University. Sociologists use four major contemporary sociological perspectives to examine social life: (1) **functionalist perspectives** assume that society is a stable, orderly system; (2) **conflict perspectives** assume that society is a continuous power struggle among competing groups, often based on **class**, **race**, **ethnicity**, or **gender**; (3) **symbolic interactionist perspectives** focus on how people make sense of their everyday social interactions; and (4) **postmodern perspectives** examine the efforts of postindustrialization consumerism and global communications in explaining social life in contemporary societies. The areas of interest and research in the social sciences, such as Sociology, Anthropology, Psychology, Economics, and Political Science, overlap in that the goal of scholars, teachers, and students is to learn more about human behavior and its causes and consequences.

LEARNING OBJECTIVES

After reading and understanding Chapter 1, you should be able to:

1. Define sociology.
2. Explain how sociology helps us to better understand our social world and ourselves.
3. Define and give examples of high, middle and low-income countries.
4. Explain how sociological theory helps us to understand social issues like consumerism.
5. Distinguish between commonsense knowledge, myths and sociological knowledge.
6. Explain what C. Wright Mills meant by the sociological imagination.
7. Define race, ethnicity, class, sex, and gender, and explain the importance of these terms to developing a sociological imagination.

8. Identify Auguste Comte, Harriet Martineau, and Herbert Spencer, and explain their unique contributions to the emergence of sociology.

9. Explain what Durkheim meant by his use of the terms *social facts* and *anomie*.

10. Relate Max Weber's concepts of rationalization and Verstehen to C. Wright Mills' Sociological Imagination.

11. Relate the global Wal-Mart Effect to other aspects of consumerism.

12. Identify and discuss the key assumptions of the Age of the Enlightenment.

13. Describe the origins of sociology in the United States and discuss the role of women in early departments of sociology and social work.

14. Distinguish between microlevel and macrolevel analyses and state which level of analysis is utilized by each of the major theoretical perspectives.

15. Define industrialization and urbanization, and explain the role of each in furthering sociological thought.

16. State the major assumptions of functionalism, conflict theory, symbolic interactionism, postmodernism, and identify the major contributors to each perspective.

17. Contrast Karl Marx's perspective on social change with that of Max Weber.

18. Compare sociology with other sciences and determine areas of overlap and important differences.

KEY TERMS

(defined at page number shown and in glossary)

KEY PEOPLE

(identified at page number shown)

CHAPTER OUTLINE

I. PUTTING SOCIAL LIFE INTO PERSPECTIVE
    A. Sociology is the systematic study of human society and social interaction.
    B. Why Study Sociology?
        1. **Sociology** helps us see the complex connections between our own lives and the larger, recurring patterns of the society and world in which we live.
            a. A **society** is a large social grouping that shares the same geographical territory and is subject to the same political authority and dominant cultural expectations.
            b. When we examine the world order, we become aware of *global interdependence* -- a relationship in which the lives of all people are intertwined closely and any one nation's problems are part of a larger global problem.
            c. Sociological research often reveals the limitations of myths associated with commonsense knowledge that guides ordinary conduct in everyday life.
    C. The Sociological Imagination
        1. According to sociologist **C. Wright Mills**, the **sociological imagination** enables us to distinguish between personal troubles and public issues.
        2. Personal troubles affect individuals and their networks and require personal solutions, while public issues affect large numbers of people which may require societal solutions.
        3. Overspending, for example, can be seen as both a personal and/or a public issue.

II. IMPORTANCE OF GLOBAL SOCIOLOGICAL IMAGINATION
   A. A *global sociological imagination* enables us to distinguish between the world's **high-income, middle-income** and **low-income countries**.
   B. Developing a personal sociological imagination requires that we take into account perspectives of people.
      1. People in today's world differ by *race* -- a term used by many people to specify groups of people distinguished by physical characteristics such as skin color -- and *ethnicity* -- cultural heritage or identity of a group, based on factors such as language or country of origin.
      2. They also differ by *class* -- the relative location of a person or group within a larger society, based on wealth, power, prestige, or other valued resources -- and by *gender* -- the meanings, beliefs, and practices associated with sex differences referred to as femininity and masculinity.
III. THE ORIGINS OF SOCIOLOGICAL THINKING
   A. Sociology and the Age of Enlightenment
      1. In this European time period, critical reasoning and experience is emphasized; religion is viewed with skepticism.
      2. The individual's pursuit of enlightened self-interest is conducive to the welfare of society as a whole; new forms of political and economic organizations, such as democracy and capitalism, emerge.
   B. Sociology and the Age of Revolution, Industrialization, and Urbanization
      1. The Enlightenment produced an *intellectual revolution*; political and economic revolutions in America and then in France.
      2. **Industrialization** -- the process by which societies are transformed from dependence on agriculture and handmade products to an emphasis on manufacturing and related industries -- and **urbanization** -- the process by which an increasing proportion of a population lives in cities rather than rural areas -- contributed to the development of sociological thinking.
IV. THE DEVELOPMENT OF MODERN SOCIOLOGY
   A. Early Thinkers: A concern with social order and stability
      1. **Auguste Comte** coined the term sociology and stressed the importance of **positivism** -- a belief that the world can best be understood through scientific inquiry. He developed the law of the three stages: the theological, metaphysical, and scientific (or positive). Some consider Comte to be the founder of sociology. Both Comte and his mentor, Saint-Simon, advocated the use of the methods of the natural sciences as models for the objective study of society.
      2. **Harriet Martineau** translated and condensed Comte's works. Her most influential work was *Society in America* in which she emphasized diversity in the United States based on race, class, and gender. She believed in racial and gender equality, enlightened reform, and creating a science of human nature and eventually a just society.

5

3. **Herbert Spencer** used an evolutionary perspective to explain stability and change in societies. He coined the term "survival of the fittest," equating this process of *natural selection* with progress and success. His view of society is known as **social Darwinism**, the idea that those species of animals, including humans, best adapted to their environment survive and prosper, whereas the poorly adapted die out.
4. According to **Emile Durkheim**, societies are built upon **social facts** which are patterned ways of acting, thinking, and feeling that exist outside any one individual and exert social control over each person. **Anomie** is a condition in which social control becomes ineffective as a result of the loss of shared values and of a sense of purpose in society. Durkheim's ideas could be applied to the analysis of credit cards as a type of social glue that creates a sense of belonging to a group.

B. Differing Views on the Status Quo: Stability versus Change
1. **Karl Marx** believed that conflict -- especially *class conflict* -- is inevitable.
   a. Class conflict is the struggle between members of the capitalist class, or *bourgeoisie* and the working class, or *proletariat*; conflict is necessary for producing social change and a better society.
   b. Exploitation of workers by capitalists results in workers' *alienation* -- a feeling of powerlessness and estrangement from other people and from oneself. When workers overthrow capitalists, a free and classless society results.
2. **Max Weber** disagreed with Marx's idea that economics is the central force in social change, emphasizing that other factors, such as religion, influence economic systems. His concern with the growth of large-scale organizations is reflected in his work on bureaucracy. According to Weber, *rational bureaucracy*, rather than class struggle, is the most significant factor in determining the social relations between people in industrial societies. By emphasizing the goal of value-free inquiry and employing *verstehen*, sociologists could then understand how others see the world.
3. **Georg Simmel** emphasized that society is best seen as a web of patterned interactions among people. By analyzing how interaction patterns vary depending on the size of the social groups and the *forms* and the content of the social interaction, Simmel provided major insights into social life. He assessed the costs of industrialization and urbanization on individuals. In one of his most important books, the *Philosophy of Money*, he examines the issue of consumerism.

C. The Beginnings of Sociology in the United States
1. The first U.S. department of sociology was at the University of Chicago. **Robert E. Park** studied the disintegrating influence of urbanization on social life, such as the influence of increases in the crime rate and racial and class antagonisms.
2. **George H. Mead** founded the symbolic perspective.

3. **Jane Addams** wrote *Hull House Maps and Papers* which was used by other Chicago sociologists for the next forty years. She received a Nobel Prize for her assistance to the underprivileged.
4. **W.E.B. Du Bois** founded the second U.S. department of sociology at Atlanta University and wrote *The Philadelphia Negro: a Social Study*, examining Philadelphia's African American community. He noted that a dual heritage creates conflict for people of color, which he referred to as *double consciousness.*

V. CONTEMPORARY THEORETICAL PERSPECTIVES
   A. A **theory** is a set of logically interrelated statements that attempts to describe, explain, and (occasionally) predict social events. Theories provide a framework or perspective -- an overall approach or viewpoint toward some subject -- for examining various aspects of social life.
   B. **Functionalist perspectives** are based on the assumption that society is a stable, orderly system characterized by societal consensus.
      1. Societies develop social structures, or institutions that persist because they play a part in helping society survive. These institutions include: the family, education, government, religion, and economy.
      2. **Talcott Parsons** stressed that all societies must make provisions for meeting social needs in order to survive. For example, a division of labor (distinct, specialized functions) between husband and wife is essential for family stability and social order.
      3. **Robert K. Merton** distinguished between intended and unintended functions of social institutions.
         a. **Manifest functions** are intended and/or overtly recognized by the participants in a social unit.
         b. **Latent functions** are unintended functions that are hidden and remain unacknowledged by participants.
         c. *Dysfunctions* are the undesirable consequences of any element of society. For example, a dysfunction of education in the United States is the perpetuation of gender, racial, and class inequalities, while a manifest function of shopping and consuming is that of creating a booming economy benefiting many.
      4. Applying **Conflict Perspectives** to Shopping and Consumption
         a. Conflict theorists focus on how inequalities based on race, sex, and income affect people's ability to acquire things they want and need.
         b. Conflict perspectives stress Veblen's concept of conspicuous consumption, or the continuous display of wealth and status through purchases.
         c. Juliet Schor discusses the consequences of conspicuous consumption for those in low-income categories.
   C. According to **conflict perspectives**, groups in society are engaged in a continuous power struggle for control of scarce resources.
      1. Along with **Karl Marx**, **Max Weber** believed that economic conditions were important in producing inequality and conflict in society; however, Weber also suggested that power and prestige are other sources of inequality.

2. **C. Wright Mills** believed that the most important decisions in the United States are made largely behind the scenes by the *power elite*, a small clique composed of the top corporate, political, and military officials. He contended that value-free sociology was impossible because social scientists must make value-related choices.
3. **Feminist perspectives** focus on patriarchy -- a system in which men dominate women. That which is considered masculine is more highly valued than that which is considered feminine.

D. Symbolic Interactionist Perspective
1. Functionalist and conflict perspectives focus primarily on **macrolevel analysis** -- an examination of whole societies, large scale social structures, and social systems. By contrast, symbolic interactionist approaches are based on a **microlevel analysis** -- an examination of everyday interactions in small groups rather than large scale social structures.
2. **Symbolic Interactionist** perspectives are based on the assumption that society is the sum of the interactions of individuals and groups. Symbols help people derive meanings from social situations.
   a. **George Herbert Mead**, a founder of this perspective, emphasized that a key feature distinguishing humans from other animals is the ability to communicate in symbols -- anything that meaningfully represents something else.
   b. Some interactionists focus on people's behavior while others focus on each person's interpretation or definition of a given situation.
   c. Others, such as **Erving Goffman**, utilize *dramaturgical analysis*, the idea that individuals go through their lives like actors performing on stage and engage in "impression management."

E. Postmodern Perspectives
1. According to **postmodern perspective**, existing theories have been unsuccessful in explaining social life in contemporary societies that are characterized by post industrialization, consumerism, and global communications.
2. Postmodern societies are characterized by an *information explosion*, the *rise of a consumer society*, and the emergence of a *global village* in which people around the world communicate with each other by electronic technologies.
3. Postmodern perspectives would focus on the shift away from production to consumerism.
4. Social life is not an objective reality waiting for us to discover how it works, but rather is how we think about it.

F. All four of these perspectives will be used in this textbook as lenses through which to view our social world

VI. COMPARING SOCIOLOGY WITH OTHER SOCIAL SCIENCES
   A. Anthropology is the study of human existence over geographic space and evolutionary time; it is divided into four main subfields: sociocultural, linguistic, archaeological, and biological anthropology.
   B. Psychology is the systematic study of behavior and mental processes—what occurs in the mind.
   C. Economics is the study of how the limited resources of a society are allocated among competing demands and is divided into two branches: macroeconomics and microeconomics.
   D. Political Science is the study of power relations, and how power is distributed, and political institutions such as the state, government, and political parties.
   E. In sum, the areas of interest and research in the social sciences overlap, the goal is to learn more about human behavior.

ANALYZING AND UNDERSTANDING THE BOXES

After reading the chapter and studying the outline, re-read the boxes and write down key points and possible questions for class discussion.

Sociology in Everyday Life: How Much Do You Know About Consumption and Credit Cards?

Key Points:

Discussion Questions:

1.

2.

3.

Sociology in Global Perspective: The Global Wal-Mart Effect? Big Box Stores and Credit Cards

Key Points:

Discussion Questions:

1.

2.

3.

Sociology and Social Policy: Online Shopping and Your Privacy

Key Points:

Discussion Questions:

1.

2.

3.

You Can Make a Difference: Dealing with Money Matters in a Material World

Key Points:

Discussion Questions:

1.

2.

3.

PRACTICE TESTS

MULTIPLE CHOICE QUESTIONS

Select the response that best answers the question or completes the statement.

1. Sociology is the systematic study of
    a. intuition and commonsense knowledge.
    b. human society and social interaction.
    c. the production, distribution, and consumption of goods and services in a society.
    d. personality and human development.

2. A _____ is a large social grouping that shares the same geographical territory and is subject to the same political authority and dominant cultural expectations.
    a. terrority
    b. culture
    c. city-state
    d. society

3. All of the following are reasons to study sociology, **except**
    a. sociology helps us gain a better understanding of ourselves and our social world.
    b. sociology utilizes scientific standards to study society.
    c. sociology confirms the accuracy of commonsense knowledge
    d. sociology helps us look beyond our personal experiences and gain insights into society.

4. According to Wright Mills, the sociological imagination refers to the ability to
    a. distinguish between personal troubles and public issues.
    b. see the relationship between preliterate and literate societies.
    c. be completely objective in examining social life.
    d. seek out one specific cause for a social problem such as suicide.

5. Widespread unemployment and massive, nationwide consumer debt are examples of
    a. public issues.
    b. personal troubles.
    c. public troubles.
    d. societal issues.

6. The ability to provide theory and research beyond one's own country enveloping countries all over the world is known as a _____ approach.
   a. global
   b. developed nation
   c. developing nation
   d. personal awareness

7. The world's _____ countries are nations with industrializing economies, particularly in urban areas, and moderate levels of national and personal income.
   a. high-income
   b. middle-income
   c. low-income
   d. subordinate-income

8. Even in high-income countries problems of personal debt threaten economic and social stability. In the U.S.A. alone, during a one year period (2006), _____ people filed for bankruptcy.
   a. 500,000
   b. 550,000
   c. 1 million
   d. 1.1 million

9. Many of the nations of Africa and Asia, particularly the People's Republic of China and India, where people typically work the land are examples of _____ countries.
   a. high-income
   b. middle-income
   c. subordinate-income
   d. low-income

10. Brazil and Mexico are examples of _____ countries.
    a. high-income
    b. high middle-income
    c. middle-income
    d. low-income

11. Femininity and masculinity are _____-related terms.
    a. sex
    b. gender
    c. biology
    d. anatomically

12. Ethnicity can be based upon the _____ of the group.
    a. cultural heritage
    b. cultural identity
    c. native language
    d. All of these choices can be used to identify ethnicity.

13. The first systematic analysis of society is found in the _____.
    a. work of Auguste Comte
    b. philosophies of the Age of Reason
    c. work of the eighteenth century social philosophers
    d. philosophies of early Greek philosophers

14. In eighteenth century France, _____ provided a place for women and men to discuss ideas and opinions.
    a. the Christian church
    b. salons
    c. the Parliament
    d. city hall

15. In France, the Enlightenment is also referred to as the _____.
    a. Industrial Revolution
    b. French Revolution
    c. Age of Discord
    d. Age of Reason

16. Two historical factors that contributed to the development of sociological thinking were
    a. industrialization and urbanization.
    b. industrialization and immigration.
    c. urbanization and centralization.
    d. urbanization and immigration.

17. In his major works, Comte drew heavily on the ideas of his mentor
    a. Karl Marx.
    b. Harriet Martineau.
    c. Saint-Simon.
    d. Frederick Engels.

18. French philosopher Auguste Comte's philosophy became known as _____ -- a belief that the world can best be understood through scientific inquiry.
    a. absolutism
    b. positivism
    c. functionalism
    d. activism

19. Auguste Comte described the *Law of the three stages*: he believed that knowledge began in the _____ -- explanations were based on religion and the supernatural.
    a. theological state
    b. metaphysical stage
    c. scientific stage
    d. positive stage

20. This British sociologist translated and condensed Comte's work and was noted for her study of social customs in Great Britain and the United States.
    a. Juliet B. Schor
    b. Jane Addams
    c. Harriet Martineau
    d. Mary Wollstonecraft

21. _____ is the belief that those species of animals, including human beings, best adapted to their environment survive and prosper, whereas those poorly adapted die out.
    a. Social Darwinism
    b. Social eugenics
    c. Social statics
    d. Social facts

22. _____ argued that conflict -- especially class conflict -- is necessary in order to produce social change and a better society.
    a. Auguste Comte
    b. Emile Durkheim
    c. Karl Marx
    d. Harriet Martineau

23. One of Simmel's most exciting studies, _____, analyzes consumerism.
    a. *The Protestant Ethic and the Spirit of Capitalism*
    b. *The Philosophy of Money*
    c. *Society in America*
    d. *The Rules of Sociological Method*

24. The first U.S. department of sociology was established at _____, where _____ was one of the best known women in the field.
    a. University of Chicago; Jane Addams
    b. Harvard University; Harriet Martineau
    c. University of California; Arlie Hochschild
    d. Princeton University; Sara McLanahan

25. Social theorist _____ observed that rapid social change can lead to a breakdown in traditional values and an increase in _____.
    a. Du Bois; subjectivity
    b. Weber; alienation
    c. Durkheim; anomie
    d. Comte; positivism

26. The word that Max Weber used to stress the need for sociologists to take into account other peoples feelings, viewpoints, or attitudes.
    a. anomie
    b. *verstehen*
    c. rationalization
    d. *gesundheit*

27. The credit card industry has contributed to the _____ process by the efficiency with which it makes loans and deals with consumers.
    a. ritualization
    b. formalization
    c. procenduralization
    d. rationalization

28. W. E. B. Du Bois referred to the identity conflict of being a black and an American as
    a. group consciousness.
    b. the American dilemma.
    c. false consciousness.
    d. double-consciousness.

29. A _____ is defined as a set of logically interrelated statements that attempts to describe, explain, and (occasionally) predict social events.
    a. fact
    b. theory
    c. hypothesis
    d. perspective

30. _____ perspectives are based on the assumption that society is a stable, orderly system.
    a. Functionalist
    b. Symbolic interactionist
    c. Conflict
    d. Feminist

31. _____ perspectives are based on the assumption that groups are engaged in a continuous power struggle for control of scarce resources.
    a. Functionalist
    b. Symbolic interactionist
    c. Conflict
    d. Postmodern

32. Some students meeting their friends at a mall to "hang out" would likely be viewed as a _____ of shopping and consumerism.
    a. latent function
    b. latent dysfunction
    c. manifest function
    d. manifest dysfunction

33. According to your text, all of the following are conflict theorists, **except**
    a. Max Weber.
    b. Talcott Parsons.
    c. Georg Simmel.
    d. C. Wright Mills

34. The _____ approach directs attention to women's experiences and the importance of gender as an element of social structure.
    a. functionalist
    b. symbolic interactionist
    c. conflict
    d. feminist

35. Social scientist Thorstein Veblen described early wealthy U.S. industrialists as engaging in _____ - the continuous public display of one's wealth and status through purchases such as expensive houses, clothing, motor vehicles, and other consumer goods.
    a. massive consumption
    b. conspicuous consumption
    c. random consumption
    d. inconspicuous consumption

36. Goffman's dramaturgical analysis represents the _____ perspective.
    a. conflict
    b. functionalist
    c. symbolic interactionist
    d. postmodern

37. Which sociological perspective would focus on the fact that our thoughts and behavior are shaped by our social interactions with others?
    a. functionalist perspective
    b. conflict perspective
    c. symbolic interactionist perspective
    d. postmodern perspective

38. Signs, gestures, written language, and shared values are all examples of
    a. symbols.
    b. psychological defense mechanisms.
    c. norms.
    d. roles.

39. The theory that best explains consumerism, global communications, and postindustrial social life is the _____ perspective.
    a. postmodern
    b. functionalist
    c. conflict
    d. symbolic interactionist

40. A _____ would more likely examine how the political process – such as the efforts of lobbyists and interest groups – affects credit card interest rates and consumer spending in the United States.
    a. sociologist
    b. economist
    c. political scientist
    d. anthropologist

TRUE/FALSE QUESTIONS

1. Global interdependence affects the lives of all people.
   T   F

2. All sociologists attempt to discover patterns or commonalties in human behavior.
   T   F

3. Industrialization first occurred in France between 1700 and 1850.
   T   F

4. The writings of Mary Wollstonecraft strongly influenced the ideas of human equality.
   T   F

5. Auguste Comte is considered by some to be the "founder of sociology."
   T   F

6. Herbert Spencer coined the term "survival of the fittest."
   T   F

7. Count Henri de Saint-Simon, as a mentor to Comte, was primarily interested in social reform.
   T   F

8. In the Marxian framework, class conflict is the struggle between the capitalist class and the working class.
   T   F

9. According to Marx, the "fetishism of commodities" describes the situation wherein the proletariat recognize that their labor gives the commodity its value.
   T   F

10. The first U.S. departments of sociology were located at the University of Chicago and at Harvard University.
    T   F

11. From a functionalist perspective, leadership, decision making, and employment outside the home to support the family all constitute expressive tasks.
    T   F

12. According to Robert K. Merton, latent functions are intended and/or overtly recognized by the participants in a social unit.
    T   F

13. The conflict perspective is one unified theory using the branches of neo-Marxism and feminism.
    T   F

14. Symbolic interaction perspectives are based on a macrolevel analysis of society.
    T   F

15. Sociology, Psychology, Economics and Political Science are all examples of social sciences.
    T   F

FILL-IN-THE-BLANK QUESTIONS

1. The first systematic analysis of society is found in the philosophies of early Greek philosophers such as _____ and _____.

2. _____ translated Comte's work; was a sociologist in her own right.

3. _____ suggested that societies are built upon social facts.

4. _____ is considered the founder of sociology.

5. _____ studied the rationalization of society.

6. _____ coined the term "survival of the fittest."

7. _____ focused on small social group patterns of interaction.

8. _____ coined the term "double-consciousness."

9. According to Marx _____ are exploited by the _____.

10. Weber's important book _____ evaluated the role of the Protestant Reformation in producing a social climate in which capitalism could exist and flourish.

11. The _____ analysis focuses on small groups.

12. The _____ perspective examines the emergence of a global village.

13. A _____ is an undesirable consequence of any element of society.

14. According to Box 1.2, *Sociology in Global Perspective*, Wal-Mart operates stores in more than 34 cities in _____.

15. According to Box 1.3, *Sociology and Social Policy*, some web sites insert _____ onto the hard drive of your computer in order to track where you go and what you do on the site.

SHORT ANSWER/ESSAY QUESTIONS

1. How might the sociological imagination help the individual personally and professionally?
2. Why were the early social thinkers concerned with social stability? Why did so many analyze society from the perspectives of the biological and physical sciences?
3. How do race, class, and gender affect, and how are they affected by, shopping and consumption?
4. How did the Age of Enlightenment and the Industrial Revolution contribute to the emergence of sociology in Europe?
5. What are the major tenets of the social sciences discussed in the text?
6. Identify the seven major early thinkers in the development of modern sociology and summarize their unique contributions to t he discipline.

STUDENT CLASS PROJECTS AND ACTIVITIES

1. Research the origins of the social sciences, other than Sociology, mentioned in this chapter. Provide answers to the following: (a) describe the development of each: how, why, and when did they develop; (b) identify the early founders or early thinkers in each discipline and describe their contributions and influence on each social science; (c) identify the major contemporary theorists in each of these social sciences and explain their contributions to each; (d) explain how most theorists define the term "social science" today. Submit your research to your professor on the due date, following any instructions given to you.
2. As discussed in your text, C. Wright Mills distinguishes between "personal troubles" and "public issues." You are to compose a list of ten personal troubles that you have recently experienced—for example, money problems, problems in a specific course, problems in interpersonal relationships, maintaining a decent grade point average, inability to find a part-time job, parental pressures, etc. You are to be specific in your listing. This information will be kept strictly confidential by your professor. Next, you are to write a paragraph about an analysis of the problems. Of these, select five from the list of ten that could possibly extend beyond your personal problems and could be considered a public issue. You are to specifically provide information that would indicate how these five do have their roots in social causes. Next, choose one of the "public problems" and briefly research that specific problem, providing some statistical data that would verify that indeed this specific problem is a social issue. (Hint: a recent divorce in your family.) You certainly can find research indicating that this one problem is certainly not only a private issue, but a public issue as well. You must provide a bibliographical reference for the one specific problem that you choose to briefly research. Submit your paper to your instructor following any other directions given to you concerning the writing of the paper.

3. Choose ten specific cartoons from various sources that illustrate some sociological concepts or ideas or any ideas of interest to sociologists as discussed in Chapter 1. Provide a copy of each of the cartoons in your paper. Type at least a paragraph per each cartoon describing what concept or idea or area of sociology the cartoon is illustrating, as well how it illustrates that concept, and why you believe it can be of interest to sociologists. Provide any other additional commentary on each cartoon. Finally, evaluate not only the cartoons, but the value/worth of this project. Submit your paper to your instructor following any other directions your instructor may have given you concerning the writing of the paper. Be certain to submit your paper on the correct date!

4. Select a specific historical or contemporary well-known sociological theorist. Your instructor may provide a list for your selection. In this project you are to: (1) investigate the background of the theorist, (2) the lifestyle of the theorist, and (3) provide any other important bibliographic information on that theorist. (4) Include in your paper the major works and ideas of your particular theorist and (5) evaluate the contribution that your theorist makes to the field of sociology. (6) Briefly evaluate the value of this project and (7) submit your paper to your instructor on the due date.

INTERNET ACTIVITIES

1. To get a better grip on what sociology is all about, explore the **American Sociological Association's web site: http://www.asanet.org/**. Click on the latest statistical fact sheet on the "Health of Sociology." Next, click on "Reasoning for Majoring in Sociology." What are the top three responses? Share these responses with your faculty and students in class. Check the "What's New" site for some of the latest research findings and some of the controversial issues in the discipline.

2. In order to introduce to the broad field of sociology, explore the **SOCIOSITE: http://www.pscw.uva.nl/sociosite/topics/index.html**. From this page you can access dozens of subject areas within the discipline. The web site has hundreds of resources organized within these subject areas.

3. **A Sociological Tour Through Cyberspace** is a web site created by Professor Michael C. Kearl: **http://www.trinity.edu/~mkearl/index.html#in**. There are many resources on this site. Read the essay on credit cards found on the site. Next, read the section on Social Science Data Resources. Discover how to recognize the differences between good data and bad, misinterpreted information.

4. As Chapter 1 points out, social thought has a very long history. Learn more about the historical development of key ideas as well as biographical information about a range of social thinkers and sociologists by going to **A Sociology Timeline from 1600: http://www.wwu.edu/~stephan/timeline.html** or locate through a search engine, find the site ED Stephan's Timeline of Sociology. Locate on the timeline the social thinkers covered in Chapter 1; then explore the historical context in which they lived.

5. To conduct a more extensive analysis of one of the social thinkers introduced in this chapter, click on to this site to learn more about Jane Addams: www.lkwdpl.org/wihohio/adda-jan.htm. Respond to these questions: Did Jane Addams graduate from college? How did Yale University honor her? What was Hull House? Why did this program serve as a model program to assist newly arrived immigrants into the United States? Was Jane Addams indeed the first American woman to receive the Nobel Peace Prize? Why did she receive this most prestigious international award? What was her role in the NAACP? In the Women's Suffrage Movement? Note the additional web sites on this Jane Addams Home page providing additional opportunities for you to advance your knowledge about this early founder in the discipline of sociology.

INFOTRAC COLLEGE EDITION EXERCISES

Visit the **InfoTrac College Edition** website at: **http://www.wadsworthmedia.com/webtutor/infotrac.htm**. You will arrive at a screen that enables you to search topics.

1. Log onto the specific journal, **Social Forces**. Then record both the titles and the numbers of research that focuses on race, class, or gender during the past two years. How many were published during the time period? Report your findings to the class.
2. Conduct a keyword search for **American Consumerism**. Using the resources found on InfoTrac, write an essay on the global impact of American culture.
3. Under the subject search for **Credit Cards**, there are a number of subdivisions. Go to the *Usage* subdivision where you will find a number of articles. Select a number of articles to bring to class and work in groups to synthesize information with reading from the text and class presentations.
4. Put together a subject search for the term **Sociology**. There are dozens of categories and over a hundred articles. Explore the discipline and bring back to class information of interest that was not presented in class or in the assigned readings.

SOLUTIONS

MULTIPLE CHOICE QUESTIONS

1. B, p. 4	15. D, p. 10	29. B, p. 23
2. D, p. 4	16. A, pp. 11-12	30. A, p. 23
3. C, p. 4	17. C, p. 13	31. C, p. 25
4. A, p. 5	18. B, p. 13	32. A, p. 24
5. A, p. 5	19. A, p. 13	33. B, p. 25
6. A, p. 8	20. D, p. 14	34. D, p. 25
7. B, p. 8	21. A, pp. 14-15	35. B, p. 26
8. D, p. 8	22. C, p. 16	36. C, p. 28
9. B, p. 8	23. B, p. 21	37. C, p. 27
10. C, p. 8	24. A, p. 22	38. A, p. 27
11. B, p. 9	25. C, p. 15	39. A, p. 30
12. D, p. 9	26. B, p. 19	40. C, p. 32
13. D, p. 9	27. D, p. 19	
14. B, p. 11	28. D, p. 22	

TRUE/FALSE QUESTIONS

1. T, p. 4	6. T, p. 14	11. F, p. 23
2. T, p. 5	7. T, p. 13	12. F, p. 23
3. F, p. 11	8. T, p. 16	13. F, p. 25
4. T, p. 11	9. F, p. 17	14. F, p. 26
5. T, p. 13	10. F, p. 22	15. T, pp. 31-32

FILL-IN-THE-BLANK QUESTIONS

1. Aristotle, p. 9; Plato, p.9
2. Harriet Martineau, p. 14
3. Emile Durkheim, p. 19
4. Auguste Comte, p. 13
5. Max Weber, p. 19
6. Herbert Spencer, p.14
7. Georg Simmel, p.20
8. W.E.B. Du Bois, p. 22
9. proletariats (workers), p.22; Bourgeoisie (owners), p.16
10. *The Protestant Ethic and The Spirit of Capitalism*, p. 18
11. microlevel, p. 26
12. postmodern, p.30
13. dysfunction, p. 24
14. China, p. 10
15. "cookies," p. 18

2

SOCIOLOGICAL RESEARCH METHODS

BRIEF CHAPTER OUTLINE

CHAPTER SUMMARY

Social research is a key part of sociology. The sociological perspective incorporates theory and research to arrive at a more accurate understanding of society and provide a factual and objective counterpoint to commonsense knowledge and ill-informed sources of information. Sociological research involves *debunking* and utilizes approaches that answer questions through the **empirical approach**, an approach that attempts to answer questions through the systematic collection and analysis of data. Theory and research form a continuous cycle that encompasses both deductive and *inductive approaches*. Many sociologists engage in *quantitative* research, which focuses on data that can be measured numerically. Other research is *qualitative*, based upon interpretive description rather than statistics. Research models are tailored to the specific problem being investigated and the focus of the researcher, which may be quantitative or qualitative. The following are steps in the conventional quantitative research: (1) select and define the research problem, (2) review previous research,

(3) formulate the hypothesis, (4) develop the research design, (5) collect and analyze the data, (6) draw conclusions and report the findings. Although the qualitative approach follows the conventional research approach, which is to: (1) formulate the problem to be studied instead of creating a hypothesis, (2) collect and analyze the data and (3) report the results, it also has several unique features. These features are: (1) the research begins with a general approach rather than a highly developed plan, (2) the researcher decides when the literature review and theory application should take place, (3) the study presents a detailed view of the topic, (4) the researcher must have access to people or other resources that can provide the necessary data, and (5) the researcher uses appropriate methods for acquiring useful qualitative data. **Research methods**—systematic techniques for conducting research—include **survey research**, **secondary analysis of existing data**, **field research**, and **experiments**. Many sociologists use *triangulation*—multiple methods—in a single study in order to gain a wider scope of data and points of view. Studying human behavior raises important ethical issues for sociologists, such as the research of Zellner and Humphreys. The challenge for social research today is to find new ways to integrate knowledge, action, and all people in the research process, in order to fill in the gaps of existing knowledge about social life.

LEARNING OBJECTIVES

After reading Chapter 2, you should be able to:

1. Describe the key steps in conducting qualitative research.
2. State the major strengths and weaknesses of secondary analysis of existing data.
3. Describe the major ethical concerns in sociological research.
4. Describe the research cycle from the deductive and inductive points of view.
5. Describe the six steps in the conventional research process.
6. Explain why validity and reliability are important considerations in sociological research.
7. Explain the concept of triangulation.
8. Describe the need for systematic research.
9. Differentiate between quantitative and qualitative research and give examples of each.
10. Distinguish between a representative sample and a random sample and explain why sampling is an integral part of quantitative research.
11. Describe the major types of surveys and indicate their major strengths and weaknesses.
12. Describe the major methods of field research and indicate when researchers are most likely to utilize each of them.
13. Describe the structure of an experiment and distinguish between laboratory and field experiments.

14. Indicate the relationship between dependent and independent variables in a hypothesis.

15. Distinguish between sociology and common sense.

KEY TERMS

(Defined at page number shown and in glossary)

KEY PEOPLE

(identified at page number shown)

CHAPTER OUTLINE

I.   WHY IS SOCIOLOGICAL RESEARCH NECESSARY?
    A. Common sense and sociological research
        1. Sociologists obtain knowledge of human behavior through research, which results in a body of information that helps to move beyond guesswork and common sense in understanding society.

B. Sociology and scientific evidence
   1. Sociology involves *debunking*—the unmasking of fallacies—in everyday and official interpretations of society.
   2. Two approaches—the *normative*, which uses religion, customs, habits, and traditions—and the attempt to answer questions concerning social life.
   3. The *empirical* approach is referred to as the conventional model, or scientific method; it utilizes two types of empirical studies: *descriptive,* which provide social facts or describe social reality and *explanatory* studies, which attempt to explain cause and effect relationships.
C. The theory and research cycle
   1. A theory is a set of logically interrelated statements that attempts to describe, explain, and (occasionally) predict social events.
   2. Research is the process of systematically collecting information for the purposes of testing an existing theory or generating a new one.
   3. The theory and research cycle consists of the *deductive approach* and *inductive approach.*
   4. Theory gives meaning to research; research helps support theory.
II. THE SOCIOLOGICAL RESEARCH PROCESS
A. The "conventional" research model
   1. *Quantitative research* is based on the goal of scientific objectivity and focuses on data that can be measured in numbers, while *qualitative research* uses interpretive descriptions (words) rather than statistics (numbers) to analyze underlying meanings and patterns of social relationships.
   2. The steps in the conventional research method include:
      a. Selecting and defining the research problem
      b. Reviewing previous research
      c. Formulating a hypothesis (if applicable)
      d. Developing research design selecting either *cross-sectional* or *longitudinal* time frames
      e. Collecting and analyzing data
      f. Drawing conclusions and reporting the findings
   3. Important concepts in the research process:
      a. A **hypothesis** is a statement of the relationship between two or more concepts.
      b. **Variables** are concepts with measurable traits or characteristics that can change or vary from one person, time, situation, or society to another.
      c. The **independent variable** is presumed to cause or determine a dependent variable.
      d. The **dependent variable** is assumed to depend on or be caused by the independent variable(s).
      e. To use a variable, sociologists create an *operational definition*—an explanation of an abstract concept in terms of observable features that are specific enough to measure the variable.
      f. An event that occurs as a result of many factors operating in combination must be explained in terms of *multiple causation.*

4. Important concepts in developing the research design:
   a. A *unit of analysis* is what or whom is being studied.
   b. *Cross-sectional studies* take place at a single point in time.
   c. *Longitudinal studies* focus on processes and social change over a period of time.
5. Important concepts in collecting and analyzing data:
   a. A **random sample** is chosen by chance. Every member of an entire population being studied has the same chance of being selected.
   b. In **probability sampling**, participants are deliberately chosen because they have specific characteristics, such as age, race, and ethnicity.
   c. **Validity** is the extent to which a study or research instrument accurately measures what it is supposed to measure.
   d. **Reliability** is the extent to which a study or research instrument yields consistent results.

B. A qualitative research model
   1. Qualitative research differs from quantitative research in several ways:
      a. Qualitative research is often used when the research question does not lend itself to numbers and statistical methods.
      b. Researchers may engage in *problem formulation* instead of creating a hypothesis.
      c. This type of research is often built on a collaborative approach in which the "subjects" are active participants in the design process, not just passive objects to be studied.
      d. Researchers tend to gather data in natural settings, such as where the person lives or works, rather than in a laboratory or other research setting.
      e. Data collection and analysis frequently occur concurrently, and the analysis draws heavily on the language of the persons studied, not the researcher.
   2. Qualitative research follows the conventional research approach, but has several unique features:
      a. *The researcher begins with a general approach rather than a highly detailed plan.*
      b. *The researcher has to decide when the literature review and theory application should take place.*
      c. *The study presents a detailed view of the topic.*
      d. *Access to people or other resources that can provide the necessary data is crucial.*
      e. *Appropriate research methods are important for acquiring useful qualitative data.*

III. RESEARCH METHODS
  A. **Research methods** are strategies or techniques for systematically conducting research.
  B. A **survey** is a poll in which researchers gather facts or attempt to determine the relationship between facts. Survey data are collected by using self-administered questionnaires, personal interviews, and/or telephone surveys.
      1. **Respondents** are persons who provide data for analysis through interviews or questionnaires.
      2. A **questionnaire** is a printed research instrument containing a series of items for the subjects' response. Questionnaires may be self-administered by respondents or administered by interviewers in face-to-face encounters or by telephone.
      3. An **interview** is a data-collection encounter in which an interviewer asks the respondent questions and records the answers. Survey research often uses *structured interviews*, in which the interviewer asks questions from a standardized questionnaire.
      4. Several strengths of survey research include the ability to describe the characteristics of the large population and enhancing the ability to do *multivariate analysis* – research involving more than two independent variables.
      5. Several weaknesses of survey research include the concern of validity, respondents may not be truthful, and the belief that the data do not always constitute "hard facts" that other analysts might use.
  C. In **secondary analysis of data**, researchers use existing material and analyze data that originally was collected by others.
      1. Existing data sources include public record, official reports of organizations or government agencies, surveys taken by researchers in universities and private corporations, books, magazines, newspapers, radio, television, and personal documents.
      2. **Content analysis** is the systematic examination of *cultural artifacts* or various forms of communication to extract thematic data and draw conclusions about social life.
      3. Several strengths of secondary analysis include: the data are readily available and inexpensive; the chance of bias reduction; and the possibility of using longitudinal data.
      4. Some weaknesses include: the data may be incomplete, unauthentic or inaccurate and *coding* the data may be difficult.
  D. **Field research** is the study of social life in its natural setting: observing and interviewing people where they live, work, or play.
      1. In **participant observation**, researchers collect systematic observations while being part of the activities of the groups they are studying.
      2. A *case study* is an in-depth, multi-faceted investigation of a single event, person, or social grouping; a *collective case study* involves multiple cases.

3. An **ethnography** is a detailed study of the life and activities of a group of people by researchers who may live with that group over a period of years.
4. An **unstructured interview** is an extended, open-ended interaction between an interviewer and an interviewee.
5. Strengths of field research include: providing a wealth of information to generate theories; enabling the study of social processes overtime; and affording the opportunity to study race, ethnicity and gender over time.
6. Some weaknesses of field research include; the question of generalization to a larger population; the data collected are descriptive, rather than precise measurement; and cause and effect relationships cannot be depicted.

E. **Experiments** – carefully designed situations in which the researcher studies the impact of certain variables on subjects' attitudes or behavior – typically require that subjects be divided into two groups:
1. The **experimental group** contains the subjects who are exposed to the independent variable (the experimental condition) to study its effect on them.
2. The **control group** contains the subjects who are not exposed to the independent variable.
3. Experiments may be conducted in a *laboratory* or *natural setting*.
4. At the conclusion of the research, the experiment and control groups are compared to see if they differ in relation to the dependent variable, and the hypothesis about the relationship of the two variables is confirmed or rejected. **Correlation** exists when two variables in an experiment are associated more frequently than could be expected by chance.
5. Some strengths of the controlled experiment include the researcher's ability to isolate the experiment and the possibility of continued replication.
6. Researchers acknowledge that experiments have several weaknesses such as: the rigid control and manipulation of variables; the artificial setting; and the problem of generalization to other groups. Another weakness is known as the **Hawthorne effect** – changes in the subject's behavior caused by the researcher's presence or by the subject's awareness of being studied.

F. Multiple Methods: Triangulation
1. Many sociologists utilize *triangulation* – the use of multiple approaches in a single study.
2. Triangulation refers to both multiple research methods and multiple data sources.

IV. ETHICAL ISSUESS IN SOCIOLOGICAL RESEARCH
A. The ASA Code of Ethics
1. The American Sociological Society (ASA) has a Code of Ethics that sets forth certain basic standards sociologists must follow in conducting research.
2. Sociologists are committed to adhering to this code and to protecting research participants; however, many ethical issues arise that cannot be resolved easily.
B. The Zellner Research
1. The research sought to determine if some automobile accidents were actually suicides.
2. To recruit respondents, he misrepresented the reasons for his study.

C. The Humphreys Research
   1. This research studied homosexual acts in public restrooms in parks.
   2. Humphreys did not ask permission of his subjects, nor did he inform them that they were being studied. His subsequent personal interviews with them were held under false pretenses.
D. Research Summary
   1. These examples of research demonstrate the difficulties of resolving ethical issues.
   2. The challenge today is to find new ways to integrate knowledge and action, and to include all people in the research process in order to fill the gaps in our existing knowledge about social life.

ANALYZING AND UNDERSTANDING THE BOXES

After reading the chapter and studying the outline, re-read the boxes and write down key points and possible questions for class discussion.

Sociology and Everyday Life: How Much Do You Know About Suicide?

Key Points:

Discussion Questions:

1.

2.

3.

Sociology in Global Perspective: Comparing Suicide from Different Nations

Key Points:

Discussion Questions:

1.

2.

3.

Framing Suicide in the Media: Sociology Verses Sensationalism

Key Points:

Discussion Questions:

1.

2.

3.

You Can Make a Difference: Responding to a Cry for Help

Key Points:

Discussion Questions:

1.

2.

3.

PRACTICE TESTS

MULTIPLE CHOICE QUESTIONS

Select the response that best answers the question or completes the statement:

1. The _____ approach involves gathering evidence to answer questions through systematic collection and analysis of data.
    a. empirical
    b. descriptive
    c. normative
    d. evaluative

2. A type of research which describes social reality or provides facts about the social world is:
    a. explanatory
    b. descriptive
    c. evaluative
    d. observational

3. In the _____ approach, the researcher begins with a theory and uses research to test the theory.
    a. inductive
    b. deductive
    c. model
    d. command

4. The first step in the conventional research process is
    a. reviewing the literature.
    b. selecting a research model.
    c. formulating the hypothesis.
    d. selecting and defining the research problem.

5. When concepts have two or more degrees or values, they are referred to as
    a. hypotheses.
    b. triangulation.
    c. variables.
    d. theories.

6. Which of the following can serve as a variable in a study?
    a. age
    b. sex
    c. ethnic background
    d. all of these choices

7. In Emile Durkheim's study of suicide, the degree of social integration was the
    a. operational definition.
    b. dependent variable.
    c. independent variable.
    d. spurious correlation.

8. In a medical study, lung cancer could be the _____ variable, while smoking could be the _____ variable.
    a. dependent, independent
    b. independent, dependent
    c. valid, reliable
    d. reliable, valid

9. In creating an operational definition, a sociologist
    a. duplicates research in a precise manner.
    b. analyzes the findings of a research.
    c. makes a concept or variable measurable.
    d. conducts field research.

10. Suppose we are investigating the primary causes of suicide in the late 1990s. Upon looking into recent cases of suicide, we find out that a number of the people had lost their jobs; that they had been unemployed off and on for the past ten years; that they had no religious affiliation; and that a number of them had been divorced within the past five years. The analysis reflects what the text terms:
    a. singular determination
    b. multivariate involvement
    c. plural association
    d. multiple causation

11. A _____ sample refers to a situation in which every member of an entire population has the same chance of being selected for a study.
    a. selective
    b. random
    c. representative
    d. longitudinal

12. _____ is the extent to which a study or research instrument accurately measures what it is supposed to measure; _____ is the extent to which a study or research instrument yields consistent results.
    a. Validity, replication
    b. Replication, validity
    c. Validity, reliability
    d. Reliability, validity

13. The national census is an example of _____ .
    a. survey research
    b. a field experiment
    c. secondary analysis
    d. participant observation research

14. Research involving more than two independent variables is known as:
    a. multivariate analysis
    b. secondary analysis
    c. random sampling
    d. field analysis

15. Researchers who use existing material and analyze data that was originally collected by others are engaged in:
    a. unethical conduct
    b. primary analysis
    c. secondary analysis
    d. survey analysis

16. Observation, ethnography, and case studies are examples of:
    a. survey research
    b. experiments
    c. secondary analysis of existing data
    d. field research

17. In an experiment, the subjects in the control group
    a. are exposed to the independent variable.
    b. are not exposed to the independent variable.
    c. are exposed to the dependent variable.
    d. are not exposed to the dependent variable.

18. A sports sociologist finds that body temperature increases as the "pumping iron" increases. In this example, body temperature would be the:
    a. dummy variable.
    b. null hypothesis.
    c. dependent variable.
    d. independent variable.

19. _____ exists when two variables change together in a predictable direction.
    a. Correlation
    b. Triangulation
    c. Multiple causation
    d. Quantification

20. According to the text, important ethical concerns were raised by all of the following researchers, **except**:
    a. Elijah Anderson
    b. William Zellner
    c. Laud Humphreys
    d. both William Zellner and Laud Humphreys

21. Sociologists use _____ to obtain knowledge of human behavior which moved beyond guesswork and common sense.
    a. research
    b. myths
    c. common sense beliefs
    d. scientific

22. In his research, Durkheim related suicide to the issue of:
    a. group cohesiveness
    b. individual acts
    c. personal social problems
    d. isolated acts

23. The _____ approach, referred to as the conventional model or the scientific method, attempts to answer questions through the systematic collection and analysis of data.
    a. explanatory
    b. descriptive
    c. empirical
    d. normative

24. Sociologists would use a(n) _____ study of suicide in asking the question why African American men over age sixty-five have a significantly lower rate of suicide than white males in the same age bracket.
   a. objective
   b. evaluative
   c. descriptive
   d. explanatory

25. _____ attempt to describe social reality or provide facts about some group, practice, or event.
   a. Explanatory studies
   b. Observational studies
   c. Descriptive studies
   d. Evaluative studies

26. With _____ research, the goal is scientific objectivity, and the focus is on data that can be measured numerically.
   a. explanatory
   b. qualitative
   c. quantitative
   d. normative

27. The first step in the "conventional" research model, which focuses on quantitative research, is to
   a. develop a research design.
   b. select and define the research problem.
   c. review previous research.
   d. collective and analyze the data.

28. In a hypothesis, the researcher assumes the _____ to be caused by another variable.
   a. independent variable
   b. dependent variable
   c. multiple variable
   d. null variable

29. In this sociology course, your professor may have specified the points needed, or the average test grades needed, to earn an "A" in the course, thus providing you a(n) _____ of a "grade" in the course.
   a. operational definition
   b. explanatory definition
   c. qualitative explanation
   d. correlated definitions

30. In _____, every member of an entire population being studied has the same chance of being selected.
    a. empirical sampling
    b. descriptive sampling
    c. probability sampling
    d. random sampling

31. When events are too complex to be caused by one variable, sociologists would explain those causes in terms of _____.
    a. pluralistic association
    b. multiple causation
    c. multiple associations
    d. multiplicity causation

32. _____ studies are based on observations that take place at a single point in time; these studies focus on behavior or responses at a specific moment.
    a. Participant
    b. Cross-sectional
    c. Longitudinal
    d. Sectional

33. In _____, participants are deliberately chosen because they have specific characteristics.
    a. probability sampling
    b. validity sampling
    c. random sampling
    d. reliable sampling

34. In his research, Emile Durkheim concluded that _____ suicides were relatively high in Protestant countries in Europe because Protestants believed in individualism and were more loosely tied to the church than were Catholics.
    a. altruistic
    b. fatalistic
    c. egoistic
    d. anomic

35. Students in this sociology class develop a research instrument to measure happiness among the students enrolled in the university. You are concerned with the issue of _____ in that you want your research instrument to accurately measure what it is suppose to measure.
    a. representation
    b. predictability
    c. reliability
    d. validity

36. Which of the following methods would most likely be used to study the attitudes of the American public on global warming?
    a. an experiment
    b. participant observation research
    c. a survey
    d. a case study

37. _____ researchers frequently attempt to study the social world from the point of view of the people they are studying.
    a. Qualitative
    b. Quantitative
    c. Reliable
    d. Experimental

38. Secondary analysis is referred as _____ research because it has no impact on the people being studied.
    a. obtrusive
    b. unobtrusive
    c. non-opinionated
    d. reliable

39. A researcher would probably use _____ in studying how gender is depicted in award-winning children's books.
    a. survey research
    b. content analysis
    c. field research
    d. an experiment

40. _____ is the study of social life in its natural setting.
    a. Field research
    b. A natural experiment
    c. Secondary analysis
    d. Case studies

TRUE/FALSE QUESTIONS

1. Emile Durkheim was one of the first sociologists to challenge the commonsense view of suicide.
   T   F

2. The normative approach answers questions through systematic collection and analysis of data.
   T   F

3. Theory gives meaning to research; research helps support theory.
   T   F

4. In the deductive approach, the researcher collects data and then generates theories from the analysis of that data.
   T   F

5. With qualitative research, the focus is on data that can be measured numerically.
   T   F

6. The first step in the conventional research model is to formulate the hypothesis.
   T   F

7. Cross-sectional studies study what has happened over a period of time.
   T   F

8. In random sampling, each member of a population has an equal chance of being selected.
   T   F

9. Validity is when a study gives consistent results to different research over time.
   T   F

10. Altruistic suicide occurs when individuals are excessively integrated into society.
    T   F

11. The research method determines how the data will be collected.
    T   F

12. In survey research, every person of that population must be interviewed.
    T   F

13. Participant observation is a quantitative research method.
    T   F

14. Sociologist William F. Whyte's classic study of *Street Corner Society* was conducted in Boston's low-income Irish neighborhood.
    T   F

15. The Hawthorne effect refers to changes in the subject's behavior caused by the researcher's presence or by the subject's awareness of being studied.
    T   F

FILL-IN-THE-BLANK QUESTIONS

1. Sociology involves _____, which challenges false or mistaken ideas or opinions.

2. A(n) _____ is a statement of the relationship between two or more concepts.

3. _____ studies compare data over a period of time.

4. In _____ sampling, participants are deliberately chosen because they have specific characteristics.

5. According to Durkheim, _____ suicide results from a lack of shared values or purpose.

6. Specific strategies for systematically conducting research are known as
   _____.

7. Persons who provide data for analysis through interviews or questions are
   called _____.

8. Researchers use _____ research to study social life in its natural setting.

9. In a classic study, sociologist _____ conducted long term participant
   observation studies in Boston's low-income Italian neighborhoods.

10. The controversial research of _____ examined "tearoom trade."

11. In conducting research, a(n) _____ makes it possible to measure a
    concept or a variable.

12. According to Box 2.1 regarding suicide, the rates of suicide are highest among
    _____ Americans.

13. According to Box 2.2 on a global perspective of suicide, the country of
    _____ has the highest suicide rate.

14. As explained in Box 2.3, "Framing Suicide in the Media," the term _____
    refers to the process by which information and entertainment are packaged by
    the mass media.

15. According to Box 2.4, "You Can Make a Difference: Responding to a Cry for
    Help," _____ is a type of depression involving withdrawing from activities
    that a person previously enjoyed, and one that should serve as a warning sign of
    suicide.

SHORT ANSWER/ESSAY QUESTIONS

1. What is the difference between qualitative and quantitative research? Give an
   example of each.
2. What is the relationship between theory and research?
3. What are the steps in the conventional research process?
4. List and describe the five basic types of research methods.
5. Identify four ethical concerns which must be considered when conducting
   research, and give three examples of controversial research discussed in your
   textbook.

STUDENT CLASS PROJECTS AND ACTIVITIES

1. Collect, over a period of two or three weeks, five articles from newspapers or
   magazines reporting the results of a public opinion poll. (Examples: NEW YORK
   TIMES, GALLUP, CNN, etc.). Evaluate the polls and respond in writing to these
   questions: (1) Were the questions fully reported? (2) What is the size of the sample,
   its composition (i.e., gender, age, race, nationality, region of country), and the
   method of selecting the sample (i.e., was it random, stratified, etc.) (3) How was the
   data collected (telephone, mail, etc.)? (4) What was the wording of the questions

asked in the poll? (5) Summarize the findings of the poll. (6) Did you find any problems of the poll? Was it scientifically designed? (7) Include any other information in your paper appropriate to this assignment. (8) Provide bibliographic reference for the polls you have selected. (9) Evaluate this project. Follow the directions of the writing of this project given to you by your instructor. Submit your paper on the due date.

2. Conduct a survey of any 20 people and ask these questions: (1) what is sociology? (2) What do sociologists do? (3) What is your opinion of sociologists? (4) Have you had a course in sociology? If so, where? (5) Have you had a course in a social science? (6) Do you know a sociologist? If so, who? (7) Include any other kind of information you may want to obtain from your subjects. (8) Include the names, gender, the ages, and, if students, the college majors of your subjects. (9) Provide a summary and an evaluation of this project in the writing of your paper. Submit your paper to your instructor following any other directions your instructor may have given you in the writing of this project. (10) Submit your paper on the correct date.

3. Recently, *Parade Magazine* published a poll of the top ten issues frustrating to Americans. The top 10 recorded in this poll were: (1) the faltering economy; (2) government misconduct and mismanagement; (3) the medical care crisis; (4) crime; (5) the abortion issue; (6) the tax burden; (7) government waste; (8) the crumbling infrastructure; (9) the learning decline; and (10) free trade. You are to conduct an informal poll, collecting data from 10 to 20 students enrolled in your institution. (1) You are to have your subjects list and briefly explain their top 10 frustrations. (2) Summarize your findings and list the top ten frustrations according to your subjects. (3) Compare and contrast this list with that of your subjects. (4) Provide an analysis of this comparison. (5) Provide the names, ages, and gender of your subjects in your written paper. (6) Provide a conclusion and an evaluation of this project in your paper. Submit your paper to your instructor following any other directions your instructor may have given you in the writing of this paper.

INTERNET ACTIVITIES

1. Learn more about the United States by accessing the **U.S. Census website**: **http://www.census.gov/acsd/www/subjects.html**. The best beginning site is the American FactFinder. This is a more user-friendly entry point to see reports that are being generated from data collected during the 2000 Census with an update through 2005. Note that several of the: subjects Index" sites have been revised as recent as August, 2007.

2. More students are using Internet sources for their class research projects. Before you join the herd, access the **Guide for Citing Electronic Information**, **http://www.wpunj.edu/wpcpages/library/citing.htm**, provided by the William Paterson University (New Jersey). The site contains guidelines for citing everything from e-mails to online journal articles. Also included are links to other related web pages.

3. The **National Opinion Research Center**, **http://www.norc.uchicago.edu/projects/gensoc.asp**, at the University of Chicago is one of the largest survey research organizations. On this site you can access information about the NORC, past and present. Next gather information on the General Social Survey. Then, go to the FAQ link on this site and bring to class a basic description of the GSS.
4. Some research organizations openly admit that their research is guided by a particular set of political values. One such example is the **Cato Institute**, **http://www.cato.org/**, a politically conservative research organization guided by Libertarian Principles. Go to their home page and click on daily commentaries and then select a topic that interests you. Can you find examples of research that is guided by conservative principles? How does this compare with the standards of objectivity that are discussed in the text?
5. When you start thinking about the difficulties of researching highly subjective social phenomena, consider accessing the **World Database of Happiness**: **http://www.eur.nl/fsw/research/happiness/**. This site is an ongoing register of scientific research related to subjective appreciation of life. Can you find out which nation is cited as the most happy? Can you discover the degree of unhappiness of nations, as measured by the rate of suicide in nations?

INFOTRAC COLLEGE EDITION EXERCISES

Visit the **InfoTrac College Edition** website at: **http://www.wadsworthmedia.com/webtutor/infotrac.htm**. You will arrive at a screen that enables you to search topics.

1. Look up the keywords **Suicide Research** on InfoTrac. Among the articles should be those that address social factors. Bring articles to class and be prepared to discuss them in small groups. Determine what kinds of research methods were used in each study.
2. The **Hawthorne Effect** is a phenomenon that you may hear referenced in other courses and reading assignments. A recent re-examination of the phenomenon and the original research contains some new surprises. Use InfoTrac to look up the keywords **Hawthorne Effect** (be sure of the spelling) and read the article. Construct a list of new facts contained in the report and bring the report to class.
3. Why are surveys chosen over other methods of collecting data? Research the keyword **Survey** and select a survey study to examine. Read the research report and then try to determine why a survey was used.

SOLUTIONS

MULTIPLE CHOICE QUESTIONS

1. A, p. 39	15. C, p. 55	29. A, p. 44
2. B, p. 39	16. D, p. 59	30. D, p. 46
3. B. p. 40	17. B, p. 63	31. B, p. 44
4. D, p. 42	18. C, p. 63	32. B, p. 45
5. C, p. 43	19. A, p. 63	33. A, p. 47
6. D, p. 44	20. A, p. 61	34. C, p. 48
7. C, p. 44	21. A, p. 38	35. D, p. 47
8. A, p. 44	22. A, p. 38	36. C, p. 52
9. C, p. 44	23. B, p. 39	37. A, p. 50
10. D, p. 44	24. D, p. 39	38. B, p. 55
11. B, p. 46	25. C, p. 39	39. B, p. 56
12. C, p. 48	26. C, p. 41	40. A, p, 57
13. A, p. 52	27. B, p. 42	
14. A, p. 55	28. B, p. 44	

TRUE/FALSE QUESTIONS

1. T, p. 38	6. F, p. 42	11. T, p. 51
2. F, p. 39	7. F, p. 45	12. F, p. 52
3. T, p. 41	8. T, p. 46	13. F, p. 57
4. F, p. 40	9. F, p. 47	14. F, p. 61
5. F, p. 41	10. T, p. 49	15. T, p. 65

FILL-IN-THE-BLANK-QUESTIONS

1. debunking, p. 39
2. hypothesis, p. 43
3. Longitudinal, p. 45
4. probability, p. 47
5. anomic, p. 49
6. research methods, p. 51
7. respondents, p. 52
8. field, p. 57
9. William F. Whyte, p. 61
10. Laud Humphreys, p. 69
11. operational definition, p. 44
12. white, p. 40
13. Lithuania, p. 47
14. media framing, p. 58
15. serious depression, p. 68

3
CULTURE

BRIEF CHAPTER OUTLINE

CHAPTER SUMMARY

Culture is the knowledge, language, values, customs, and material objects that are passed from person to person and from one generation to the next. At the macrolevel, culture can be a stabilizing force or a source of discord, conflict, and even violence. At the microlevel, culture is essential for individual survival. Sociologists distinguish between **material culture** -- the physical creations of society -- and **nonmaterial culture** -- the abstract or intangible human creations of society (such as **symbols**, **language**, **values**, and **norms**). According to the **Sapir-Whorf hypothesis**, language shapes our perception of reality. For example, language may create and reinforce inaccurate perceptions based on gender, race, ethnicity, or other human attributes. Cultural change and diversity are intertwined. In the United States, diversity is reflected through race, ethnicity, age, sexual orientation, religion, occupation, and so forth. **Culture shock** refers to the anxiety people experience when they encounter cultures radically different from their own. **Ethnocentrism** -- a belief based on the assumption that one's own culture is superior to others -- is counterbalanced by **cultural relativism** -- the belief that the behaviors and customs of a society must be examined within the context of its own culture. While it is assumed that **high culture** appeals primarily to elite audiences, **popular culture** is believed to appeal to members of the middle and working classes. Political and religious leaders in some cultures fear and oppose **cultural imperialism**, the extensive infusion of one nation's cultures into other cultures, while others point to numerous cross-cultural influences. Sociologists regard culture as a central ingredient in human behavior and share a similar purpose; however, they typically use different theoretical perspectives in their research of culture. A functional analysis of culture assumes that a common language and shared values help produce consensus and harmony. According to some conflict theorists, culture may be used by certain groups to maintain their privilege and exclude others from society's benefits. Symbolic interactionists suggest that people create, maintain, and modify their culture as they go about their everyday activities. Postmodern theorists suggest that there are many cultures within the United States alone. In order to grasp a better understanding of how popular culture may simulate reality rather than being reality, postmodernists believe that we need a new way of conceptualizing culture and society. As we look toward even more diverse and global cultural patterns in the future, it is important to apply the sociological imagination not only to our society, but to the entire world as well.

LEARNING OBJECTIVES

After reading Chapter 3, you should be able to:

1. Define culture.
2. List and briefly explain ten core values in U.S. society.
3. State the definition of norms and distinguish between folkways, mores, and laws.
4. Distinguish between high culture and popular culture and between fads and fashion.
5. Describe how culture can be both a stabilizing force and a source of conflict in societies.
6. Describe subcultures and countercultures; give examples of each.
7. Describe the importance of culture in determining how people think and act on a daily basis.
8. Use your experiences with food as a symbolic representation of culture.
9. Describe the importance of language and relate to the Sapir-Whorf hypothesis.
10. Contrast ideal and real culture and give examples of each.
11. State the definitions for culture shock, ethnocentrism, and cultural relativism, and explain the relationship between these three concepts.
12. Describe the functionalist, conflict, symbolic interactionist, and postmodernist perspectives on culture.
13. Compare several of the forms that popular culture takes.
14. Distinguish between discovery, invention, and diffusion as means of cultural change.
15. Explain why the rate of cultural change is uneven.

KEY TERMS

(defined at page number shown and in glossary)

counterculture, p. 91
cultural imperialism, p. 95
cultural lag, p. 87
cultural relativism, p. 93
cultural universals, p. 78
culture, p. 73
culture shock, p. 92
diffusion, p. 88
discovery, p. 87
ethnocentrism, p. 92
folkways, p. 86
invention, p. 87
language, p. 79

laws, p. 86
material culture, p. 76
mores, p. 86
nonmaterial culture, p. 78
norms, p. 85
popular culture, p. 93
sanctions, p. 86
Sapir-Whorf hypothesis, p. 80
subculture, p. 90
symbol, p. 79
taboos, p. 86
technology, p. 77
values, p. 84

KEY PEOPLE

(identified at page number shown)

Jean Baudrillard, p. 99
Pierre Bourdieu, p. 96
Napoleon Chagnon, p. 92
Stephen M. Fjellman, p. 100
Marvin Harris, p. 93
Bronislaw Malinowski, p. 95

George Murdock, p. 78
William F. Ogburn, p. 87
Edward Sapir and Benjamin Whorf, p. 80
Georg Simmel, p. 98
Ann Swidler, p. 76
Robin M. Williams, p. 84

CHAPTER OUTLINE

I. CULTURE AND SOCIETY IN A CHANGING WORLD
    A. **Culture** is the knowledge, language, values, customs, and material objects that are passed from person to person and from one generation to the next in a human group or society.
    B. Culture is essential for our individual survival and for our communication with other people.
    C. As our society becomes more diverse and communication among members of international cultures more frequent, the need to appreciate diversity and to understand how people in other cultures view their world increases.

D. **Material culture** consists of the physical or tangible creations that members of a society make, use, and share while **nonmaterial culture** consists of the abstract or intangible human creations of society that influence people's behavior.

E. According to anthropologist George Murdock, **cultural universals** are customs and practices that occur across all societies. Examples include appearance, activities, social institutions, and customary practices.

II. COMPONENTS OF CULTURE

A. A **symbol** is anything that meaningfully represents something else.

B. **Language** is defined as a set of symbols that express ideas and enable people to think and communicate with one another.

1. Language and social reality

a. According to the **Sapir-Whorf hypothesis**, language shapes the view of reality of the speaker.

b. Most sociologists contend that language may *influence* but does not *determine* social reality.

2. Language and gender

a. Examples of situations in which the English language ignores women include using the masculine gender to refer to human beings in general, and nouns that show the gender of the person we expect in a particular occupation.

b. Words have positive connotations when relating to male power, prestige, and leadership; when related to women, they convey negative overtones of weakness, inferiority, and immaturity.

3. Language, race, and ethnicity

a. Language may create and reinforce our perceptions about race and ethnicity by transmitting preconceived ideas about the superiority of one category of people over another.

b. The "voice" of verbs may devalue contributions of members of some racial-ethnic groups.

c. Adjectives can have different meanings when used in certain contexts.

C. **Values** are collective ideas about what is right or wrong, good or bad, and desirable or undesirable in a particular culture.

1. Ten U.S. core values are:

a. Individualism

b. Achievement and success

c. Activity and work

d. Science and technology

e. Progress and material comfort

f. Efficiency and practicality

g. Equality

h. Morality and humanitarianism

i. Freedom and liberty

j. Racism and group superiority

2. *Value contradictions* are values that conflict with one another or are mutually exclusive (achieving one makes it difficult to achieve another).
3. *Ideal culture* refers to the values and standards of behavior that people in a society profess to hold; *real culture* refers to the values and standards of behavior that people actually follow.

D. **Norms** are established rules of behavior or standards of conduct.
1. **Folkways** are everyday customs that may be violated without serious consequences within a particular culture.
2. Mores are strongly held norms that may not be violated without serious consequences within a particular culture. **Taboos** are mores so strong that their violation is considered to be extremely offensive.
3. **Laws** are formal, standardized norms that have been enacted by legislatures and are enforced by formal sanctions.

III. TECNOLOGY, CULTURAL CHANGE, AND DIVERSITY
A. Cultural change is continual in societies, and these changes are often set in motion by these processes:
1. **Technology** refers to the knowledge, techniques, and tools that allow people to transform resources into a usable form and the knowledge and skills required to use what is developed.
2. **Cultural lag** is a gap between the technical development (material culture) of a society and its moral and legal institutions (nonmaterial culture).
3. **Discovery** is the process of learning about something previously unknown or unrecognized.
4. **Invention** is the process of combining existing cultural items into a new form.
5. **Diffusion** is the transmission of cultural items or social practices from one group or society to another.

B. Cultural diversity
1. *Cultural diversity* refers to the wide range of cultural differences found between and within nations.
   a. Homogeneous societies include people having a common culture.
   b. Heterogeneous societies include people who are dissimilar.
2. A **subculture** is a group of people who share a distinctive set of cultural beliefs and behaviors that differ in some significant way from that of the larger society. Examples include Old Order Amish and Chinatowns.
3. A **counterculture** is a group that strongly rejects dominant societal values and norms and seeks alternative lifestyles. Examples include skinheads and members of some paramilitary militias.

C. **Culture shock** is the disruption that people feel when they encounter cultures radically different from their own, and they believe they cannot depend on their own taken-for-granted assumptions about life.

D. **Ethnocentrism** is the assumption that one's own culture and way of life are superior to all others. **Cultural relativism** is the belief that the behaviors and customs of a society must be viewed and analyzed within the context of its own culture.

IV. A GLOBAL POPULAR CULTURE?
  A. *High culture* consists of activities usually patronized by elite audiences while **popular culture** consists of activities, products, and services which are assumed to appeal primarily to members of the middle and working class.
  B. *Culture capital theory* is based on the assumption that high culture is a device used by the dominant class to exclude the subordinate classes.
  C. Forms of popular culture:
    1. A *fad* is a temporary but widely copied activity followed enthusiastically by large numbers of people.
    2. A *fashion* is a currently valued style of behavior, thinking, or appearance that is longer lasting and more widespread than a fad.
    3. **Cultural imperialism** is the extensive intrusion of one nation's culture into other nations, and is viewed as a threat with many countries becoming westernized.
    4. Others argue that if a global culture comes into existence it will include many components from many societies and cultures.
V. SOCIOLOGICAL ANALYSIS OF CULTURE
  A. According to functionalist theorists, societies where people share a common language and core values are more likely to have consensus and harmony.
  B. Conflict theorists suggest that values and norms help create and sustain the privileged position of the powerful in society. According to Karl Marx, ideas are *cultural relations* of a society's most powerful members who use *ideology* to maintain their positions of dominance.
  C. According to symbolic interactionist theorists, people create, maintain, and modify culture as they go about their everyday activities; symbols make communication with others possible by providing shared meanings.
  D. According to postmodern theorists, no single perspective can grasp the complexity and diversity of the social world; we should speak of *cultures* rather than *culture*. Reality may not be what it seems to be; we should therefore deconstruct existing beliefs and theories about culture in hopes of gaining new insights.
VI. CULTURE IN THE FUTURE
  A. In future decades, the issue of cultural diversity will increase in importance.
  B. Some of the most important changes in cultural patterns may include: (1) television and radio, films and videos, and electronic communications including computers and cyberspace will continue to accelerate the flow of information; however, (2) most of the world's population will not participate in this technological revolution.
  C. The study of culture helps us understand not only our own "tool kit" of symbols, stories, rituals, and world views, but to expand our insights to include those of other people of the world who also seek strategies for enhancing their own quantity and quality of life.

ANALYZING AND UNDERSTANDING THE BOXES

After reading the chapter and studying the outline, re-read the boxes and write down key points and possible questions for class discussion.

Sociology and Everyday Life: How Much Do You Know About Global Food and Culture?

Key Points:
Discussion Questions:

1.

2.

3.

You Can Make a Difference: Understanding People from Other Cultures

Key Points:

Discussion Questions:

1.

2.

3.

Sociology in Global Perspective: The Malling of China: What Part Does Culture Play?

Key Points:

Discussion Questions:

1.

2.

3.

Framing Culture in the Media: You are What you Eat?

Key Points:

Discussion Questions:

1.

2.

3.

PRACTICE TESTS

MULTIPLE CHOICE QUESTIONS

Select the response that best answers the question or completes the statement.

1. _____ consists of knowledge, language, values, customs, and material objects.
   a. Social structure
   b. Society
   c. Culture
   d. Social organization

2. All of the following statements regarding culture are true, **except**:
   a. culture is essential for our survival
   b. culture is essential for our communications with other people
   c. culture is fundamental for the survival of societies
   d. culture reflects human instincts

3. In terms of their functions, cultural universals are
   a. useful because they ensure the smooth and continued operation of society.
   b. the result of attempts by a dominant group to impose its will on a subordinate group.
   c. independent from functional necessities.
   d. very similar in form from one group to another and from one time to another within the same group.

4. How a person is dressed demonstrates
   a. the use of language to construct social reality.
   b. how symbols affect our thoughts about class.
   c. cultural universals.
   d. ideal cultural norms.

5. Regarding the relationship between language and gender, the text points out that
   a. the pronouns *he* and *she* are seldom used in everyday conversation by most people.
   b. the English language largely has been purged of sexist connotations.
   c. the English language ignores women by using the masculine form to refer to human beings in general.
   d. words in the English language typically have positive connotations when relating to female power, prestige, and leadership.

6. From a(n) _____ perspective, a shared language is essential to a common culture.
   a. functionalist
   b. conflict
   c. feminist
   d. interactionist

7. The most frequently spoken language in U.S. homes other than English is:
   a. Italian
   b. Spanish
   c. French
   d. German

8. Which of the following hypothetical statements does **not** express a core U.S. value?
   a. "How well does it work?"
   b. "Is this a realistic thing to do?"
   c. "My freedom is important to me."
   d. "It is good to be lazy."

9. All of the following are examples of U.S. folkways, **except**:
   a. using underarm deodorant
   b. brushing our teeth
   c. avoiding sexual relationships with siblings
   d. wearing appropriate clothing for specific occasions

10. The most common type of formal norms is:
    a. folkways
    b. mores
    c. sanctions
    d. laws

11. Laws that deal with public safety and well-being are classified as:
    a. criminal law
    b. civil law
    c. state ways
    d. federal law

12. When one part of a culture changes at a faster pace than another, _____ occurs.
    a. cultural diffusion
    b. cultural lag
    c. cultural shock
    d. cultural diversity

13. _____ is the process of reshaping existing cultural items into a new form.
    a. Discovery
    b. Diffusion
    c. Invention
    d. Restoration

14. According to the text, all of the following are examples of countercultures **except**:
    a. the Old Order Amish
    b. neo-Nazi skinheads
    c. beatniks and flower children
    d. the Nation of Islam

15. The disorientation that people feel when they encounter cultures radically different from their own is referred to as:
    a. cultural diffusion
    b. cultural relativism
    c. cultural disorientation
    d. culture shock

16. National anthems, pledges to a flag, and nationalism are forms of positive _____.
    a. ethnocentrism
    b. cultural relativism
    c. cultural diffusion
    d. cultural indifference

17. Anthropologist Marvin Harris has pointed out that the Hindu taboo against killing cattle is very important to the economic system in India. This exemplifies:
    a. ethnocentrism
    b. cultural relativism
    c. cultural diffusion
    d. cultural indifference

18. Some view the widespread infusion of the English language into non-English speaking countries as a form of
    a. cultural relativism.
    b. cultural imperialism.
    c. cultural diversity.
    d. cultural lag.

19. Which theoretical perspective suggests that ideas are used by agents of the ruling class to maintain their positions of dominance in a society?
    a. symbolic interactionist
    b. postmodernist
    c. conflict
    d. structural functionalism

20. Which theoretical perspective suggests that reality in a culture may not be what it seems?
    a. symbolic interaction
    b. postmodernist
    c. conflict
    d. structural functionalism

21. An example of a(n) _____ is a spider building a web because of basic biological needs.
    a. instinct
    b. impulse
    c. drive
    d. reflex

22. A(n) _____ is an unlearned, biologically-determined, involuntary response to some physical stimuli.
    a. instinct
    b. reflex
    c. impulse
    d. drive

23. Biologically determined impulses common to all members of a species that satisfy needs such as sleep, food, water, and sexual gratification are known as:
    a. instincts
    b. drives
    c. reflexes
    d. reactions

24. According to sociologist Ann Swindler, _____ is/are a "tool kit of symbols, stories, rituals, and world views, which people may use in varying configurations to solve different kinds of problems."
    a. instincts
    b. society
    c. reflexes
    d. culture

25. _____ consists of the physical or tangible creations that members of a society make, use, and share.
    a. Technology
    b. Nonmaterial culture
    c. Cultural symbols
    d. Material culture

26. Sociologists define _____ as the knowledge, techniques, and tools that make it possible for people to transform resources into usable forms, and the knowledge and skills required to use them after they are developed.
    a. nonmaterial culture
    b. technology
    c. material culture
    d. normative culture

27. _____ consists of the abstract or intangible human creations of society that influence people's behavior.
    a. Nonmaterial culture
    b. Material culture
    c. A symbol
    d. Technology

28. A central component of _____ culture is _____, which are the mental acceptance or conviction that certain things are true or real.
    a. material, values
    b. nonmaterial, beliefs
    c. nonmaterial, values
    d. material, beliefs

29. Anthropologist George Murdock compiled a list of over seventy _____ -- customs and practices that occur across all societies.
    a. morals
    b. beliefs
    c. cultural universals
    d. symbols

30. All of the following statements regarding cultural universals are true, **except:**
    a. Cultural universals include appearance, activities, social institutions, and customary practices.
    b. Some customs and practices are found in all cultures.
    c. The specific forms of cultural universals vary from one group to another and from one time to another within the same group.
    d. Sociologists are in agreement that cultural universals are the result of functional necessity.

31. A _____ is anything that meaningfully represents something else.
    a. symbol
    b. belief
    c. value
    d. norm

32. The _____ suggests that language not only expresses our thoughts and perceptions but also influences our perception of reality.
    a. Murdock-Williams hypothesis
    b. Sapir-Whorf hypothesis
    c. Ogburn-Mead hypothesis
    d. Sumner-Spencer hypothesis

33. Recent data gathered by the U.S. Census Bureau indicate that approximately _____ percent of the people in this country speak a language other than English at home.
    a. 5
    b. 10
    c. 15
    d. 20

34. _____ theorists view language as a source of power and social control; it perpetuates inequalities between people and between groups because words are used to "keep people in their place."
    a. Symbolic interactionist
    b. Postmodern
    c. Structural functionalist
    d. Conflict

35. Lationo/as in New Mexico and south Texas employ _____, which are sayings that are unique to the Spanish language, as a means of expressing themselves and as a reflection of their cultural heritage.
    a. dichos
    b. machos
    c. notas
    d. memos

36. According to sociologist Robin Williams, _____ is an American core value that emphasizes helpfulness, personal kindness, aid in mass disasters, and organized philanthropy.
    a. equality
    b. achievement and success
    c. democracy
    d. morality and humanitarianism

37. "The United States stands for equal opportunity for all." This statement illustrates _____ culture
    a. ideal
    b. real
    c. diverse
    d. universal

38. Jon, a college student believes in the idea of success, but does not study or attend classes as regularly as he could in order to achieve a high grade point average. His behavior illustrates _____ cultures.
    a. ideal
    b. real
    c. material
    d. universal

39. Chinese Americans in San Francisco, Korean Americans and Puerto Ricans in New York, and Mexican Americans in San Antonio are examples of _____.
    a. countercultures
    b. majority subcultures
    c. ethnic subcultures
    d. popular cultures

40. According to _____ theorists, much of what has been written around the world is Eurocentric.
    a. postmodern
    b. conflict
    c. symbolic interaction
    d. structural functionalism

TRUE/FALSE QUESTIONS

1. Culture is composed of people, whereas a society is composed of ideas.
   T   F

2. Culture is essential for individual, but not as fundamental for the survival of societies.
   T   F

3. Humans have a number of basic instincts.
   T   F

4. Drives are learned, biologically-determined involuntary responses to some physical stimuli.
   T   F

5. The subject matter of jokes is thought to be a cultural universal.
   T   F

6. As a result of their significance, symbols can simultaneously produce loyalty and animosity.
   T   F

7. According to the Sapir-Whorf hypothesis, language shapes the view of reality of its speakers.
T   F

8. Functionalist theorists view language as a source of power and social control that perpetuates inequalities in society.
T   F

9. According to sociologist Robin M. Williams, racism and group superiority is a core value in the United States.
T   F

10. Unlike folkways, mores have strong moral and ethical connotations that may not be violated without serious consequences.
T   F

11. Popular culture is the U.S.A.'s second largest export to other countries.
T   F

12. Fads tend to be longer lasting and more widespread than fashions.
T   F

13. An example of cultural imperialism would be if most world cultures became westernized.
T   F

14. Chinatowns and other ethnic subcultures help first generation immigrants adapt to abrupt cultural change.
T   F

15. When Napoleon Chagnon visited the Yanomamo, he found that they live in a culture very similar to that of the United States.
T   F

16. The term "chair" is considered to be an example of a "genderless" term.
T   F

17. Sanctions are informal norms that may be violated without serious consequences within a particular culture.
T   F

18. According to Box 3.1, giving round-shaped foods to the parents of new babies is considered to be lucky in some cultures.
T   F

19. According to Box 3.3, the world's largest shopping mall is located in the United States.
T   F

20. According to Box 3.4, pesticides are never used in growing organic foods.
T   F

FILL-IN-THE-BLANK QUESTIONS

1. The physical or tangible creations that members of a society make, use, and share are examples of _____.
2. A(n) _____ is anything that meaningfully represents something else.
3. The _____ hypothesis argues that language shapes the view of reality of its speakers.
4. Collective ideas about what is right or wrong, good or bad are known as _____.
5. _____ culture refers to the values that people profess to hold, while _____ culture refers to values and standards that people actually follow.
6. _____ sanctions are rewards for appropriate behavior.
7. Mores so strong that their violation is considered to be extremely offensive are known as _____.
8. Societies wherein people are dissimilar in "social characteristics" such as religion, income, or race/ethnicity are referred to as _____ societies.
9. A group that strongly rejects dominant societal values within a culture are known as a(n) _____.
10. A(n) _____ includes clothing, behavior, thinking, or appearance that is longer lasting and more widespread than a fad.
11. _____ culture consists of classical music, opera, ballet, and live theater usually patronized by the elite.
12. _____ culture in the United States is thought to be "home grown."
13. According to Pierre Bourdieu's _____ theory, high culture is a device used by the dominant class to exclude the subordinate class.
14. A(n) _____ is a category of people who share some distinguishing attribute, belief, values and/or norms that set them apart from the dominant culture.
15. According to Jean Baudrillard, the world of culture is based on _____, not reality itself.

SHORT ANSWER/ESSAY QUESTIONS

1. What are the major types of norms in our society? Give an example of each.
2. Define the concept and then explain how the following could reflect diversity within our society: (1) counterculture; (2) subculture; (3) non-material culture; (4) fads; (5) fashions; and (6) symbols.
3. Explain the relationship between: (a) language and social reality; (b) language and gender; and (c) language, race, and ethnicity.
4. What is culture? How do sociologists distinguish between "culture" and "society"?
5. Which values of those discussed in this chapter seems to be the most important and enduring in the United States today? Which would seem to be the most important in the future?
6. What does the "Lived Experience" introduction at the beginning of this chapter say about the importance of food in becoming "an American"?

STUDENT CLASS PROJECTS AND ACTIVITIES

1. Examine the lyrics to ten popular contemporary songs that have to do with identity and the problems of teenagers in America. Country-western, rock, folk, rap, and Christian rock movement are some possible sources. Write your response to the following: (1) Analyze the content of the lyrics in terms of culture; (2) Examine the norms that the lyrics promote; (3) What are the suggestions made by the lyrics? (4) What are the themes of the lyrics? (5) What type of music accompanies the lyrics, and what are some of the impacts of the musical sounds? Write up your paper according to any given instructions and submit your paper at the appropriate time.

2. Write a paper on this issue: "Are non-industrial cultures an 'endangered species' in the world today?" In this paper, discuss the localizing and globalizing forces of all cultures today. Examine why, in previous obscure cultures, the native cultures appear to be eroding and becoming "acculturated" into the norms of the industrialized world. Is this globalizing force welcomed by the few remaining aboriginal cultures today? Why? Why not? Is it functional for them? Why? Why not? Select a specific culture where this process may or may not be occurring. For example, you could research the Australian aborigine; the Yanomamo; some of the native groups in Venezuela and other South American cultures; the Mayan Indians of Mexico; the Bushmen in Africa; the Zulu in South Africa; the IK in Uganda, Kenya and the Sudan; the Wajos, Bugis, and Makassarese of Sulawesi, the Navaho or Hopi in the United States, etc. Provide bibliographic references for your data and any other information that you may find relevant to your topic. In your paper, provide a conclusion and a personal evaluation of this project. Be sure to submit your paper on the proper due date!

3. Write a paper on one event that is considered to be of the top ten most important cultural events worldwide in the history of humankind. You can choose from one of the top ten of the following list, or you may choose another event, if you prefer. The top ten, according to the staff of TIME MAGAZINE: (1) the Crusades; (2) the signing of the Magna Carta; (3) the Black Death; (4) the Renaissance; (5) the discovery of the Americas; (6) the Reformation; (7) the Industrial Revolution; (8) American and French Revolutions; (9) World War II; and (10) the Rise and Fall of Communism. Explain why you chose a specific event, and fully describe, discuss and evaluate that event. You must provide in your paper a historical background to that specific event. You must include in your discussion why this specific event was most influential in worldwide history. In your writing of this project, you must include bibliographic references and an evaluation of this project. Submit your paper according to specifications of your instructor on the assigned due date.

4. Since the 9-11-01 terrorist attacks in the United States, the word *jihad* has sparked debates all over the world. Research the origin of the word, the definition of the word, and various uses of the word. Explain how some acts of world terrorism are related to the term and specifically how/why terrorism is directed against the United States and the western world. Does the term evoke acts and thoughts of ethnocentrism? Explain. Next, conduct a survey of 20-25 people either on your campus or in your community. Ask these questions: (1) What does *jihad* mean to you? (2) what do you think this word means to most people in the United States? (3) What do you think this word means, or implies, to Muslims in the United States?

To Muslims in the Middle East and in other various parts of the world?
(4) Summarize your results, following any directions given to you by your instructor.

INTERNET ACTIVITIES

1. To read about current issues on postmodern culture, go to the leading electronic journal of interdisciplinary thought on contemporary cultures published by The Johns Hopkins University, **Postmodern Culture** (P M C), **http://www3.iath.virginia.edu/pmc/contents.all.html**, and click on the current issue. Explore the recently published articles. Select one to read, critique it, and share your report in class.
2. *The Voice of the Shuttle*, is one of the most extensive collections of cultural studies resources. Using a reliable search engine, type in the words "The Voice of the Shuttle Cultural Studies." Click on <u>VOS: Cultural Studies</u>. This site designates the intersection between cultural criticism/theory and selective resources in sociology, media studies, postcolonial studies, economics, literature, and other fields chosen to represent the alignments that now signify "culture" for the contemporary humanities. This is a great research site for further information on cultural studies and will link you to a number of sociology sites that might be useful in future research projects.
3. Find out who lives in the *Global Village*: **http://www.empowermentresources.com/info2/theglobalvillage.html**. It answers the question, "What would the world look like if it contained just 1000 people?" The responses are indeed provocative.
4. This is the first site that you should surf if you are interested in **cultural studies**: **http://www.culturalstudies.net/**. It is an academic site where you can find journals, articles, papers, bibliographies, reading lists, theorists and critics. This is a good site for resource material for papers and reports.

INFOTRAC COLLEGE EDITION EXERCISES

Visit the **InfoTrac College Edition** website at: **http://www.wadsworthmedia.com/webtutor/infotrac.htm**. You will arrive at a screen that enables you to search topics.

1. Go to InfoTrac and look up the concept **social norms**. You should first view the encyclopedia excerpt. From the text of the definition, jump to the definition of other related concepts (blue colored links). Follow the related concepts and develop a string of at least five branching off of the social norms definition where you began. Cut and paste these encyclopedia definitions and create a single document that provides a broader perspective on social norms.
2. Discover the meaning of the concept **ethnocentrism** from reading its use in the periodical literature found on InfoTrac. Could you explain why this phenomenon develops in a group of sixth graders?
3. Look up the subject **technology culture** on InfoTrac. Scan several of the articles. Create a list of problems that appear to be related to technological aspects of culture.

SOLUTIONS

MULTIPLE CHOICE QUESTIONS

1. C, p. 73
2. D, p. 74
3. A, p. 78
4. B, p. 79
5. C, p. 80
6. A, p. 83
7. B, p. 83
8. D, p. 84
9. C, p. 86
10. D, p. 86
11. A, p. 86
12. B, p. 87
13. C, p. 87
14. A, p. 90

15. D, p. 92
16. A, p. 92
17. B, p. 93
18. B, p. 95
19. C, p. 95
20. B, p. 99
21. A, p. 75
22. B, p. 75
23. B, p. 75
24. D, p. 76
25. D, p. 76
26. B, p. 77
27. A, p. 78
28. B, p. 78

29. C, p. 78
30. D, p. 79
31. A, p. 79
32. B, p. 80
33. D, p. 83
34. D, p. 83
35. A, p. 83
36. D, p. 84
37. A, p. 85
38. B, p. 85
39. C, p. 91
40. A, p. 99

TRUE/FALSE QUESTIONS

1. F, p. 73
2. F, p. 75
3. F, p. 75
4. F, p. 75
5. F, p. 78
6. T, p. 79
7. T, p. 80

8. F, p. 83
9. T, p. 85
10. T, p. 86
11. T, p. 94
12. F, p. 94
13. T, p. 94
14. T, p. 91

15. F, p. 92
16. T, p. 80
17. F, p. 86
18. T, p. 76
19. F, p. 96
20. F, p. 98

FILL-IN-THE-BLANK QUESTIONS

1. material culture, p. 76
2. symbol, p. 79
3. Sapir-Whorf, p. 80
4. values, p. 84
5. ideal, p. 85; real, p. 85
6. positive, p. 86
7. taboos, p. 86
8. heterogeneous, p. 88
9. counterculture, p. 91
10. fashion, p. 94
11. high, p. 93
12. popular, p. 93
13. cultural capital, p. 93
14. subculture, p. 90
15. simulation, p. 99

4
SOCIALIZATION

BRIEF CHAPTER OUTLINE

CHAPTER SUMMARY

Socialization is the lifelong process through which individuals acquire a self identity and the physical, mental, and social skills needed for survival in society. Socialization is essential for the individual's survival and for human development; it also is essential for the survival and stability of society. People are a product of two forces: heredity and social environment. Most sociologists agree that while biology dictates our physical makeup, the social environment largely determines how we develop and behave.

Humans need social contact to develop properly. Cases of isolated children have shown that people who are isolated during their formative years fail to develop their full emotional and intellectual capacities and that social contact is essential in developing a self. A variety of psychological and sociological theories have been developed to explain child abuse but also to describe how a positive process of socialization occurs. Psychological theories focus primarily on how the individual personality develops. Sigmund Freud developed the psychoanalytic perspective, while Erik Erikson, drawing from Freud's theory, identified eight psychosocial stages of development. Jean Piaget pioneered the field of cognitive development, which emphasizes the intellectual development of children. The stages of moral development were developed by Lawrence Kohlberg and later criticized and corrected by Carol Gilligan. Sociological perspectives examine how people develop an awareness of self and learn about their culture. This **self-concept** is not present at birth; it arises in the process of social experience. Charles Horton Cooley developed the image of the **looking-glass self** to explain how people see themselves through the perceptions of others. George Herbert Mead linked the idea of self concept to **role-taking** and to learning the rules of social interaction. Ecological perspectives emphasize cultural or environmental influences on human development. When children do not have a positive environment in which to develop a self concept, it becomes difficult to form a healthy social self. According to sociologists, **agents of socialization**—including families, schools, peer groups, and mass media—teach us what we need to know in order to participate in society. Gender, racial-ethnic, and social class are all determining factors in the life long socialization process. **Anticipatory socialization**—the process by which knowledge and skills are learned for future roles—often occurs before achieving a new status. **Resocialization**—the process of learning new attitudes, values, and behaviors, either voluntarily or involuntarily—sometimes takes place in **total institutions**. In today's global society, because of the rapid pace of technological change, we must learn how to anticipate and consider the consequences of the future.

LEARNING OBJECTIVES

After reading Chapter 4, you should be able to:

1. Define socialization and explain why this process is essential for the individual and society.
2. Explain Freud's views on the conflict between individual desires and the demands of society,
3. Present the stages of psychosocial development proposed by Erikson.
4. Outline the stages of cognitive development as set forth by Piaget.
5. State the major agents of socialization and describe their effects on children's development.
6. Describe Mead's concept of the generalized other and explain socialization as an interactive process.
7. Describe the major strengths and weaknesses of Erikson's developmental theory.
8. Compare and contrast the moral development theories of Kohlberg and Gilligan.

9. Explain what is meant by gender socialization and racial socialization.
10. Demonstrate some of the symbolic uses of the cell phone among teenagers.
11. Describe the process of resocialization and explain why it often takes place in a total institution.
12. Distinguish between sociological and sociobiological perspectives on the development of human behavior.
13. Outline the stages of the life course and explain how each stage varies based on gender, race/ethnicity, class, and positive or negative treatment.
14. Compare the key components of the human development theories of Cooley and Mead.
15. Explain why cases of isolated children are important to understanding of the socialization process.
16. Evaluate the contribution of Cooley and Mead to our understanding of the socialization process

KEY TERMS

(defined at page number shown and in glossary)

KEY PEOPLE

(identified at page number shown)

CHAPTER OUTLINE

I. WHY IS SOCIALIZATION IMPORTANT AROUND THE GLOBE?
   A. **Socialization** is the lifelong process of social interaction through which individuals acquire a self-identity and the physical, mental, and social skills needed for survival in society.
   B. Human Development: Biology and Society
      1. Every human being is a product of biology, society, and personal experiences, or heredity and environment.
      2. **Sociobiology** is the systematic study of how biology affects social behavior.
   C. Problems Associated with Social Isolation and Maltreatment
      1. Social environment is a crucial part of an individual's socialization; people need social contact with others in order to develop properly.
      2. Researchers have attempted to demonstrate the effects of social isolation on nonhuman primates that are raised without contact with others of their own species.
      3. Isolated children illustrate the importance of socialization.
      4. The most frequent form of child maltreatment is child neglect.
II. SOCIAL PSYCHOLOGICAL THEORIES OF HUMAN DEVELOPMENT
   A. Freud's Psychoanalytic Perspective suggested that human behavior and personality originate from unconscious forces within individuals.
      1. The **id** is the component of personality that includes all of the individual's basic biological drives and needs that demand immediate gratification.
      2. The **ego** is the rational, reality oriented component of personality that imposes restrictions on the innate pleasure seeking drives of the id.
      3. The **superego** consists of the moral and ethical aspects of personality. When a person is well-adjusted, the ego successfully manages the opposing forces of the id and the superego.
   B. **Erik H. Erikson** identified eight psychosocial stages of development, each of which is accompanied by a crisis or potential crisis that involves transitions in social relationships:
      1. Trust versus Mistrust (birth to age 1).
      2. Autonomy versus Shame and Doubt (1-3 years).
      3. Initiative versus Guilt (3-5 years).
      4. Industry versus Inferiority (6-11 years).
      5. Identity versus Role Confusion (12-18 years).
      6. Intimacy versus Isolation (18-35 years).
      7. Generativity versus Self absorption (35-55 years).
      8. Integrity versus Despair (maturity and old age).
   C. **Jean Piaget**'s theory of cognitive development relates to changes over time in how people think. Piaget was interested in how children obtain, process, and use information. He believed that children experience four stages of cognitive development:
      1. *Sensorimotor Stage* (birth to age 2) – children understand the world only though sensory contact and immediate action because they cannot engage in symbolic thought or use language.

2. *Preoperational Stage* (ages 2-7) – children begin to use words as mental symbols and to develop the ability to use mental images.
3. *Concrete Operational Stage* (ages 7-11) – children think in terms of tangible objects and actual events; they also can draw conclusions about the likely physical consequences of an action without always having to try it out.
4. *Formal Operational Stage* (age 12 through adolescence) – adolescents are able to engage in highly abstract thought and understand places, things, and events they have never seen. Beyond this point, changes in thinking are a matter of changes in degree rather than in the nature of their thinking.

D. **Lawrence Kohlberg** classified moral reasoning into three sequential levels:
1. *Preconventional Level* (ages 7-10) – children give little consideration to the views of others.
2. *Conventional Level* (age 10 through adulthood) – children initially believe that behavior is right if it receives wide approval from significant others, including peers, and then a law and order orientation, based on how one conforms to rules and laws.
3. *Postconventional Level* (few adults reach this stage) – people view morality in terms of individual rights. At the final stage of moral development, "moral conduct" is judged by principles based on human rights that transcend government and laws.

E. Gilligan's View on Gender and Moral Development
1. One of the major critics of Kohlberg's work was psychologist **Carol Gilligan**, who noted that Kohlberg's model was based solely on male responses.
2. She noted that, because of differences in socialization and life experiences, men and women approach moral issues differently.
3. Her research addressed male bias in Kolberg's research.
4. Her research identified three states in female development.

III. SOCIOLOGICAL THEORIES OF HUMAN DEVELOPMENT
A. Cooley, Mead, and Symbolic Interactionist Perspectives
1. Without social contact, we cannot form a **self-concept** – the totality of our beliefs and feelings about ourselves.
2. According to **Charles Horton Cooley's looking-glass self**, a person's sense of self is derived from the perceptions of others through a three step process:
   a. We imagine how our personality and appearance will look to other people.
   b. We imagine how other people judge the appearance and personality that we think we present.
   c. We develop a self concept.
3. **George Herbert Mead** linked the idea of self concept to **role-taking** – the process by which a person mentally assumes the role of another person in order to understand the world from that person's point of view.
   a. **Significant others** are those persons whose care, affection, and approval are especially desired and who are most important in the development of the self; these individuals are extremely important in the socialization process.

b. Mead divided the self into the "I" – the subjective element of the self that represents the spontaneous and unique traits of each person – and the "me" – the objective element of the self, which is composed of the internalized attitudes and demands of other members of society and the individual's awareness of those demands.

c. Mead outlined three stages of self development:
   i. *preparatory stage* – children largely imitate the people around them;
   ii. *play stage* (from about age 3 to 5) – children learn to use language and other symbols, thus making it possible for them to pretend to take the roles of specific people;
   iii. *game stage* – children understand not only their own social position but also the positions of others around them. At this time, the child develops a **generalized other** – an awareness of the demands and expectations of the society as a whole or of the child's subculture.

4. Some recent symbolic interactionist research emphasize that socialization is a collective process in which *children* are active and creative agents, not passive recipients in the socialization process.
   a. Children are capable of actively constructing their own shared meanings as they acquire language skills and accumulate interactive experiences.
   b. Some researchers believe that the peer group is the most significant public realm for children.

5. Ecological perspectives emphasize cultural or environmental influences on human development.
   a. *Ecological systems* theory focuses on the interactions a child has with other people and situations.
   b. The ecological systems are as follows: (1) the *microsystem*, (2) the *mesosystem*, (3) the *exosystem*, and (4) the *macrosystem*.

IV. AGENTS OF SOCIALIZATION
   A. **Agents of socialization** are the persons, groups, or institutions that teach us what we need to know in order to participate in society. These are the most pervasive agents of socialization in childhood.
   B. The family is the most important agent of socialization in all societies.
      1. Functionalists emphasize that families are the primary agents for the procreation and socialization of children, as well as the primary source of emotional support.
      2. Conflict theorists stress that socialization reproduces the class structure in the next generation.
      3. The social construction/symbolic interactionist perspective suggests that children affect their parents' lives and change the overall household environment.
   C. The school has played an increasingly important role in the socialization process as the amount of specialized technical and scientific knowledge has expanded rapidly.
      1. Schools teach specific knowledge and skills; they also have a profound effect on a child's self image, beliefs, and values.

2. From a functionalist perspective, schools are responsible for: (1) socialization – teaching students to be productive members of society; (2) transmission of culture; (3) social control and personal development; and (4) the selection, training, and placement of individuals on different rungs in the society.

3. According to conflict theorists such as Samuel Bowles and Herbert Gintis, much of what happens in school amounts to a hidden curriculum – the process by which children from working-class and lower-income families learn to be neat, to be on time, to be quiet, to wait their turn, and to remain attentive to their work – attributes that are important for later roles in the work force.

4. Symbolic interactionists might focus on how daily interactions and practices in school affect the construction of students' beliefs, practices, and values.

D. A **peer group** is a group of people who are linked by common interests, equal social position, and (usually) similar age.

1. Peer groups function as agents of socialization by contributing to our sense of "belonging" and our feelings of self worth.

2. Individuals must earn their acceptance with their peers by meeting the group's demands for a high level of conformity to its own norms, attitudes, speech, and dress codes.

E. The **mass media** is an agent of socialization that has a profound impact on both children and adults.

1. The media function as socializing agents in several ways: (1) they inform us about events; (2) they introduce us to a wide variety of people; (3) they provide an array of viewpoints on current issues; (4) they make us aware of products and services that, if we purchase them, supposedly will help us to be accepted by others; and (5) they entertain us by providing the opportunity to live vicariously (through other people's experiences).

2. Recent studies have shown that, on average, U.S. children spend more time per year in front of television sets, computers, and video games than they spend per year in school.

a. Television has been praised for offering children numerous positive, educational, and prosocial experiences; but blamed for children having lower grades in school, reading fewer books, exercising less and becoming overweight.

b. The influence of the Internet on young people continues to be a concern; young people make up the fastest- growing segment of Internet users.

c. Cultural studies scholars and some postmodern theorists believe that "media culture" has in recent years dramatically changed the sociological process for very young children.

V. GENDER AND RACIAL-ETHNIC SOCIALIZATION

A. **Gender socialization** is the aspect of socialization that contains specific messages and practices concerning the nature of being female or male in a specific group or society.

1. Families, schools, and sports tend to reinforce traditional roles through gender socialization.

2. From an early age, media, including children's books, television programs, movies, and music provide subtle and not-so-subtle messages about masculine and feminine behavior.

B. **Racial socialization** is the aspect of socialization that contains specific messages and practices concerning the nature of one's racial or ethnic status as it relates to: (1) personal and group identity; (2) intergroup and interindividual relationships, and (3) position in the social hierarchy.

VI. SOCIALIZATION THROUGH THE LIFE COURSE

A. Socialization is a lifelong process: each time we experience a change in status, we learn a new set of rules, roles, and relationships.

1. Even before we enter a new status, we often participate in **anticipatory socialization** – the process by which knowledge and skills are learned for future roles.

2. The most common categories of age are infancy, childhood, adolescence, and adulthood (often subdivided into young adulthood, middle adulthood, and older adulthood).

B. During infancy and early childhood, family support and guidance are crucial to a child's developing self concept; however, some families reflect the discrepancy between cultural ideals and reality – children grow up in a setting characterized by fear, danger, and risks that are created by parental neglect, emotional maltreatment, or premature economic and sexual demands.

C. Anticipatory socialization for adult roles is often associated with adolescence; however, some young people may plunge into adult responsibilities at this time.

D. In early adulthood (usually until about age forty), people work toward their own goals of creating meaningful relationships with others, finding employment, and seeking personal fulfillment. Wilbert Moore divided workplace, or occupational, socialization into four phases:

1. career choice
2. anticipatory socialization
3. conditioning and commitment
4. continuous commitment

E. Between the ages of 40 and 65, people enter middle adulthood, and many begin to compare their accomplishments with their earlier expectations.

F. In older adulthood, some people are quite happy and content; others are not. Difficult changes in adult attitudes and behavior may occur in the last years of life when people experience decreased physical ability and **social devaluation** – when a person or group is considered to have less social value than other groups.

G. It is important to note that everyone does not go through certain passages or stages of a life course at the same age and that life course patterns are strongly influenced by race, ethnicity, gender, and social class.

VII. RESOCIALIZATION

A. **Resocialization** is the process of learning a new and different set of attitudes, values, and behaviors from those in one's previous background and experience.

1. *Voluntary resocialization* occurs when we enter a new status of our own free will (e.g., medical or psychological treatment or religious conversion).

2. *Involuntary resocialization* occurs against a person's wishes and generally takes place within a **total institution** – a place where people are isolated from the rest of society for a set period of time and come under the control of the officials who run the institution. Examples include military boot camps, prisons, concentration camps, and some mental hospitals.

VIII. SOCIALIZATION IN THE FUTURE
    A. Families are likely to remain the institution that most fundamentally shapes and nurtures personal values and self identity.
    B. However, parents increasingly may feel overburdened by this responsibility, especially without societal support – such as high quality, affordable child care – and more education in parenting skills.
    C. A central issue facing parents and teachers as they socialize children is the growing dominance of the media and other forms of technology.
    D. With the rapid pace of technological change in the socialization process, we must anticipate – and consider the consequences of – the future.

ANALYZING AND UNDERSTANDING THE BOXES

After reading the chapter and studying the outline, re-read the boxes and write down key points and possible questions for class discussion.

Sociology and Everyday Life: How Much Do You Know About Early Socialization and Child Care?

Key Points:

Discussion Questions:

1.

2.

3.

Sociology and Social Policy: Who Should Pay for Child Care?

Key Points:

Discussion Questions:

1.

2..

3.

Sociology in Global Perspective: The Youthful Cry Heard Around the world: "Everybody Else Has a Cell Phone! Why Can't I Have One?"

Key Points:

Discussion Questions:

1.

2.

3.

You Can Make a Difference: Helping a Child Reach Adulthood

Key Points:

Discussion Questions:

1.

2.

3.

PRACTICE TESTS

MULTIPLE CHOICE QUESTIONS

Select the response that best answers the question or completes the statement.

1. _____ is the lifelong process of social interaction through which individuals acquire a self identity.
   a. Human development
   b. Socialization
   c. Behavior modification
   d. Imitation

2. The systematic study of how biology affects social behavior is known as:
   a. sociophysiology
   b. sociobiology
   c. sociology
   d. social psychology

3. Harry and Margaret Harlow's experiment with rhesus monkeys demonstrated that
   a. food was more important to the monkeys than warmth, affection, and physical comfort.
   b. monkeys cannot distinguish between a nonliving "mother substitute" and their own mother.
   c. socialization is not important to rhesus monkeys; their behavior is purely instinctive.
   d. without socialization, young monkeys do not learn normal social or emotional behavior.

4. In discussing child maltreatment, the text points out that:
   a. many types of neglect constitute child maltreatment.
   b. the extent of this problem has been exaggerated by the media.
   c. most sexual abuse perpetrators are punished by imprisonment.
   d. most child maltreatment occurs in families living below the poverty line.

5. According to Sigmund Freud, the _____ consists of the moral and ethical aspects of personality.
   a. id
   b. ego
   c. superego
   d. libido

6. Which of the following is not one of Jean Piaget's stages of cognitive development?
   a. preoperational
   b. concrete operational
   c. formal operational
   d. post operational

7. The stages of moral development were initially set forth by _____ and then criticized by _____.
    a. Charles Horton Cooley, George Herbert Mead
    b. Lawrence Kohlberg, Carol Gilligan
    c. Jean Piaget, Lawrence Kohlberg
    d. Erik Erikson, Jean Piaget

8. All of the following are components of the self concept, **except**:
    a. the physical self
    b. the active self
    c. the functional self
    d. the psychological self

9. The theories of Charles Horton Cooley and George Herbert Mead can best be classified as _____ perspectives.
    a. symbolic interactionist
    b. functionalist
    c. conflict
    d. feminist

10. According to Charles Horton Cooley, we base our perception of who we are on how we think other people see us and on whether this seems good or bad to us. He referred to this perspective as the:
    a. self-fulfilling prophecy
    b. generalized other
    c. looking-glass self
    d. significant other

11. All of the following are stages in Mead's theory of self development, **except** the
    a. anticipatory stage.
    b. game stage.
    c. play stage.
    d. preparatory stage.

12. The _____ refers to the child's awareness of the demands and expectations of society as a whole or of the child's subculture.
    a. looking-glass self
    b. id
    c. ego
    d. generalized other

13. All of the following are components of the ecological systems **except**:
    a. ecosystem
    b. microsystem
    c. mesosystem
    d. exosystem

14. Sociologist Melvin Kohn has suggested that _____ is one of the strongest influences on what and how parents teach their children.
    a. race/ethnicity
    b. religion
    c. social class
    d. age

15. Currently, about _____ percent of all U.S. preschool children are in day care of one kind or another.
    a. 15
    b. 25
    c. 60
    d. 75

16. As agents of socialization, peer groups are thought to "pressure" children and adolescents because
    a. individualism is encouraged and rewarded in these groups.
    b. individuals must earn their acceptance with their peers by conforming to the group's norms.
    c. individuals are encouraged to put friendship above material possessions.
    d. individuals are discouraged from making long term friends in the peer group.

17. According to Patricia Hill Collins, _____ play an important role in the gender socialization and motivation of African-American children, especially girls.
    a. "other mothers"
    b. substitute mothers
    c. aunts
    d. siblings

18. The process by which knowledge and skills are learned for future roles is known as:
    a. resocialization
    b. anticipatory socialization
    c. cybersocialization
    d. expectant socialization

19. Social devaluation is most likely to be experienced during this stage of the life course:
    a. infancy and childhood
    b. adolescence
    c. early adulthood
    d. older adulthood

20. All of the following are examples of voluntary resocialization, **except**:
    a. becoming a student
    b. going to prison
    c. becoming a Buddhist
    d. joining Alcoholics Anonymous

21. The example of actress Drew Barrymore's description of childhood maltreatment in her family is used at the beginning of the chapter on socialization to illustrate that
    a. large numbers of children experience maltreatment at the hands of family members.
    b. large numbers of children experience maltreatment at the hands of caregivers.
    c. maltreatment is of interest to sociologists because it impacts a child's social growth, behavior, and self-image
    d. all of the above

22. All of the following statements regarding socialization are TRUE, **except:**
    a. Socialization is essential for the individual's survival and for human development.
    b. Socialization is essential for the survival and stability of society.
    c. Socialization enables a society to "reproduce" itself by passing on cultural content from one generation to the next.
    d. Socialization is a learning process that takes place primarily during childhood.

23. _____ is considered to be the pioneer of sociobiology.
    a. Sigmund Freud
    b. E.O. Wilson
    c. Kingsley David
    d. Urie Bronfenbrenner

24. Sociologist Kingsley David was interested in the case of Anna, a child who was kept in an attic-like room, because
    a. he was studying parental attitudes of young, unmarried mothers.
    b. he wanted to know more about what happens when a child is raised in isolation.
    c. he was attempting to determine how children develop a generalized other.
    d. he was examining the relationship between sexual motives and human behavior.

25. The case of Genie, an isolated child, illustrates that
    a. with proper therapy, children who have been isolated can become a part of the mainstream.
    b. children who have experienced extreme isolation do not live long enough to reach adulthood.
    c. children who experience social isolation and neglect may be defined as "retarded" when they reach adulthood.
    d. isolated children can recover from any physical damages.

26. Drawing from Freud's theory, _____ identified eight psychosocial stages of development, reason that each stage is accompanied by a crisis or potential crisis that involves transitions in social relationships.
    a. Erik H. Erikson
    b. Charles Horton Cooley
    c. Jean Piaget
    d. George Herbert Mead

27. In Erik Erikson's theory, during the _____ state of psychosocial development, children gain a feeling of control over their behavior, develop a variety of physical and mental abilities, and begin to assert their independence.
    a. trust versus mistrust
    b. autonomy versus shame and doubt
    c. initiative versus guilt
    d. industry versus inferiority

28. Swiss psychologist Jean Piaget was a pioneer in the field of _____, which relates to changes over time in how people think.
    a. psychoanalysis
    b. psychosocial development
    c. cognitive development
    d. interactionism

29. According to Carol Gilligan, the key weakness of Lawrence Kohlberg's work is that it
    a. underestimates human potential for immorality.
    b. was based on only male subjects.
    c. ignores key social psychological insights.
    d. overemphasizes the subconscious mind.

30. The sum total of perceptions and feelings that an individual has of being a distinct, unique person represents the:
    a. personality
    b. psyche
    c. self
    d. individual orientation

31. According to the text's discussion on the looking-glass self,
    a. our looking-glass self is who we actually are.
    b. is based upon our own perception of self.
    c. is based on how we think people see us.
    d. is based on our correct perceptions.

32. The "I" is the _____ element of self.
    a. subjective
    b. reflective
    c. objective
    d. conjective

33. _____ refers to the process by which a person mentally assumes the role of another person in order to understand the world from that person's point of view.
   a. Role exploration
   b. Role assumption
   c. Role-searching
   d. Role-taking

34. Urie Bronfenbrenner's _____ theory focuses on the overall context in which child development occurs.
   a. conflict
   b. postmodern
   c. ecological systems
   d. symbolic interaction

35. According to sociologist Melvin Kohn, one of the strongest influences on <u>what</u> and <u>how</u> parents teach their children is
   a. social class.
   b. religious background.
   c. racial and/or ethnic background.
   d. the media.

36. According to George Herbert Mead, when children between the ages of 3 and 5 play "house" or "firefighter," they are in the _____ stage of self-development.
   a. game
   b. play
   c. preparatory
   d. "me"

37. According to Mead, "selves"
   a. are present at birth.
   b. reflect only significant others.
   c. can never change.
   d. can only exist in relation to other selves.

38. Research on the influence of television on children has concluded that children who spend considerable time watching television often
   a. have lower grades in school.
   b. read fewer books.
   c. exercise less.
   d. all of the above

39. Sociologists use the concept of _____ to refer to the aspect of socialization that contains specific messages and practices concerning the nature of being female or male in a specific group or society.
   a. sex role socialization
   b. gender socialization
   c. feminism and masculine
   d. peer group socialization

40. The most important aspects of one's racial identity are
    a. influenced by the media.
    b. influenced by one's peers.
    c. influenced by one's neighbors.
    d. influenced by one's family.

TRUE/FALSE QUESTIONS

1. Around the globe, the process of socialization is essential for the survival of both the society and the individual.
   T   F

2. After-school programs have greatly reduced the number of children who are home alone after school.
   T   F

3. Unlike humans, nonhuman primates such as monkeys and chimpanzees do not need social contact with others of their species in order to develop properly.
   T   F

4. Extensive therapy used in a program to socialize Genie proved to be highly successful.
   T   F

5. The cases of "Anna" and "Genie" make us aware of the importance of the socialization process because they show the detrimental effects of social isolation and neglect.
   T   F

6. Sigmund Freud theorized that our personalities are largely unconscious – hidden away outside our normal awareness.
   T   F

7. Erik Erikson suggested that if infants receive good care and nurturing from their parents, they will develop a sense of trust that is important to their future psychosocial development.
   T   F

8. According to George H. Mead, our looking-glass self is based on our perception of how other people think of us.
   T   F

9. According to Cooley, the self develops only through contact with others.
   T   F

10. The family is the most important agent of socialization in all societies.
    T   F

11. Conflict theorists assert that students have different experiences in the school system, depending on their class, racial ethnic background, gender, and the neighborhood in which they live.
   T   F

12. Peer pressure does not affect preschool children.
   T   F

13. Gender socialization contributes to people's beliefs about what the "preferred" sex of a child should be and in influencing our beliefs about acceptable behaviors for males and females.
   T   F

14. Scholars have found that ethnic values begin to crystallize among children at about the age of adolescence.
   T   F

15. Occupational socialization tends to be most intense immediately after a person makes the transition from school to the workplace.
   T   F

16. Military boot camps, jails and prisons are all examples of total institutions.
   T   F

17. According to Box 4.1, all states have reporting requirements for child maltreatment.
   T   F

18. According to Box 4.2, paying for child care is a large concern in most other high-income countries, as in the United States.
   T   F

19. Cell phones, according to Box 4.3, have become a major socializing agent for teenagers.
   T   F

20. If someone has reason to believe that child maltreatment is occurring, the first step that should be taken is to report it to authorities, according to Box 4.4.
   T   F

FILL-IN-THE-BLANK QUESTIONS

1. The lifelong process of social interaction through which individuals acquire a self-identity is defined as _____.

2. The systematic study of how biology affects social behavior is _____.

3. According to Freud, the _____ consists of the moral and ethical aspects of personality.

4. The _____ is the totality of our beliefs and feelings about ourselves.

5. According to Cooley, the _____ refers to the way in which a person's sense of self is derived from the perceptions of others.

6. According to Mead, the _____ refers to the child's awareness of the demands and expressions of the society as a whole or of the child's subculture.

7. A _____ group is a group of people who are linked by common interest, equal social position and similar age.

8. Gender _____ is important in determining what we think the "preferred" sex of a child should be and our ideas about acceptable behaviors for females and males.

9. The process by which knowledge and skills are learned for future roles is termed _____.

10. The process wherein a person or group is considered to have less social value than other persons or groups is known as _____.

11. _____ are the persons, groups, or institutions that teach us what we need to know in order to participate in society.

12. A(n) _____ is a place where people are isolated from the rest of society for a set period of time and are under the control of the officials who run the institution.

13. _____ are those persons whose care, affection and approval are especially desired and who are most important in the development of the self.

14. _____ is the aspect of socialization that contains specific messages and practices concerning the nature of one's racial or ethnic status.

15. _____ is the process of learning a new and different set of attitudes, values and behaviors from those in one's background and previous experiences.

SHORT ANSWER/ESSAY QUESTIONS

1. What do the cases of isolated children suggest about socialization and human development?
2. What are the differences between the sociological theories of human development and the psychological theories of human development?
3. How do we develop a self-concept according to sociologists?
4. Why, according to Anne Kasper, are Mead's ideas about the social self more applicable to men than women?
5. What is Lawrence Kohlberg's perspective on moral development and what is Carol Gilligan's criticism of his perspective?

STUDENT CLASS PROJECTS AND ACTIVITIES

1. Think back to your childhood and list the 4 or 5 televisions programs that you watched most often. Next, list the 4 or 5 programs you watch most often today. Respond in writing to these questions: (1) What values were demonstrated in each of the respected programs–yesteryear and today? (2) Were there, or are there, any stereotypes presented via the programs? If so, what were/are they? (3) Were there/are there any values presented that contradict your parents' values? If so, what were/are they? (4) How do you personally evaluate each of these television programs? Include any other information pertinent to this project. Submit your paper on the due date, following any other specific instructions provided by your professor.

2. Write a five to ten page biography of your life, with special emphasis on your own socialization processes. Include responses to the following questions: (1) Identify the major agencies of socialization in your life; (2) Who, specifically (by name and status) were the "significant others" in your life? (3) What type of values did your significant others advocate? (4) Who were/are the generalized others in your life? Why? What impact did they have? (5) In what way did any of the above shape your self-image and goals? Provide any additional information in the writing of your biography. Evaluate this project and submit your paper at the appropriate time, following any specific instructions provided by your instructor.

INTERNET ACTIVITIES

1. To read and gain more insight on two of the most influential sociologists in their concepts of "the self"—Charles Horton Cooley's concept of the "looking-glass self" and George Herbert Mead's concept of the "genesis of self," and "The 'I' and the 'Me'—use this search engine: **http://www.google.com**. Type in the name of Charles Horton Cooley first; select the section on the collection of his articles and books. Note in Cooley's own words, "the looking-glass self" theory. Then, select the section, Biography of Charles Horton Cooley in order to more fully understand the major societal forces which shaped his life. Next, using the same search engine, type in the name, George Herbert Mead. Surf the many sites available on Mead, ranging from the various biographies to his major writings. Note when and how his interest in the works of Cooley and John Dewey influenced the ideas and writing of Mead. Click on the section listing MEAD'S collection of articles and read about his concept of the "genesis of self." Note, also, Mead's theory of role-taking, developmental and process by which the self is constructed and refined.

2. Michael Kearl's fabulous **sociology web gateway**, **http://www.trinity.edu/mkearl/fam-kids.html**, addresses parents and children. There are connections to pages that directly address socialization. This is an excellent resource when you are looking for more information for a paper/presentation or want to bring additional facts to a class discussion.

3. An enormous sociological gateway site, the **SOCIOSITE**, **http://www.pscw.uva.nl/sociosite/TOPICS/index.html**, has a topical index that will introduce you to dozens of subjects. Once you access this address, use the topical index and search for sites in the social psychology category. Most of the websites in this category are actually resource or gateway sites. You can construct a *Social Psychology Web Site Fact Sheet*. Related web sites can be listed, categorized, and summarized. Pay special attention to those sites related to aspects of socialization.

4. This web site, **http://www.public.asu.edu/~zeyno217/365/notes1.html**, contains information about the mass media and society. It is a concise presentation of some of the most important concepts. Read the section on "Mass media and Socialization." How does the media influence the "Sociological Imagination"? Why is "television one of the primary socializing agents of today's society?"

INFOTRAC COLLEGE EDITION EXERCISES

Visit the **InfoTrac College Edition** website at: **http://www.wadsworthmedia.com/webtutor/infotrac.htm**. You will arrive at a screen that enables you to search topics.

1. Bring to class an article that addresses a subcategory of socialization.

Gender	Racial/Ethnic	Cross-cultural	Family	School
Media	Life Course	Peers	Self concept	Role taking

2. You can use InfoTrac to conduct a search about **Carol Gilligan**, a pioneer in the field of feminist-oriented developmental psychology. Articles include several reviews of her book, *Between Voice and Silence: Women and Girls, Race and Relationship*. Compare the reviews from a number of divergent sources.

3. Use InfoTrac to locate biographical sketches of the major psychologists and sociologists introduced in this chapter. InfoTrac provides access to encyclopedia excerpts containing biographical information. Put together a one-page sketch of each theorist. Portraits can be exported from the Internet and inserted in the document. If you are taking psychology and sociology courses in the future, resources such as these might prove useful.

SOLUTIONS

MULTIPLE CHOICE QUESTIONS

1. B, p. 106	15. C. p. 122	29. B, p. 115
2. B, p. 107	16. B, p. 123	30. C, p. 115
3. D, p. 108	17. A, p. 128	31. C, p. 116
4. A, p. 110	18. B, p. 129	32. A, p. 117
5. C, p. 111	19. D, p. 131	33. D, p. 117
6. D, pp. 113-114	20. B, p. 132	34. C, p. 120
7. B, p. 115	21. D, p. 105	35. A, p. 121
8. C, p. 115	22. D, p, 106	36. B, p. 116
9. A, p. 116	23. B, p. 107	37. D, p. 119
10. C, p. 116	24. B, p.109	38. D, p. 120
11. A, p. 118	25. C, p. 110	39. B, p. 128
12. D, p. 119	26. A, pp. 111-112	40. D, p. 128
13. A, p. 120	27. B, p. 112	
14. C, p. 121	28. C, p. 113	

TRUE/FALSE QUESTIONS

1. T, p. 106	8. F, p. 116	15. T, p. 131
2. F, pp. 107-108	9. T, p. 116	16. T, p. 132
3. F, p. 108	10. T, p. 120	17. T, p. 108
4. F, p. 110	11. T, p. 122	18. F, p. 124
5. T, pp. 109-110	12. F, p. 123	19. T, pp. 126-127
6. T, p. 111	13. T, p. 128	20. T, p. 123
7. T, p. 112	14. F, p. 128	

FILL-IN-THE-BLANK QUESTIONS

1. socialization, p. 106
2. sociobiology, p. 107
3. superego, p. 111
4. self-concept, p. 115
5. looking-glass self, p. 117
6. generalized other, p. 119
7. peer, p. 123
8. socialization, p. 107
9. anticipatory socialization, p. 129
10. social devaluation, p. 131
11. agents of socialization, p. 121
12. total institution, p. 132
13. significant others, p. 117
14. racial socialization, p. 129
15. resocialization, p. 131

5

SOCIETY, SOCIAL STRUCTURE, AND INTERACTION

BRIEF CHAPTER OUTLINE

CHAPTER SUMMARY

Social structure and interaction are critical components of everyday life. At the microlevel, **social interaction** – the process by which people act toward or respond to other people – is the foundation of meaningful relationships in society. At the macrolevel, **social structure** is the stable pattern of social relationships that exist within a particular group or society. This structure includes **status**, **roles**, **groups**, and **social institutions**. Social researchers have identified five types of societies based on various levels of subsistence technology: **hunting** and **gathering**, **horticultural** and **pastoral**, **agrarian**, **industrial**, and **postindustrial** societies. Changes in social structure may dramatically affect individuals and groups, as demonstrated by Durkheim's concepts of **mechanical** and **organic solidarity** and Tönnies' *gemeinschaft* and *gesellschaft*. In examining the relationship between social structure and homelessness, several suggestions for homelessness are offered; however the answers derived are often based upon perceptions of reality in a society. Social interaction within a society is guided by certain shared meanings across situations. The microlevel perspective focuses on the social interactions among individuals, especially face-to-face encounters. Not everyone interprets social interaction situations in the same way. Race/ethnicity, gender, and social class often influence perceptions of meaning. The **social construction of reality** refers to the process by which our perception of reality is shaped by the subjective meaning we give to an experience. **Ethnomethodology** is the study of the commonsense knowledge that people use to understand the situations in which they find themselves. **Dramaturgical analysis** is the study of social interaction that compares everyday life to a theatrical presentation. **Presentation of self** refers to efforts to present our own self to others in ways that are most favorable to our own interests or image. Feeling rules shape the appropriate emotions for a given role or specific situation. Social interaction also is marked by **nonverbal communication**, which is the transfer of information between people without the use of speech. In the future, macrolevel and microlevel analyses are essential in determining how our social structures should be shaped so that they can respond to pressing social needs.

LEARNING OBJECTIVES

After reading Chapter 5, you should be able to:

1. State the definition of social structure and explain why it is important for individuals and society.

2. State the definition of status and distinguish between ascribed and achieved statuses.

3. Define role expectation, role performance, role conflict, and role strain, and give an example of each.

4. Identify and describe the five major types of societies.

5. Describe Goffman's dramaturgical analysis and explain what he meant by presentation of self.

6. Explain what is meant by the sociology of emotions and describe sociologist Arlie Hochschild's contribution to this area of study. Explain what is meant by master status and give three examples.

7. Define formal organization and explain why many contemporary organizations are known as "people-processing" organizations.

8. Define nonverbal communication and personal space and explain how these concepts relate to our interactions with others.

9. Describe the process of role exiting.

10. Explain the difference in primary and secondary groups.

11. State the definition for social institution and name the major institutions found in contemporary society.

12. Compare Emile Durkheim's typology of mechanical and organic solidarity with Ferdinand Tönnies' *gemeinschaft* and *gesellschaft*.

13. Describe the social construction of reality according to the theories of symbolic interactionism.

14. Summarize the most significant ways that homelessness is experienced.

15. Evaluate functionalist and conflict perspectives on the nature and purpose of social institutions.

16. Describe ethnomethodology and note its strengths and weaknesses.

KEY TERMS

(defined at page number shown and in glossary)

achieved status, p. 141
agrarian societies, p.151
ascribed status, p. 141
dramaturgical analysis, p. 164
ethnomethodology, p. 163
formal organization, p. 148
Gemeinschaft, p. 155
Gesellschaft, p. 156
horticultural societies, p. 150
hunting and gathering
   societies, p. 150
impression management
   (presentation of self) , p. 164
industrial societies, p.152
master status, p. 141
mechanical solidarity, p. 155
nonverbal communication, p. 166
organic solidarity, p. 155
pastoral societies, p. 150

personal space, p. 168
postindustrial societies, p. 154
primary group, p. 147
role, p. 144
role conflict, p. 145
role exit, p. 146
role expectation, p. 144
role performance, p. 144
role strain, p. 145
secondary group, p. 147
self-fulfilling prophecy, p. 163
social construction of reality, p. 162
social group, p. 147
social institution, p. 148
social interaction, p. 137
social structure, p. 137
status, p. 140
status symbol, p. 143

KEY PEOPLE

(identified at page number shown)

CHAPTER OUTLINE

I. SOCIAL STRUCTURE: THE MACROLEVEL PERSPECTIVE
    A. **Social interaction** is the process by which people act toward or respond to other people.
    B. **Social structure** is the complex framework of social institutions (such as the economy) and the social practices (such as rules and social roles) that make up a society and that organize and establish limits on people's behavior.
       1. This structure is essential for the survival of society and for the well-being of individuals.
       2. This structure provides for a social web for familial support and social relationships that connects each person to the larger society.
    C. Social structure creates boundaries that define which persons or groups will be the "insiders" and which will be the "outsiders."
       1. *Social marginality* is the state of being part insider and part outsider in the social structure. Social marginality results in stigmatization.
       2. A *stigma* is any physical or social attribute or sign that so devalues a person's social identity that it disqualifies that person from full social acceptance.
II. COMPONENTS OF SOCIAL STRUCTURE
    A. A **status** is a socially defined position in a group or society characterized by certain expectations, rights, and duties.
       1. A **status set** is comprised of all the statuses that a person occupies at a given time.
       2. Ascribed and achieved statuses:
          a. An **ascribed status** is a social position conferred at birth or received involuntarily later in life. Examples include race/ethnicity, age, and gender.
          b. An **achieved status** is a social position a person assumes voluntarily as a result of personal choice, merit, or direct effort. Examples of achieved statuses include occupation, education and income. Ascribed statuses have a significant influence on the achieved statuses we occupy.

3. A **master status** is the most important status a person occupies; it dominates all of the individual's other statuses and is the overriding ingredient in determining a person's general social position (e.g., being poor or rich is a master status).

4. **Status symbols** are material signs that inform others of a person's general social position. Examples include a wedding ring or a Rolls-Royce automobile.

B. A **role** is a set of behavioral expectations associated with a given status.

1. **Role expectation** – a group's or society's definition of the way a specific role ought to be played – may sharply contrast with **role performance** – how a person actually plays the role.

2. **Role conflict** occurs when incompatible role demands are placed on a person by two or more statuses held at the same time (e.g., a woman whose roles include full time employee, mother, wife, caregiver for an elderly parent, and community volunteer).

3. **Role strain** occurs when incompatible demands are built into a single status that a person occupies (e.g., married women in the workforce having many responsibilities at home). Sexual orientation, age, and occupation frequently are associated with role strain.

4. **Role exit** occurs when people disengage from social roles that have been central to their self identity (e.g., ex convicts, ex nuns, retirees, and divorced women and men).

C. A **social group** consists of two or more people who interact frequently and share a common identity and a feeling of interdependence.

1. A **primary group** is a small, less specialized group in which members engage in face to face, emotion based interactions over an extended period of time (e.g., one's family, close friends, and school or work related peer groups).

2. A **secondary group** is a larger, more specialized group in which the members engage in more impersonal, goal-oriented relationships for a limited period of time (e.g., schools, churches, the military, and corporations).

3. A **social network** is a series of social relationships that link an individual to others.

4. A **formal organization** is a highly structured group formed for the purpose of completing certain tasks or achieving specific goals (e.g., colleges, corporations, and the government).

D. A **social institution** is a set of organized beliefs and rules that establish how a society will attempt to meet its basic social needs. Examples of social institutions include the family, religion, education, the economy, the government, mass media, sports, science and medicine, and the military.

III. SOCIETIES, TECHNOLOGY, AND SOCIOCULTURAL CHANGE

A. **Hunting** and **gathering societies**, preindustrial societies, use simple technology for hunting animals and gathering vegetation; their technology is limited to tools and weapons used for basic subsistence.

B. **Horticultural societies** and **pastoral societies**, as examples of preindustrial societies, have existed throughout the world at some time.
   1. **Pastoral societies** are based on technology that supports the domestication of large animals to provide food, and emerged in mountainous regions and areas with low amounts of annual rainfall. They typically remain nomadic as they seek new grazing lands and water sources for their animals.
   2. **Horticultural societies** are based on technology that supports the cultivation of plants to provide food. These societies emerged in more fertile areas that were better suited for growing plants through the use of hand tools.
C. **Agrarian societies** use the technology of large-scale farming including animal-drawn or energy-powered plows and equipment, to produce their food supply; most of the world's population lives in agrarian societies that are in various stages of industrialization.
D. **Industrial societies** are based on technology that mechanizes production; this mode of production transforms predominantly rural and agrarian societies into urban and industrial societies.
E. **Postindustrial societies** are societies in which technology supports a service-and-information-based economy; knowledge is viewed as the basic source of innovations and policy formulation; education becomes an important social institution and scientific research becomes institutionalized.

IV. STABILITY AND CHANGE IN SOCIETIES
A. Sociologists Emile Durkheim and Ferdinand Tönnies developed typologies to explain how stability and change occur in the social structure of societies.
B. Durkheim's Typology:
   1. **Mechanical solidarity** refers to the social cohesion in preindustrial societies where there is minimal division of labor and people feel united by shared values and common social bonds.
   2. **Organic solidarity** refers to the social cohesion found in industrial societies in which people perform very specialized tasks and feel united by their mutual dependence.
C. Tönnies' Typology:
   1. According to Ferdinand Tönnies, the *Gemeinschaft* is a traditional society in which social relationships are based on personal bonds of friendship and kinship and on intergenerational stability. Relationships are based on ascribed statuses.
   2. The *Gesellschaft* is a large, urban society, in which social bonds are based on impersonal and specialized relationships, with little long term commitment to the group or consensus on values. Relationships are based on achieved statuses.
D. Social Structure and Homelessness - in examining the relationship between social structure and homelessness, several explanations are provided for *Gesellschaft* societies.
   1. The homeless have made bad decisions which eventually led to homelessness; they are responsible for the consequences of their actions.
   2. Homelessness is rooted in poverty; it is a result of steady, across the board lowering of the standard of living of the working class and the lower class – a social class phenomena.

V. SOCIAL INTERACTION: THE MICROLEVEL PERSPECTIVE
  A. Social interaction within a given society has certain shared meanings across situations; however, everyone does not interpret social interaction rituals in the same way.
  B. The **social construction of reality** is the process by which our perception of reality is shaped largely by the subjective meaning that we give to an experience. Our definition of the situation can result in a **self-fulfilling prophecy** – a false belief or prediction that produces behavior that makes the original false belief come true.
  C. **Ethnomethodology** is the study of the commonsense knowledge that people use to understand the situations in which they find themselves.
    1. This approach challenges existing patterns of conventional behavior in order to uncover people's background expectancies, that is, their shared interpretation of objects and events, as well as the actions they take as a result.
    2. To uncover people's background expectancies, ethnomethodologists frequently conduct breaching experiments in which they break "rules" or act as though they do not understand some basic rule of social life so that they can observe other people's responses.
  D. **Dramaturgical analysis** is the study of social interaction that compares everyday life to a theatrical presentation.
    1. This perspective was initiated by Erving Goffman, who suggested that day to day interactions have much in common with being on stage or in a dramatic production.
    2. Most of us engage in **impression management**, or **presentation of self** – people's efforts to present themselves to others in ways that are most favorable to their own interests or image.
    3. Social interaction, like a theater, has a front stage – the area where a player performs a specific role before an audience – and a back stage – the area where a player is not required to perform a specific role because it is out of view of a given audience.
  E. The Sociology of Emotions
    1. Arlie Hochschild suggests that we acquire a set of *feeling rules*, which shape the appropriate emotions for a given role or specific situation.
    2. Emotional labor occurs when employees are required by their employers to feel and display only certain carefully selected emotions.
    3. Gender, class, and race are related to the expression of emotions necessary to manage one's feelings.
  F. **Nonverbal communication** is the transfer of information between persons without the use of speech (e.g. facial expressions, head movements, body positions, and other gestures). Personal space is the immediate area surrounding a person that the person claims as private. Age, gender, kind of relationship, and social class are important factors in allocation of personal space. Power differentials between people are reflected in personal space and privacy.

VI. FUTURE CHANGES IN SOCIETY, SOCIAL STRUCTURE, AND INTERACTION

A. The social structure in the U.S. has been changing rapidly in the past decades (e.g., more possible statuses for persons to occupy and roles to play than at any other time in history).

B. Ironically, at a time when we have more technological capability, more leisure activities and types of entertainment, and vast quantities of material goods available for consumption, many people experience high levels of stress, fear for their lives because of crime, and face problems such as homelessness.

C. While some individuals and groups continue to show initiative in trying to solve some of our pressing problems, the future of this country rests on our collective ability to deal with major social problems at both the macrolevel (structural) and the microlevel of society.

ANALYZING AND UNDERSTANDING THE BOXES

After reading the chapter and studying the outline, re-read the boxes and write down key points and possible questions for class discussion.

Sociology and Everyday Life: How Much Do You Know About Homeless Persons?

Key Points:

Discussion Questions:

1.

2.

3.

Framing Homelessness in the Media: Thematic and Episodic Framing

Key Points:

Discussion Questions:

1.

2.

3.

Sociology and Social Policy: Homeless Rights Versus Public Space

Key Points:

Discussion Questions:

1.

2.

3.

You Can Make a Difference: Offering a Helping Hand to Homeless People

Key Points:

Discussion Questions:

1.

2.

3.

PRACTICE TESTS

MULTIPLE CHOICE QUESTIONS

Select the response that best answers the question or completes the statement.

1. The stable patterns of social relationships that exist within a particular group or society are referred to as:
   a. social structure
   b. social interaction
   c. social dynamics
   d. social constructions of reality

2. According to the text, a recently arrived immigrant often experiences:
   a. social depletion.
   b. social marginality.
   a. social disintegration.
   b. social misappropriation.

3. A _____ is a socially defined position in a group or society.
   a. location
   b. set
   c. status
   d. role

4. Maxine is an attorney, a mother, a resident of California, and a Jewish American. All of these socially defined positions constitute her
   a. role pattern.
   b. role set.
   c. ascribed statuses.
   d. status set.

5. A master status
   a. historically has been held only by men.
   b. is comprised of all of the statuses that a person occupies at a given time.
   c. is the most important status a person occupies.
   d. is a social position a person always assumes voluntarily.

6. According to the text, a wedding ring and a Rolls-Royce automobile are both examples of:
   a. conspicuous consumption
   b. status symbols
   c. status markers
   d. master status indicators

7. _____ is/are the dynamic aspect of a status.
    a. Role
    b. Norms
    c. Groups
    d. People

8. In her study of women athletes in college sports programs, sociologist Tracey Watson found role _____ in the traditionally incongruent identities of being a woman and being an athlete.
    a. strain
    b. distancing
    c. conflict
    d. confusion

9. According to the text, lesbians and gay men often experience role _____ because of the pressures associated with having an identity heavily stigmatized by the dominant cultural group.
    a. strain
    b. distancing
    c. conflict
    d. confusion

10. When a homeless person is able to become a domiciled person, sociologists refer to the process as
    a. role disengagement.
    b. role exit.
    c. role engulfment.
    d. role relinquishment.

11. All of the following are examples of a primary group, **except**:
    a. family
    b. close friends
    c. peer groups
    d. students in a lecture hall

12. Which of the following statements **best** describes the characteristics of a secondary group?
    a. a small, less specialized group in which members engage in impersonal, goal-oriented relationships over an extended period of time
    b. a small, less specialized group in which members engage in face to face, emotion-based interactions over an extended period of time
    c. a larger, more specialized group in which members engage in face to face, emotion-based interactions over an extended period of time
    d. a larger, more specialized group in which members engage in impersonal, goal-oriented relationships over an extended period of time

13. The National Law Center on Homelessness and Poverty and other caregiver groups that provide services for the homeless and others in need are examples of:
    a. primary groups
    b. formal organizations
    c. informal organizations
    d. social networks

14. Sociologists use the term _____ to refer to a set of organized beliefs and rules that establish how a society will attempt to meet its basic social needs.
    a. social structure
    b. social expectations
    c. social networking
    d. social institutions

15. Which one of the following types of societies is more egalitarian?
    a. hunting and gathering
    b. horticultural and pastoral
    c. agrarian
    d. industrial

16. Most of the world's population lives in _____ societies.
    a. horticultural and pastoral
    b. postindustrial
    c. agrarian
    d. industrial

17. Emile Durkheim referred to the social cohesion found in industrial societies as
    a. organic solidarity.
    b. mechanical solidarity.
    c. *Gemeinschaft.*
    d. *Gesellschaft.*

18. A young person who has been told repeatedly that she or he is not a good student and, as a result, stops studying and receives failing grades is an example of a(n)
    a. selective perception.
    b. objective deduction.
    c. subjective reality.
    d. self-fulfilling prophecy.

19. Ethnomethodologist Harold Garfinkel assigned different activities to his students to see how breaking the unspoken rules of behavior created confusion. His research involved a series of
    a. shared expectancies.
    b. breaching experiments.
    c. dramaturgical analyses.
    d. impression managements.

20. According to sociologist Arlie Hochschild, people acquire a set of _____, which shape the appropriate emotions for a given role or specific situation.
    a. role expectations
    b. emotional experiences
    c. feeling rules
    d. nonverbal cues

21. In _____ societies, social inequality is in the highest of all preindustrial societies in terms of both class and gender.
    a. hunting and gathering
    b. horticultural and pastoral
    c. agrarian
    d. industrial

22. In recent years, _____ have accounted for 40 percent of the homeless population. This is the fastest growing category of homeless persons in the United States.
    a. families with children
    b. divorced females
    c. families without children
    d. single African American males

23. In 2004, _____ made up 49 percent of the homeless population in the United States.
    a. Asian Americans
    b. Latino/as (Hispanics)
    c. whites (Caucasians)
    d. African Americans

24. Not all of the homeless are unemployed. About _____ percent hold full or part-time jobs, but earn too little to find an affordable place to live.
    a. 13
    b. 18
    c. 22
    d. 33

25. _____ theorists maintain that, in capitalistic societies where a few people control the labor of many, the social structure reflects a systems of relationships of domination among categories of people, such as owner-worker and employer-employee.
    a. Postmodern
    b. Symbolic interactionist
    c. Functionalist
    d. Conflict

26. A convicted criminal, wearing a prison uniform, is an example of a person who has been _____; the uniform says that the person has done something wrong and should not be allowed unsupervised outside the prison walls.
    a. stigmatized
    b. marginalized
    c. dislabeled
    d. alienated

27. Sociologist Robert Park coined the term _____, to refer to persons (such as immigrants) who simultaneously share the life and traditions of two distinct groups.
    a. alienation
    b. stigmatization
    c. social marginality
    d. anomie

28. _____ is a group's or society's definition of the way a specific role *ought* to be played.
    a. Status expectation
    b. Collective judgment
    c. Role expectation
    d. Role performance

29. _____ is how a person actually plays the role.
    a. Role expectation
    b. Role performance
    c. Role behavior
    d. Status set

30. Groups are an important component of social structure. A(n) _____ consists of two or more people who interact frequently and share a common identity and a feeling of interdependence.
    a. social group
    b. acting group
    c. primary group
    d. secondary group

31. Many of us build _____ that involve our personal friends in primary groups and our acquaintances in secondary groups. They are a social web that links an individual to others.
    a. cultural bonds
    b. cohesive structures
    c. social networks
    d. cliques

32. In _____ societies, the division of labor between men and women in the middle and upper classes became more distinct: men were responsible for being "breadwinners"; women were seen as "homemakers."
   a. hunting and gathering
   b. industrial
   c. agrarian
   d. horticultural and pastoral

33. _____ are characterized by an information explosion and an economy in which large numbers of people either provide or apply information or are employed in service jobs (such as fast-food server or health care worker).
   a. Industrial societies
   b. Agrarian societies
   c. Horticultural and pastoral societies
   d. Postindustrial societies

34. Sociologist Emile Durkheim believed that people in preindustrial societies feel a more or less _____.
   a. organic solidarity.
   b. cohesive bonding.
   c. direct bonding.
   d. mechanical solidarity.

35. _____ theorists emphasize that social institutions exist because they perform five essentials tasks: replacing members; teaching new members; producing, distributing, and consuming goods and services; preserving order; and providing and maintaining a sense of purpose.
   a. Premodern
   b. Conflict
   c. Functionalist
   d. Symbolic interactionist

36. _____ use the technology of large-scale farming, including animal-drawn or energy-powered plows and equipment, to produce their food supply.
   a. Hunting and gathering societies
   b. Industrial societies
   c. Horticultural and pastoral societies
   d. Agrarian societies

37. Some symbolic interactionists believe that there is very little shared reality beyond that which is socially created. They refer to this as the _____ – the process by which our perception of reality is largely shaped by the subjective meaning that we give to an experience.
   a. social construction of reality
   b. objectification of social reality
   c. subjective assessment of reality
   d. self-fulfilling prophecy

38. According to _____, interaction is based on assumption of shared expectations. For example, when you are talking with someone, what expectations do you have that you will take turns?
    a. conflict theorists
    b. ethnomethodologists
    c. functionalists
    d. rational choice theorists

39. A/an _____ is a false belief or prediction that produces behavior that makes the originally false belief come true.
    a. definition of the situation
    b. interactionist dialogue
    c. reality impression
    d. self-fulfilling prophecy

40. Physical space is an important component of nonverbal communication. Anthropologist Edward Hall asserted that _____ is the immediate area surrounding a person that the person claims as private.
    a. intimate space
    b. social space
    c. public space
    d. personal space

TRUE/FALSE QUESTIONS

1. Most homeless persons are on the streets by choice or because they were deinstitutionalized by mental hospitals.
   T   F

2. Sociologists use the term "status" to refer only to high level positions in society.
   T   F

3. Historically, the most common master statuses for women have related to positions in the family, such as daughter, wife, and mother.
   T   F

4. Role performance does not always match role expectation.
   T   F

5. Role conflict occurs when incompatible role demands are placed on a person by two or more statuses held at the same time.
   T   F

6. Role distancing occurs when social identities are inconsistent with one's own identity.
   T   F

7. Role exit usually does not require a creation of a new identity.
   T   F

8. Close friends would be an example of a primary group.
   T   F

9. Social solidarity exists when a group remains cohesive when facing obstacles.
   T   F

10. Social networks work the same way for men and women and for people from different racial ethnic groups.
    T   F

11. Agrarian societies are based on technology that supports the cultivation of plants to provide food.
    T   F

12. In an industrialized society, the family decreases in significance, as the economic, educational and political institutions grow in size.
    T   F

13. Relationships in *Gemeinschaft* societies are based on achieved rather than ascribed status.
    T   F

14. Race/ethnicity, gender, and social class play a part in the way that social interaction is interpreted.
    T   F

15. Age, gender, and social class are strong influences in the allocation of personal space.
    T   F

16. An achieved status is a social condition conferred at birth.
    T   F

17. According to Box 5.1, most homeless people choose to be homeless.
    T   F

18. According to Box 5.2, "Framing Homelessness in the Media," episodic framing refers to news stories that focus primarily on statistics about the homeless population and recent trends in homelessness.
    T   F

19. According to Box 5.3, "public space protection" has become an issue not only in the United States, but also in other countries.
    T   F

20. According to Box 5.4, most Americans rarely see homeless individuals on street corners and elsewhere in everyday activities.
    T   F

FILL-IN-THE-BLANK QUESTIONS

1. The process by which people act toward or respond to other people is defined as
    _____.

2. A status gained as a result of personal ability is defined as a/an _____ status.

3. _____ societies are based on technology that supports the cultivation of
    plants to provide food.

4. Societies based on technology that supports the domestication of large animals to
    provide food are known as _____ societies.

5. The process by which our perception of reality is largely shaped by the subjective
    meaning that we give to the experience is known as the _____.

6. The _____ means that people analyze a social context in which they find
    themselves, determine what is in their best interest, and adjust their attitudes and
    actions accordingly.

7. The study of social interaction that compares everyday life to a theatrical
    presentation is known as _____.

8. _____ refers to people's efforts to present themselves to others in ways that
    are most favorable to their own interests or image.

9. The transfer of information between persons without the use of words is called
    _____.

10. The immediate area surrounding a person that the person claims as private is
    defined as _____.

11. _____ is the study of the commonsense knowledge that people use to
    understand the situations in which they find themselves.

12. Sociologist _____ used the term _____ in referring to the area where a
    player performs a specific role before an audience.

13. Sociologist _____ referred to the social cohesion in preindustrial societies
    where there is minimal division of labor and people feel united by shared values as a
    _____.

14. The _____ field focuses on certain social rules for the appropriate expressions
    of feelings and thoughts in a society.

15. According to Goffman, the _____ is the area where a player is not required to
    specific role because it is out of the view of a given audience.

SHORT ANSWER/ESSAY QUESTIONS

1. What are the major components of social structure? Give examples of each.
2. What are the five major components of societies? List some of the major characteristics of each type. In which pre-industrial society is the social inequality the highest? The lowest? Why? Which type of society brought about some of the greatest innovations in all of human history?
3. How is homelessness related to the social structure of a society? In what type of society do you think there might be more homelessness? Why?
4. What do we mean by the social construction of reality? Give some examples.
5. What are some of the projected future changes in the social structure in the United States? How should our society react to these changes?
6. According to the photo essay "Things We Take for Granted" section of this chapter, how were the social structures and institutions affected by Hurricane Katrina, and simultaneously how were the people in this area directly affected? Why are social institutions and social interaction so important in defining who we are and what we want to become?

STUDENT CLASS PROJECTS AND ACTIVITIES

1. Utilizing the concept of social networks as discussed in the text, trace your own social networks. Construct diagrams linking yourself to your family and other relatives, and to your friends and to other acquaintances. After constructing the diagrams, answer the following questions/statements: (1) describe the members in your network (who are they; and what is their gender, approximate age, and current employment, if applicable?). (2) How and why are they linked to you (3) Summarize the impact of each person in your network; are the ties weak or strong? Why? (4) What persons in your network are most important now? Why? (5) What persons in your network will be especially important to you in future years? (6) Is there any potential for anyone in your network to harm you in any fashion? What can/should you do about this? (7) Is there any potential for anyone in your network to provide you with special advantages for a potential career? Explain in detail your responses to the above questions. Write your paper in the manner described by your instructor and submit your paper on the due date.
2. As mentioned in the text, changes in social structure have a dramatic impact on individuals, groups, and societies. Many nations in the world today are experiencing radical changes as they attempt to modernize; while many contemporary societies are experiencing great change as they become more complex with the introduction of new technology. You are to select any country that is experiencing such change utilizing at least one of the typologies mentioned in the text (that of Durkheim or Tönnies). Countries that could illustrate the former are the People's Republic of China; any of the 15 Republics of the U. S. S. R.; any of the countries of the former Yugoslavia; many of the countries of South America, such as Brazil; many of the countries of Africa, such as Zaire; many of the countries of Southeast Asia or the Pacific Rim, such as Thailand, South Korea, Malaysia, Laos; Iran; or India. Countries

that could illustrate the former are: Japan, the United States, Canada, France, Germany, or Britain. You should examine the changes from both a microlevel and a macrolevel perspective. Trace the changes in (1) the statuses, (2) the roles, (3) groups, (4) social institutions, and (5) in the society for both the individual and the country as a whole. You should also explain the effects these changes will have on the future options and opportunities of people living in that particular country. Submit your paper to your instructor, following any specific guidelines for your finished paper.

3. The text uses homelessness as a concrete example to help you understand the difficult concept of social structure. Providing effective solutions to this problem is even more difficult. Many politicians suggest that one reason for the increase in homelessness is a direct result of the shortage of available low-income housing. You are to research and write an issue paper on this topic: "Should more low-income housing be created to help the homeless?" In your paper, provide the following: (1) a background of this problem; (2) factual information germane to homelessness; (3) the positions of opposing viewpoints; (4) arguments for and against the issue; (5) your own position and resolution of homelessness. Suggested note: you may attempt a synthesis of opposing positions; leave the issue essentially unresolved; or provide a new position; or a completely different, innovative resolution.

INTERNET ACTIVITIES

1. ***Emotions: Observing the Fusion Between Body, Mind and Society***: **http://www.trinity.edu/mkearl/spsy-emo.html**. This is a fantastic web site that will help you understand the connections between sociology, psychology and physiology as they address the topic of emotions. The site contains links to research articles, survey data, and an online bibliography.

2. The Center for Housing Policy at the University of York in the United Kingdom provides information on a wide-range of housing issues, such as homelessness, housing needs of older people and street homelessness. Locate this site at **www.york/ac.uk/inst/chp/**

3. Civil Practices Network (CPN) is a collaborative and non-partisan project dedicated to bringing together diverse social organizations in every community and institutional settings in the United States in order to revitalize and strengthen major infrastructures throughout the country. From the website, **www.cpn.org**, discover some of these rebuilding strategies.

4. You can read a section from Erving Goffman: "The Presentation of Self in Everyday Life", by Adam D. Barnhart on this site: **http://www.hewett.norfolk.sch.uk/CURRIC/soc/goffman.htm**. This site provides a detailed description and analysis of Goffman's major works. How do people handle "impression management?" What does Goffman mean by "front stage" and "back stage?"

INFOTRAC COLLEGE EDITION EXERCISES

Visit the **InfoTrac College Edition** website at**:**
http://www.wadsworthmedia.com/webtutor/infotrac.htm. You will arrive at a screen
that enables you to search topics.

1. Search for articles using the subject guide **Social Structure**. You will get 76
   periodical references. Choose the article entitled: *PATHWAYS TO ADULTHOOD IN
   CHANGING SOCIETIES: Variability and Mechanisms in Life Course Perspective.*
   This would make a great resource for a small group discussion. The article would
   help you to connect the curriculum content to the lives of your students. The author,
   Michael J. Shanahan, suggests new lines of inquiry that focus on the interplay
   between agency and social structures in the shaping of lives.
2. Look for research projects related to the study of **emotions**. Bring in copies of
   articles of interest. This becomes an effective way to discuss empirical research
   which uses categories such as emotions. You can emphasize research methods,
   science, and the sociology of emotions.
3. These three articles were accessed using InfoTrac. Each is a good beginning point
   for a discussion on **social structure and interaction**. Each can inspire the
   sociological imagination. Find other research articles that help you understand the
   powerful influence of society on you individual life.
   - *Restoring American cultural institutions.* Jerry L. Martin. Society. Jan-Feb
     1999 v36 i2 p35(6).
   - *Responses to humiliation.* (The Decent Society) Arthur Ripstein. Social
     Research. Spring 1997 v64 n1 p90 (22).
   - *On moral issues Americans prefer persuasion.* (The Communication
     Network survey) Society, Nov-Dec 1997 v35 n1 p2 (1).

SOLUTIONS

MULTIPLE CHOICE QUESTIONS

1. A, p. 137
2. B, p. 139
3. C, p. 140
4. D, p. 141
5. C, p. 141
6. B, p. 142
7. A, p. 144
8. C, p. 145
9. A, p. 145
10. B, p. 146
11. D, p. 147
12. D, p. 147
13. B, p. 148
14. D, p. 148

15. A, p. 149
16. C, p. 152
17. A, p. 155
18. D, p. 163
19. B, p. 163
20. C, p. 165
21. C, p. 152
22. A, p. 140
23. D, p. 137
24. C, p. 138
25. D, p. 139
26. A, p. 139
27. C, p. 139
28. C, p. 144

29. B, p. 144
30. A, p. 147
31. C, p. 147
32. B, p. 153
33. D, p. 154
34. D, p. 155
35. C, p. 149
36. D, p. 151
37. A, p. 162
38. B, p. 163
39. D, p. 163
40. D, p. 168

TRUE/FALSE QUESTIONS

1. F, p. 140
2. F, p. 140
3. T, p. 141
4. T, p. 144
5. T, p. 145
6. T, p. 145
7. F, p. 146

8. T, p. 147
9. T, p. 147
10. F, p. 148
11. F, p. 151
12. T, p. 153
13. F, p. 156
14. T, p. 159

15. T, p, 169
16. F, p. 141
17. F, p. 140
18. F, p. 143
19. T, p. 158
20. F, p. 170

FILL-IN-THE-BLANK QUESTIONS

1. social interaction, p. 137
2. achieved, p. 140
3. Horticultural; p. 150
4. pastoral, p. 150
5. social construction of reality, p. 162
6. definition of the situation, p. 163
7. dramaturgical analysis, p. 164
8. impression management, p. 164
9. nonverbal communication, p. 166
10. personal space, p. 168
11. Ethnomethodology, p. 163
12. Erving Goffman; front state, p. 165
13. Ferdinand Tönnies, *Gemeinschaft*, p. 155
14. sociology of emotions, p. 165
15. back stage, p. 165

6

GROUPS AND ORGANIZATIONS

BRIEF CHAPTER OUTLINE

CHAPTER SUMMARY

Groups are a key element of our social structure and much of our social interaction takes place within them. A **social group** is a collection of two or more people who interact frequently, share a sense of belonging, and have a feeling of interdependence. Social groups may be either **primary groups** – small, personal groups in which members engage in emotion based interactions over an extended period – or **secondary groups** – larger, more specialized groups in which members have less personal and more formal, goal oriented relationships. All groups set boundaries to indicate who does and who does not belong: an **ingroup** is a group to which we belong and with which we identify; an **outgroup** is a group we do not belong to or perhaps feel hostile toward. The size of a group is one of its most important features. The smallest groups are **dyads** – groups composed of two members – and **triads** – groups of three. In order to maintain ties with a group, many members are willing to conform to norms established and reinforced by group members. Three types of **formal organizations** – highly structured secondary groups formed to achieve specific goals in an efficient manner – are normative, coercive, and utilitarian organizations. A **bureaucracy** is a formal organization characterized by hierarchical authority, division of labor, explicit procedures, and impersonality in personnel concerns. The **iron law of oligarchy** refers to the tendency of organizations to become a bureaucracy ruled by the few. A recent movement to humanize bureaucracy has focused on developing human resources. The best organizational structure for the future is one that operates humanely and that includes opportunities for all, regardless of race, gender, or class.

LEARNING OBJECTIVES

After reading Chapter 6 you should be able to:

1. State definitions for ingroup, outgroup, and reference group and describe the significance of these concepts in everyday life.

2. Summarize Max Weber's perspective on rationality and outline his ideal characteristics of bureaucracy.

3. Define the iron law of oligarchy.

4. Describe the informal structure in bureaucracies and list its positive and negative aspects.

5. Distinguish between aggregates, categories, and groups from a sociological perspective.

6. Distinguish between the two functions of leadership and the three major styles of group leadership.

7. Compare normative, coercive, and utilitarian organizations and describe the nature of membership in each.

8. Describe dyads and triads and explain the phenomena of changes in interaction patterns.

9. Explain what is meant by groupthink and elaborate on reasons why it can be dangerous for organizations.

10. Describe the ways that online social networks enhance communication.

11. Distinguish between primary and secondary groups and explain how people's relationships differ in each.

12. Contrast functionalist and conflict perspectives on the purposes of groups.

13. Explain the relationship between bureaucratic hierarchies and oligarchies.

14. Explain the contributions of Solomon Asch and Stanley Milgram to our understanding about group conformity and obedience to authority.

15. Discuss the major shortcomings of bureaucracies and their effects on workers, clients or customers, and levels of productivity.

16. Evaluate U.S. and Japanese models of organization.

KEY TERMS

(defined at page number shown and in glossary)

aggregate, p. 176
authoritarian leaders, p.183
bureaucracy, p. 191
bureaucratic personality, p. 197
category, p. 176
conformity, p. 184
democratic leaders, p. 183
dyad, p. 181
expressive leadership, p. 183
goal displacement, p. 197
groupthink, p. 188
ideal type, p. 192

informal side of a bureaucracy, p. 195
ingroup, p. 177
instrumental leadership, p. 183
iron law of oligarchy, p. 198
laissez-faire leaders, p. 184
network, p. 179
outgroup, p. 177
rationality, p. 192
reference group, p. 179
small group, p. 181
triad, p. 182

KEY PEOPLE

(identified at page number shown)

Solomon Asch, p.184
James S. Coleman, p. 190
Charles H. Cooley, p. 176
Amitai Etzioni, p. 190
Irving Janis, p. 188

Robert Michels, p. 198
Stanley Milgram, p. 185
Georg Simmel, p. 181
William Graham Sumner, p. 177
Max Weber, p. 192

CHAPTER OUTLINE

I. SOCIAL GROUPS
    A. Groups, Aggregates, and Categories
        1. A **social group** is a collection of two or more people who interact frequently with one another, share a sense of belonging, and have a feeling of interdependence.
        2. An **aggregate** is a collection of people who happen to be in the same place at the same time but share little else in common.
        3. A **category** is a number of people who may never have met one another but share a similar characteristic.
    B. Types of Groups
        1. Primary and Secondary Group
            a. According to Charles H. Cooley, a **primary group** is a small group whose members engage in face-to-face, emotion-based interactions over an extended period of time.

b. A **secondary group** is a larger, more specialized group in which the members engage in more impersonal, goal-oriented relationships for a limited period of time.

2. Ingroups and Outgroups
   a. According to William Graham Sumner, an **ingroup** is a group to which a person belongs and with which the person feels a sense of identity.
   b. An **outgroup** is a group to which a person does not belong and toward which the person may feel a sense of competitiveness or hostility.

3. Reference Groups
   a. A **reference group** is a group that strongly influences a person's behavior and social attitudes, regardless of whether that individual is an actual member.
   b. Reference groups help us explain why our behavior and attitudes sometimes differ from those of our membership groups; we may accept the values and norms of a group with which we identify rather than one to which we belong.

4. Networks
   a. A **network** is a web of social relationships that link one person with other people and, through them, with other people they know.
   b. Networks extend far beyond to not only persons that are known, but also the people that know of you.

II. GROUP CHARACTERISTICS AND DYNAMICS
A. Group Size
   1. A **small group** is a collectivity small enough for all members to be acquainted with one another and to interact simultaneously.
   2. According to Georg Simmel, small groups have distinctive interaction patterns which do not exist in larger groups.
      a. In a **dyad** – a group composed of two members – the active participation of both members is crucial for the group's survival and members have a more intense bond and a sense of unity not found in most larger groups.
      b. When a third person is added to a dyad, a **triad** – a group composed of three members – is formed, and the nature of the relationship and interaction patterns change; for example two members may unite to create a *coalition* – an alliance created in an attempt to reach a shared objective or goal, which sometimes can subject the third member to group pressure to conform.
   3. As group size increases, members tend to specialize in different tasks, and communication patterns change.
B. Group Leadership
   1. Leaders are responsible for directing plans and activities so that the group completes its task or fulfills its goals.
   2. Leadership functions:
      a. **Instrumental leadership** is goal- or task-oriented; if the underlying purpose of a group is to complete a task or reach a particular goal, this type of leadership is most appropriate.

b. **Expressive leadership** provides emotional support for members; this type of leadership is most appropriate when harmony, solidarity, and high morale are needed.
3. Leadership styles:
    a. **Authoritarian leaders** make all major group decisions and assign tasks to members.
    b. **Democratic leaders** encourage group discussion and decision making through consensus building.
    c. **Laissez-faire leaders** are only minimally involved in decision making and encourage group members to make their own decisions.
4. The research of John Pryor suggested a relationship between group conformity and sexual harassment demonstrating that sexual harassment is more likely to occur when it is encouraged (or not actively discouraged) by others.
C. Group Conformity
1. **Conformity** is the process of maintaining or changing behavior to comply with the norms established by a society, subculture, or other group.
2. In a series of experiments, Solomon Asch found that the pressure toward group conformity was so great that participants were willing to contradict their own best judgment if the rest of the group disagreed with them, thus demonstrating the power that groups have in producing *compliance*.
3. Stanley Milgram (a former student of Asch's) conducted a series of controversial experiments and concluded that people's obedience to authority may be more common than most of us would like to believe.
D. Groupthink
1. Irving Janis coined the term **groupthink** to describe the process by which members of a cohesive group arrive at a decision that many individual members privately believe is unwise.
2. The tragic launch of the space shuttle *Columbia* is an example of this process.
E. Social Exchange/Rational Choice Theories
1. Social Exchange/Rational Choice theories focus on the process by which actors settle on one optimal outcome out of a range of possible choices.
2. Social Exchange theories are based on the assumption of *self-interest* as the basic motivating factor.
3. Key elements of rational choice theories include *actors* and *resources*.
III. FORMAL ORGANIZATIONS IN GLOBAL PERSPECTIVE
A. A *formal organization* is a highly structured secondary group formed for the purpose of achieving specific goals in the most efficient manner (e.g., corporations, schools, and government agencies).
B. Types of Formal Organizations
1. Amitai Etzioni classified formal organizations into three categories based on the nature of membership.

2. We voluntarily join *normative* organizations when we want to pursue some common interest or to gain personal satisfaction or prestige from being a member. Examples include political parties, religious organizations, and college social clubs.
3. People do not voluntarily become members of *coercive* organizations – associations people are forced to join. Examples include total institutions, such as boot camps and prisons.
4. We voluntarily join *utilitarian* organizations when they can provide us with a material reward we seek. Examples include colleges and universities, and the workplace.

C. Bureaucracies
1. **Bureaucracy** is an organizational model characterized by a hierarchy of authority, a clear division of labor, explicit rules and procedures, and impersonality in personnel matters.
2. According to Max Weber, bureaucracy is the most "rational" and efficient means of attaining organizational goals because it contributes to coordination and control. **Rationality** is the process by which traditional methods of social organization, characterized by informality and spontaneity, gradually are replaced by efficiently administered formal rules and procedures.
3. An **ideal type** is an abstract model which describes the recurring characteristics of some phenomenon; the ideal characteristics of bureaucracy include:
   a. Division of Labor – each member has a specific status with certain assigned tasks to fulfill.
   b. Hierarchy of Authority – a chain of command that is based on each lower office being under the control and supervision of a higher one.
   c. Rules and Regulations – standardized rules and regulations establish authority within an organization and usually are provided to members in a written format.
   d. Qualifications-Based Employment – hiring of staff members and professional employees is based on specific qualifications; individual performance is evaluated against specific standards; and promotions are based on merit as spelled out in personnel policies.
   e. Impersonality – interaction is based on status and standardized criteria rather than personal feelings or subjective factors.
4. An organization's **informal structure** is composed of those aspects of participants' day to day activities and interactions that ignore, bypass, or do not correspond with the official rules and procedures of the bureaucracy.
   a. The informal structure has also been referred to as work culture because it includes the ideology and practices of workers on the job; workers create this *work culture* in order to conform, resist, or adapt to the constraints of their jobs as well as to guide and interpret social relations on the job.
   b. There are positive and negative aspects of thoughts about informal structure in organizations; one emphasizes control (or eradication) of informal groups; the other suggests that they should be nurtured.

D. Shortcomings of Bureaucracy
  1. Inefficiency and Rigidity
     a. **Goal displacement** occurs when the rules become an end in themselves (rather than a means to an end), and organizational survival becomes more important than achievement of goals
     b. The term **bureaucratic personality** is used to describe those workers who are more concerned with following correct procedures than they are with getting the job done correctly.
  2. Resistance to Change
     a. Once bureaucratic organizations are created, they tend to resist change.
     b. Often bureaucracies tend to promote people from within the bureaucracy who may not be qualified for the positions.
  3. Perpetuation of Race, Class, and Gender Inequalities
     a. Some bureaucracies perpetuate inequality of race, class, and gender.
     b. This form of organizational structure created a specific type of work or learning environment that historically was typically created for middle and upper-middle class white men.
E. Bureaucracy and Oligarchy
  1. Max Weber believed that bureaucracy was a necessary evil because it achieved coordination and control and thus efficiency in administration; however, he believed such organizations stifled human initiative and creativity.
  2. Bureaucracy generates an enormous degree of unregulated and often unperceived social power in the hands of a very few leaders.
  3. According to Robert Michels, this results in the **iron law of oligarchy** – a bureaucracy ruled by a few people.
IV. ALTERNATIVE FORMS OF ORGANIZATION
A. "Humanizing" bureaucracy includes: (1) less rigid, hierarchical structures and greater sharing of power and responsibility; (2) encouragement of participants to share their ideas and try new approaches; and (3) efforts to reduce the number of people in dead end jobs and to help people meet outside family responsibilities while still receiving equal treatment inside the organization.
B. The Japanese model of organization has been widely praised for its innovative structure, which (until recently) has included:
  1. Long-term employment and company loyalty – Workers were guaranteed permanent employment after an initial probationary period.
  2. Quality Circles – small workgroups that meet regularly with managers to discuss the group's performance and working conditions.
  3. People-oriented – having respect for employees and emphasizing long-term goals appear to have made Japanese automobile manufacturers more successful than U.S. companies.
V. NEW ORGANIZATIONS IN THE FUTURE
A. There is a lack of consensus among organizational theorists about the "best" model of organization; however, some have suggested a *horizontal* model in which both hierarchy and functional or departmental boundaries largely would be eliminated.

B. In the horizontal structure, a limited number of senior executives would still exist in support roles (such as finance and human resources); everyone else would work in multidisciplinary teams that would perform core processes (e.g., product development or sales).

C. It is difficult to determine what the best organizational structure for the future might be; however, everyone can benefit from humane organizational environments that provide opportunities for all people regardless of race, gender, or class.

ANALYZING AND UNDERSTANDING THE BOXES

After reading the chapter and studying the outline, re-read the boxes and write down key points and possible questions for class discussion.

Sociology and Everyday Life: How Much Do You Know About Privacy in Groups and Organizations?

Key Points:

Discussion Questions:

1.

2.

3.

Framing Community in the Media: "Virtual Communities" on the Internet

Key Points:

Discussion Questions:

1.

2.

3.

Sociology and Social Policy: Computer Privacy in the Workplace

Key Points:

Discussion Questions:

1.

2.

3.

You Can Make a Difference: Developing Invisible (but Meaningful) Networks on the Web

Key Points:

Discussion Questions:

1.

2.

3.

PRACTICE TESTS

MULTIPLE CHOICE QUESTIONS

Select the response that best answers the question or completes the statement.

1. According to the "lived experience" introductory discourse in this chapter, FaceBook is
    a. a "fun" way to get to know other people.
    b. a way to avoid actual human contact.
    c. a way to join online groups with similar interests.
    d. all of the above

2. A(n) _____ is a collection of people who happen to be in the same place at the same time while a(n) _____ is a number of people who may never have met one another but share a similar characteristic.
   a. aggregate , category
   b. category, aggregate
   c. social group, aggregate
   d. category, social group

3. John had thought about trying out for the football team, but because he is not very athletic, he decides to hang out with the "stoners" at his school. He harbors some feelings of hostility for the football team. For John, the football team represents a(n)
   a. reference group.
   b. secondary group.
   c. ingroup.
   d. outgroup.

4. A _____ is an alliance created in an attempt to reach a shared objective or goal.
   a. triad
   b. coalition
   c. dyad
   d. reference group

5. _____ leadership is goal- or task-oriented; _____ leadership provides emotional support for members.
   a. Authoritarian, democratic
   b. Authoritarian, laissez-faire
   c. Expressive, instrumental
   d. Instrumental, expressive

6. A group leader at a business-related seminar is only minimally involved with the decisions made by the group and encourages members to make their own choices. This illustrates a(n) _____ leader.
   a. expressive
   b. democratic
   c. laissez-faire
   d. authoritarian

7. In one of Solomon Asch's experiments, a subject was asked to compare the length of lines printed on a series of cards without knowing that all other research "subjects" actually were assistants to the researcher. The results of Asch's experiments revealed that:
   a. 85 percent of the subjects routinely chose the correct response regardless of the assistants' opinions.
   b. 33 percent of the subjects routinely chose to conform to the opinion of the assistants by giving an incorrect answer.
   c. 10 percent of the subjects routinely chose to conform to the opinion of the assistants by giving an incorrect answer.
   d. the opinion of the assistants had no influence on the subjects' opinions.

115

8. In reference to the experiments by Stanley Milgram, the text points out that
    a. obedience to authority may be less common than most people would like to believe.
    b. many of the subjects questioned the ethics of the experiment.
    c. research such as this raises important questions concerning research ethics.
    d. most of the subjects were afraid to conform because the use of electrical current was involved.

9. According to the text, the tragic loss of the space shuttle *Columbia* is an example of:
    a. groupthink
    b. obedience to authority
    c. compliance
    d. the iron law of oligarchy

10. All of the following are types of formal organizations, **except** _____ organizations.
    a. normative
    b. anomic
    c. coercive
    d. utilitarian

11. Membership in _____ organizations is involuntary.
    a. normative
    b. anomic
    c. coercive
    d. utilitarian

12. According to Max Weber, rationality refers to:
    a. the process by which bureaucracy is gradually replaced by alternative types of organization such as quality circles.
    b. the process by which traditional methods of social organization are gradually replaced by bureaucracy.
    c. the level of sanity (or insanity) of people in an organization.
    d. the logic used by organizational leaders in decision making.

13. All of the following are ideal type characteristics of bureaucratic organizations, as specified by Max Weber, **except**:
    a. coercive leadership
    b. impersonality
    c. hierarchy of authority
    d. division of labor

14. In higher education, student handbooks, catalogs, and course syllabi are examples of:
    a. informal structures in bureaucracy
    b. impersonality
    c. hierarchy of authority
    d. rules and regulations

15. Networking consists of:
    a. relatives and close friends
    b. classmates and professors
    c. between 500 and 2,500 acquaintances
    d. all of the above

16. A group with two members is a _____, while a group with three members is a _____.
    a. triad, dyad
    b. dyad, large group
    c. dyad, triad
    d. triad, small group

17. All of the following are shortcomings of bureaucracy, **except**:
    a. impersonality
    b. inefficiency and rigidity
    c. resistance to change
    d. perpetuation of race, class, and gender inequalities

18. George Ritzer's "McDonaldization" includes all of the following **except**:
    a. efficiency
    b. predictability
    c. multiple employees
    d. control through nonhuman technologies

19. People join _____ organizations when they want to pursue a common interest.
    a. normative
    b. global
    c. coercive
    d. utilitarian

20. According to Robert Michels, all organizations encounter
    a. the Peter Principle.
    b. the iron law of oligarchy.
    c. Murphy's Law.
    d. groupthink.

21. Sociologist Charles H. Cooley used the term _____ to describe a small, less specialized group in which members engage in face-to-face, emotion-based interactions over an extended period of time.
    a. secondary group
    b. significant group
    c. tertiary group
    d. primary group

22. We have primary relationships with other individuals in our primary groups – that is, with our _____, who frequently serve as role models.
    a. generalized others
    b. significant other
    c. familial others
    d. fictive others

23. People who are the same age, race, and gender, and who share the same educational level constitute a(n):
    a. category.
    b. aggregate.
    c. formal organization.
    d. social group.

24. According to sociologist James Coleman, the key elements of the _____ theories are actors (which may include individuals, groups, corporations, and societies) and resources (which comprise the things over which actors have control and in which they have some interests).
    a. feminist
    b. social exchange
    c. postmodern
    d. rational choice

25. According to Max Weber, _____ is the process by which traditional methods of social organization, characterized by informality and spontaneity, gradually are replaced by efficiently administered formal rules and procedures.
    a. rationality
    b. division of labor
    c. top-down
    d. bottom-up

26. A _____ is an organizational model characterized by a hierarchy of authority, a clear division of labor, explicit rules and procedures, and impersonality in personnel matters.
    a. democracy
    b. bureaucracy
    c. monarchy
    d. formal organization

27. In his study of bureaucracies, sociologist _____ relied on an ideal type analysis, which he adapted from the field of economics.
    a. Irving Janis
    b. Charles Cooley
    c. Karl Marx
    d. Max Weber

28. For a person who strongly believes in the value of human rights and equal opportunity, which of the following would not constitute a reference group?
    a. the Ku Klux Klan
    b. the American Civil Liberties Union
    c. the National Organization for Women
    d. the council on Racial Equality

29. A(n) _____ is a web of social relationships that links one person with other people and, through them with other people they know.
    a. informal organization
    b. network
    c. primary group
    d. reference group

30. According to functionalists, _____ needs are particularly important for self-expression and support from family, friends, and peers.
    a. expressive
    b. instrumental
    c. group
    d. individual

31. The text points out that ingroup and outgroup distinctions
    a. may encourage social cohesion among group members.
    b. serve to prevent racism, sexism, and ageism.
    c. discourage feelings of group superiority.
    d. are less likely to exist in urban societies than in rural societies.

32. _____ size refers to the number of potential members a group has.
    a. Relative
    b. Power
    c. Absolute
    d. Potential

33. A(n) _____ is a group to which a person does not belong and toward which the person may feel a sense of competitiveness or hostility.
    a. ingroup
    b. outgroup
    c. reference group
    d. secondary group

34. _____ leadership is goal- or task-oriented.
    a. Instrumental
    b. Expressive
    c. Primary
    d. Secondary

35. _____ leaders encourage group discussion and decision making through consensus building.
    a. Democratic
    b. Authoritarian
    c. Laissez-faire
    d. Socialistic

36. _____ is a form of compliance in which people follow direct orders from someone in a position of authority.
    a. Cohesion
    b. Obedience
    c. Learning
    d. Socialization

37. Thorstein Veblen used the term _____ to characterize situation in which workers have become so highly specialized, or have been given such fragmented jobs to do, that they are unable to come up with creative solutions to problems.
    a. trained incapacity
    b. bureaucratic personality
    c. bureaucratic formation
    d. organization men

38. The tendency of a bureaucracy to be ruled by a very limited number of people is referred to as the
    a. iron cage.
    b. Peter Principal.
    c. iron law of oligarchy.
    d. absolute regime.

39. Eric works for the All-Star Baseball Card Company. Bi-weekly, Eric and about twelve of his co-workers meet with the shift supervisor to discuss ways to improve the work setting and the company's product. This group is an example of a(n):
    a. quality circle
    b. primary group
    c. bureaucratic subdivision
    d. assessment committee

40. In his study of sexual harassment, psychologist John Pryor found that when the "trainers" were lead to believe that sexual harassment was condoned and were then left alone with the women, they took full advantage of the situation in _____ percent of the experiments.
    a. 30
    b. 50
    c. 70
    d. 90

TRUE/FALSE QUESTIONS

1. People in aggregates share a common purpose.
   T   F

2. Although formal organizations are secondary groups, they also contain many primary groups within them.
   T   F

3. According to functionalists, people form groups to meet instrumental and expressive needs.
   T   F

4. If one member withdraws from a dyad, the group ceases to exist.
   T   F

5. Laissez-faire leaders encourage group discussion and decision making through consensus building.
   T   F

6. Stanley Milgram's research focused on people's obedience to authority.
   T   F

7. The research of Solomon Asch demonstrated that few people actually will bow to social pressure in small-group settings.
   T   F

8. Recent research on sexual harassment by psychologist John Pryor suggests that there is a relationship between group conformity and harassment.
   T   F

9. The explosion of the space shuttle *Columbia* has been cited as an example of groupthink.
   T   F

10. In the utilitarianism doctrine of rational choice theory, the purpose of all action should be to bring about the greatest happiness to the greatest number of people.
    T   F

11. Over the past century, the number of formal organizations increased dramatically in the United States.
    T   F

12. An ideal type is an abstract model that describes the recurring characteristics of some phenomenon.
    T   F

13. The formal structure of an organization has been referred to as "bureaucracy's other face."
    T   F

14. The "grapevine" informal structure of a bureaucracy spreads information much slower than the "official" channels, formal channels of communication.
T   F

15. Women and people of color who are employed in positions traditionally held by white men often experience categorical exclusion from the informal structures.
T   F

16. A category is a number of people who may never have met one another, but share a similar characteristic.
T   F

17. According to Box 6.1, a fast food restaurant can require all employees under the age of twenty-one to submit to periodic, unannounced drug testing.
T   F

18. Sociologists Robyn Bateman Driskell and Larry Lyon, according to Box 6.2 "Framing Community in the Media: Virtual Communities on the Internet," suggest that true communities can be established in the digital environment of cyberspace.
T   F

19. According to Box 6.3, "Computer Privacy in the Workplace" employees argue that they have the right to monitor everything that their employees do on company-owed computers.
T   F

20. Box 6.4, "Developing Invisible (But Meaningful) Networks on the Web," suggests that establishing social spaces almost always leads to creating greater social and individual problems.
T   F

FILL-IN-THE-BLANK QUESTIONS

1. According to _____, small groups have distinctive interaction patterns that do not exist in larger groups.

2. Sociologist _____ coined the terms "ingroup" and "outgroup."

3. The term "primary group" was coined by _____.

4. A series of controversial experiments dealing with people's obedience to authority were conducted by _____.

5. The term "groupthink" was coined by _____.

6. The theory of a _____ has been developed to explain how social class distinctions are perpetuated through different types of employment.

7. The term "iron law of oligarchy" was coined by _____.

8. Many U.S. corporations use _____ to achieve better product quality and to lower productive costs.

9. Officials in a bureaucracy practice _____ when they emphasize rules and procedures more than the achievement of the goals.

10. The term _____ describes workers more concerned with following correct procedures than they are with getting the job done correctly.

11. In the early 1980s in the United States, there was a movement to _____ bureaucracy with the major goal being to establish an organizational environment that develops rather than impedes human resources.

12. For several decades, the _____ model of organization has been widely praised for its innovative structure.

13. In analyzing group characteristics and dynamics, the key factors of _____ theories are actors and resources.

14. Social exchange theories are based upon the assumption that _____ is the basic motivating factors in people's interaction.

15. The tragic 2003 explosion of the space shuttle *Columbia* while preparing to land has been cited as an example of _____.

SHORT ANSWER/ESSAY QUESTIONS

1. According to the social exchange/rational choice theory, what might some men gain from sexually harassing workers? What are some possible losses or penalties? What do women gain or lose by reporting incidents of sexual harassment?
2. Define and provide an example of each of the following three theories: (1) Parkinson's law; (2) Peter Principle; and (3) Michel's Iron Law of Oligarchy.
3. What is bureaucracy? What are some of the strengths and weaknesses of bureaucracies?
4. What is groupthink? How might one affect or be affected by groupthink?
5. What were the insights gained from Stanley Milgram's research? Do these insights outweigh the personal costs to the subjects of the research?

STUDENT CLASS PROJECTS AND ACTIVITIES

1. Prior to taking office, the U.S. President-elect announces to the public his/her choice of all the cabinet and chief advisory appointments. These appointments profoundly shape many of the policies and directions of the U.S. Federal Government, which is one of the largest bureaucracies in the country. As a required written project, select ten of these positions. Describe the position, in terms of the purpose of the office; the requirements, if any, for the position; and the length of the term of that position. Name and briefly describe the credentials of the person who occupies that office. Some examples of cabinet posts are: U.S. Attorney General, Agriculture Secretary, Secretary of the Interior, Transportation Secretary, Secretary of State, Secretary of Defense, Education Secretary, Energy Secretary, Secretary of Veterans Affairs, Commerce Secretary, Labor Secretary, Secretary of the Treasury, etc. Some

examples of non-cabinet appointments are: Office of Management and Budget, White House Chief of Staff, White House Council of Economic Advisers, Environmental Protection Agency Administrator, etc. Submit your paper at the designated time according to the format and instructions provided by your instructor.

2. One example of a formal organization is your college or university. Gather information about the characteristics of the organization. Is it constructed according to the specific characteristics of a bureaucracy as described by Max Weber? Explain your response. You may gather information from the college catalog, available organizational charts, and/or via interviews with personnel who work in the organization. Draw and organizational chart to demonstrate the characteristics.

3. Visit a factory with an assembly line, preferably a large factory. Obtain or construct an organizational chart of the factory. Try to speak with several employees who occupy different positions in the factory. Carefully record all responses. The questions are: (1) What is their specific job description? (2) Why do they work there? (3) What is their specific responsibility? (4) Are they satisfied with their position? Why? Why not? (5) What about the formal and informal aspect of the work environment? What is the degree of friendship with other workers or management? (6) Do they have any control over their responsibilities? Explain. (7) Are there any improvements they would make? If so, what specifically? Submit your papers at the appropriate date, following directions provided by your professor.

4. When Hurricane Katrina (August 29, 2005) hit the Gulf Coast, the failure of local, state, and federal government agencies to respond immediately and the subsequent dysfunctions that followed created national and international debates and resultant programs of assistance. Research this disaster in various publications from that time. Provide a description of what happened, what areas and cities of the United States were affected, and the resulting effects of this devastating hurricane. Research the *present* situation of the people, the cities and the areas affected. Provide statistical as well as socio-cultural descriptions of the hurricane and its major impact on the major social institutions: the family, education, religious, economic, and political institutions. Provide sketches of the people affected and their personal stories. Provide any pictures, statistical tables, and any other information pertinent to this topic. Submit your paper following your professor's directions of the writing of your paper/project.

INTERNET ACTIVITIES

1. Go to **Verstehen**, **http://www.faculty.rsu.edu/~felwell/Theorists/Weber/Whome.htm**, the Max Weber web site and read about Bureaucracy, Rationality, Oligarchy and Ideal Types.
2. Here are a number of *McDonaldization* sites. There are even more "out there" on the World Wide Web. As you begin to study this theory, take a look at the WWW and ask for more information. Hold a "show and tell" with other class members and relate ideas from these sites to what you have learned about groups and organizations and Weber's theories:
   http://www.mcdonaldization.com/main.shtml;
   http://www.geocities.com/mcdonaldization/; and,
   http://www.umsl.edu/~rkeel/010/mcdonsoc.html

3. Read what Thich Naht Hanh, a nominee for the Nobel Peace Prize, writes in his recent best seller book, *The Art of Power,* offering suggestions for effective and instrumental leadership.

4. The sites below provide information on the devastating impact of Hurricane Katrina. From the information at the sites describe how the victims of the hurricane are coping What has been the federal government's response to the crisis? Locate current additional sites to gather the most recent information and present your written report in class: **http://www.msnbc.msn.com/id/9107338/**; **http://www.hurricaneadvisories.com/**

5. To learn more about which people volunteer most, access the homepage of a special project by Lester M. Salamon and associates at **http://www.ihu.edu/~cnp/**

INFOTRAC COLLEGE EDITION EXERCISES

Visit the **InfoTrac College Edition** website at: **http://www.wadsworthmedia.com/webtutor/infotrac.htm**. You will arrive at a screen that enables you to search topics.

1. Look for articles on **Social Isolation** that compare the consequences of web interaction rather than face-to-face interaction in a group. Note that much of the research to date finds negative consequences(loneliness and other psychological consequences). After looking at this research, why do you think people miss interacting with a group face to face?

2. Look for research on **conformity**. What types of situations seem to increase or decrease conformity? What types of group membership is more likely to increase conformity (e.g., gangs, cults, etc.)?

3. This exercise is intended for the professor in the classroom. Have your students search for the keyword **leadership style** in the title, abstract or citation of articles. Instruct them to record information on effective leadership styles and to bring that information to class. Divide your students into small groups and ask them to work in small groups creating a forum on leadership. Instruct them to use information from any of these articles and from this chapter to construct a brochure on "How to Be a Successful Leader." You could then have the students in your class to hold a university/college-wide forum on leadership styles, and offer the brochure to all participants.

4. There are 425 periodical references listed under the category of **bureaucracy**. Try to discover the meaning of this common word. Bring to class examples of these articles. From these articles, construct a definition of the concept along with its essential dimensions.

SOLUTIONS

MULTIPLE CHOICE QUESITONS

1. D, p. 175	15. D, p. 180	29. B, p. 179
2. A. p. 176	16. C, p. 181	30. A, p. 180
3. D, p. 177	17. A, p. 197	31. A, p. 178
4. B, p. 182	18. C, p. 174	32. A, p. 182
5. D, p. 183	19. A, p. 190	33. B, p. 177
6. C, p. 184	20. B, p. 198	34. A, p. 180
7. B, p. 185	21. A, p. 176	35. A, p. 183
8. C, p. 187	22. B, p. 176	36. B, p. 185
9. A, p. 188	23. A, p. 176	37. A, p. 197
10. B, p. 190	24. D, p. 190	38. C, p. 198
11. C, p. 191	25. A, p. 192	39. A, p. 200
12. B, p. 192	26. B, p. 192	40. D, p. 187
13. A, p. 192	27. D, p. 192	
14. D, p. 193	28. A, p. 179	

TRUE/FALSE QUESITONS

1. T, p. 176	8. T, p. 188	15. T, p. 198
2. T, p. 176	9. T, p. 188	16. T, p. 176
3. T. p. 180	10. T, p. 188	17. T, p. 178
4. T. p. 181	11. T, p. 190	18. F, p. 181
5. F, p. 184	12. T, p. 192	19. T, p. 196
6. T, p. 185	13. F, p. 194	20. T, p. 200
7. F, p. 185	14. F, p. 195	

FILL-IN-THE-BLANK QUESTIONS

1. Georg Simmel, p. 181
2. William Graham Sumner, p. 177
3. Charles H. Cooley, p. 176
4. Stanley Milgram, P. 185
5. Irving Janis, P. 188
6. dual labor market, p. 198
7. Robert Michels, p. 198
8. quality circles, p. 200
9. goal displacement, p. 197
10. bureaucratic personality, p. 197
11. humanize, p. 199
12. Japanese, p. 199
13. rational choice, p. 190
14. self-interest, p. 188
15. groupthink, p. 188

7

DEVIANCE AND CRIME

BRIEF CHAPTER OUTLINE

CHAPTER SUMMARY

By definition, **deviance** is any behavior, belief, or condition that violates significant social norms in the society or group in which it occurs. Crime is a form of deviant behavior that violates criminal law and is punishable by fines, jail terms, and other sanctions. A subcategory, **juvenile delinquency**, refers to a violation of a law or the commission of a status offense by young people. All societies create norms in order to define, reinforce, and to help teach acceptable behavior. They also have various mechanisms of **social control**, systematic practices developed by social groups to encourage conformity and to discourage deviance. What causes deviance, and why is it functional for society? **Functionalists** suggest that deviance is inevitable in all societies and serves several functions: it clarifies rules, unites groups, and promotes social change. Functionalists use **strain theory** and opportunity theory, such as **access to illegitimate opportunity structures** and to argue that socialization into the core value of material success without the corresponding legitimate means to achieve that goal accounts for much of the crime committed by people from lower income backgrounds, especially when a person's ties to society are weakened or broken. **Conflict theorists** suggest that people with economic and political power define as criminal any behavior that threatens their own interests and are able to use the law to protect their own interests. **Symbolic Interactionists** use **differential association theory**, **differential reinforcement theory**, **rational choice theory of deviance**, **social bonding theory** and **labeling theory** to explain how a person's behavior is influenced and reinforced by others. **Postmodernists'** perspectives on deviance examine the intertwining nature of knowledge, power, and technology on social control and discipline. While the law classifies crime into felonies and misdemeanors based on the seriousness of the crime, sociologists categorize crimes according to how they are committed and how society views them. Some general categories of crime include: **violent, property, public order, occupational and corporate crime, organized crime and political crime.** Studies show that many more crimes are committed than are reported in official crime statistics. In explaining terrorism and crime, the rational choice approach suggests that terrorists are rational actors who constantly calculate the gains and losses of participation in violent—and often suicidal—acts against others. Gender, age, class, and race are key factors in official crime statistics. The criminal justice system includes the police, the courts, punishment, and corrections. These agencies often have considerable discretion in dealing with deviance. The death penalty, or capital punishment, has been used in the United States as an appropriate and justifiable response to very serious crimes and continues to be controversial. As we move into the future, we need new approaches for dealing with crime and delinquency. Equal justice under the law needs to be guaranteed, regardless of race, class, gender, or age. Global crime – the networking of powerful criminal organizations and their associates in shared activities around the world – has expanded rapidly in the era of global communications and rapid transportation networks. Reducing global crime will require a global response, including the cooperation of law enforcement agencies around the world.

LEARNING OBJECTIVES

After reading Chapter 7, you should be able to:

1. Explain the nature of deviance and describe its most common forms.

2. Describe the types of behavior included in conventional crimes.

3. Define the criminal justice system.

4. Describe the underground economy and the ways it enables criminal networks.

5. Discuss the concept of global crime, including the major types of global crime.

6. Discuss the functions of deviance from a functionalist perspective and outline the principal features of strain, opportunity, and control theories.

7. Discuss conflict perspectives on deviance.

8. Identify and distinguish the three varieties of feminist approaches to deviance and crime.

9. Distinguish between legal and sociological classifications of crime.

10. Differentiate between occupational and corporate crime and explain "criminals."

11. Describe organized crime and political crime and explain how each may weaken social control in a society.

12. Compute the official crime statistics for your city.

13. Describe the key components of differential association theory, differential reinforcement theory, social control theory, rational choice theory, and labeling theory.

14. Explain how police, courts, and prisons practice considerable discretion in deal with offenders.

15. State the four functions of punishment and explain how disparate treatment of the poor, all people of color, and white women is evident in the U.S. prison system.

16. Explain why official crime statistics may not be an accurate reflection of actual crime.

KEY TERMS

(defined at page number shown and in glossary)

KEY PEOPLE

(identified at page number shown)

CHAPTER OUTLINE

I.   WHAT IS DEVIANCE?
   A. By definition, **deviance** is any behavior, belief or condition that violates significant social norms in the society or group in which it occurs.
      1. Deviance is relative and it varies in its degree of seriousness: some forms of deviant behavior are officially defined as a crime.
      2. **Crime** is a form of deviant behavior that violates criminal law and is punishable by fines, jail terms, and other negative sanctions.
      3. **Juvenile delinquency** refers to a violation of law or the commission of a status offense by young people.
   B. What is social control?
      1. All societies have norms that govern acceptable behavior and mechanisms of social control, systematic practices developed by social groups to encourage conformity and discourage deviance.
      2. **Criminology** is the systematic study of crime and the criminal justice system, including police, courts, and prisons.
II. FUNCTIONALIST PERSPECTIVES ON DEVIANCE
   A. What causes deviance, and why is it functional for society?
      1. Emile Durkheim regarded deviance as a natural and inevitable part of all societies.

2. Deviance is universal because it serves three important functions:
   a. Deviance clarifies rules.
   b. Deviance unites a group.
   c. Deviance promotes social change.
3. Functionalists acknowledge that deviance also may be dysfunctional for society; if too many people violate the norms, everyday existence may become unpredictable, chaotic, and even violent.
B. Strain Theory: Goals and Means to Achieve Them
   1. According to **strain theory**, people feel strain when they are exposed to cultural goals that they are unable to obtain because they do not have access to culturally-approved means of achieving those goals. Robert Merton identified five ways in which people adapt to cultural goals and approved ways of achieving them:
      a. Conformity
      b. Innovation
      c. Ritualism
      d. Retreatism
      e. Rebellion
C. Opportunity Theory: Access to Illegitimate Opportunities
   1. According to Richard Cloward and Lloyd Ohlin, for deviance to occur, people must have access to **illegitimate opportunity structures** – circumstances that provide an opportunity for people to acquire through illegitimate activities what they cannot achieve through legitimate channels.
III. CONFLICT PERSPECTIVES ON DEVIANCE
A. Deviance and Power Relations
   1. According to conflict theorists, people in positions of power maintain their advantage by using the law to protect their own interests.
B. Deviance and Capitalism
   1. According to the critical approach, the way laws are made and enforced benefits the capitalist class by ensuring that individuals at the bottom of the social class structure do not infringe on the property or threaten the safety of those at the top.
C. Feminist Approaches
   1. While there is no single feminist perspective on deviance and crime, three schools of thought have emerged:
      a. Liberal feminism is based on the assumption that women's deviance and crime is a rational response to gender discrimination experienced in work, marriage, and interpersonal relationships.
      b. Radical feminism is based on the assumption that women's deviance and crime is related to patriarchy (male domination over females) that keeps women more tied to family, sexuality, and home, even if women also have full time paid employment.
      c. Socialist feminism is based on the assumption that women's deviance and crime is the result of women's exploitation by capitalism and patriarchy (e.g., their overrepresentation in relatively low wage jobs and their lack of economic resources).

2. Feminist scholars of color have pointed out that these schools of feminist thought do not include race and ethnicity in their analyses.
3. Approaches Focusing on Race, Class, and Gender
    a. According Arnold's research, criminal behavior of women in her study was linked to class, gender, and racial oppression, which the women experienced daily in their families and at work and school.
    b. According to feminist sociologists and criminologists, research on women as both victims and perpetrators is long overdue.
    c. The research of Martin and Jurik found that women in the criminal justice system experience significant barriers in justice occupations.

IV. SYMBOLIC INTERACTIONIST PERSPECTIVES ON DEVIANCE
    A. Differential Association Theory and Differential Reinforcement Theory
        1. **Differential association theory** states that individuals have a greater tendency to deviate from societal norms when they frequently associate with persons who are more favorable toward deviance than conformity.
        2. *Differential reinforcement* theory suggests that both deviant behavior and conventional behavior are learned through the same social processes.
    B. Rational Choice Theory
        1. The **rational choice theory of deviance** states that deviant behavior occurs when a person weighs the costs and benefits of nonconventional or criminal behavior and determines that the benefits will outweigh the risks involved in such actions.
    C. Control Theory: **Social Bonding**
        1. **Social bond** theory holds that the probability of deviant behavior increases when a person's ties to society are weakened or broken.
    D. Labeling Theory
        1. **Labeling theory** states that deviants are those people who have been successfully labeled as such by others.
        2. **Primary deviance** is the initial act of rule breaking.
        3. **Secondary deviance** occurs when a person who has been labeled a deviant accepts that new identity and continues the deviant behavior.
        4. **Tertiary deviance** occurs when a person labeled a deviant seeks to normalize the behavior by relabeling it as non-deviant.

V. POSTMODERNIST PERSPECTIVES ON DEVIANCE
    A. Postmodernist's perspective on deviance examines the intertwining nature of knowledge, power, and technology on social control and discipline.
    B. Technologies make widespread surveillance and disciplinary power possible in many settings.
        1. These settings include state-police network, factories, schools, and hospitals.
        2. Discipline would not sweep uniformly through society due to opposing forces that would use their power to oppose such surveillance.

VI. CRIME CLASSIFICATIONS AND STATISTICS
    A. How the Law Classifies Crime
        1. Crimes are divided into felonies and misdemeanors based on the seriousness of the crime.

B. Other Crime Categories
  1. The **Uniform Crime Report** is the major source of information on crimes and is complied by the Federal Bureau of Investigation.
      a. **Violent crime** consists of actions – murder, forcible rape, robbery, and aggravated assault – involving force or the threat of force against others.
      b. **Property crime** – includes burglary, motor vehicle theft, larceny theft, and arson.
      c. **Public order crime** – involves an illegal action voluntarily engaged in by participants, such as prostitution or illegal gambling.
      d. **Occupational** and **corporate crime** involves illegal activities committed by people in the course of their employment or financial affairs.
      e. **Organized crime** is a business operation that supplies illegal goods and services for profit.
      f. **Political crime** refers to illegal or unethical acts involving the usurpation of power by government officials, or illegal/unethical acts perpetrated against the government by outsiders seeking to make a political statement, undermine the government, or overthrow it.
C. Crime Statistics
  1. Official crime statistics, such as those found in the *Uniform Crime Report*, provide important information on crime; however, the data reflect only those crimes that have been reported to the police.
  2. The National Crime Victimization Survey and anonymous self reports of criminal behavior have made researchers aware that the incidence of some crimes, such as theft, is substantially higher than reported in the UCR.
  3. Crime statistics do not reflect many crimes committed by persons of upper socioeconomic status in the course of business because they are handled by administrative or quasi judicial bodies.
D. Terrorism and Crime
  1. In the twenty-first century, the United States and other nations are confronted with world terrorism and crime.
  2. The nebulous nature of the "enemy" and other problems have resulted in a "war on terror" launched by the government.
  3. **Terrorism** is the calculated, unlawful use of physical force or threats of violence against persons or property in order to intimate or coerce a government, organization or individual for the purpose of gaining some political, religious, economic or social objective.
E. Street Crimes and Criminals
  1. Gender and Crime
      a. The three most common arrest categories for both men and women are driving under the influence of alcohol or drugs (DUI), larceny, and minor or criminal mischief types of offenses.

    b. Liquor law violations (such as underage drinking), simple assault, and disorderly conduct are middle range offenses for both men and women, and the rate of arrests for murder, arson, and embezzlement are relatively low for both men and women.

    c. There is a proportionately greater involvement of men in major property crimes and violent crime.

2. Age and Crime

    a. Arrest rates for index crimes are highest for people between the ages of 13 and 25, with the peak being between ages 16 and 17.

    b. Rates of arrest remain higher for males than females at every age and for nearly all offenses.

3. Social Class and Crime

    a. Individuals from all social classes commit crimes; they simply commit different kinds of crime.

    b. Persons from lower socioeconomic backgrounds are more likely to be arrested for violent and property crimes; only a very small proportion of individuals who commit white collar or elite crimes will ever be arrested or convicted.

4. Race and Crime

    a. In 2005, whites (including Latino/as) accounted for about 70 percent of all arrests for index crimes; arrest rates for whites were higher in nonviolent property crimes such as fraud and larceny theft, but were lower than the rates for African Americans in violent crimes such as robbery and murder.

    b. In 2005, African Americans made up about 12 percent of the U.S. population but accounted for almost 28 percent of all arrests.

    c. In 2005, Latino/as made up about 13 percent of the U.S. population and accounted for about 13 percent of all arrests; over two-thirds of their offenses were for nonindex crimes such as alcohol and drugs-related offenses and disorderly conduct.

    d. In 2005, about 1 percent of all arrests were Asian-Americans or Pacific Islanders.

    e. In 2005, about 1 percent of all arrests were Native Americans, designated in the UCR as "American Indian" or "Alaskan Native."

    f. Arrest records tend to produce over generalizations about who commits crime because arrest statistics are not an accurate reflection of the crimes actually committed in our society; white collar and elite criminals are not always arrested.

    g. Arrests should not be equated with guilt: being arrested does not mean that a person is guilty of the crime.

5. Crime Victims

    a. Men are more likely to be victimized by crime although women tend to be more fearful of crime, particularly those directed toward them, such as forcible rape.

b. The elderly also tend to be more fearful of crime, but are the least likely to be victimized. A Justice Department study found that native Americans are more likely to be victims of violent crimes. Young men of color between the ages of 12 and 24 have the highest criminal victimization rates.

c. The burden of robbery victimization falls more heavily on males than females, African Americans more than whites, and young people more than middle aged and older persons.

VII.   THE CRIMINAL JUSTICE SYSTEM

A. *The criminal justice system* includes the police, the courts, and correctional facilities. This system is a collection of bureaucracies that has considerable discretion in the use of personal judgment regarding whether to take action on a situation and, if so, what kind of action to take.

B. The **police** are responsible for crime control and maintenance of order.
1. *Racial profiling* remains a highly charged issue.
2. The current emphasis of *community-oriented policing* has enhanced the image of police departments.

C. The courts determine the guilt or innocence of those accused of committing a crime.
1. Defendants may be urged to plea bargain.
2. More serious crimes are more likely to proceed to trail.
3. Sentencing involves various sanctions such as fines, probation, incarceration, house arrest, electronic monitoring and capital punishment.

D. **Punishment** is any action designed to deprive a person of things of value (including liberty) because of something the person is thought to have committed. **Corrections** is often used in place of punishment.
1. Historically, punishment has had four major goals: *retribution*, *general deterrence*, *incapacitation*, and *rehabilitation*.
2. Newer approaches, such as *restoration*, promote peacemaking, rather than punishment for offenders.
3. *Corrections* is a major activity today; more than one out of 20 males, one out of 100 females are under some form of correctional control.
4. Some argue that a *determinate sentence* with *mandatory sentencing guidelines* would help reduce disparities in corrections based on race, ethnicity, and class.

E. The Death Penalty
1. For many years, capital punishment, or the **death penalty**, has been used in the United States as a justifiable response to serious crimes.
2. Scholars have documented race and class biases in the imposition of the death penalty in this country.
3. Questions that remain today regarding capital punishment include the execution of:
a. Possible innocent individuals
b. Juvenile offenders
c. Those not having effective legal counsel
d. Those believed to be insane

4. Two landmark supreme court cases ruled on issues of capital punishment;
    a. the mentally retarded cannot be executed.
    b. juries, not judges, must decide if a convicted murderer should receive the death sentence.
VIII. DEVIANCE AND CRIME IN THE UNITED STATES IN THE FUTURE
    A. Although many people in the United States agree that crime is one of the most important problems facing this country, they are divided over what to do about it.
    B. The best approach for reducing delinquency and crime ultimately is prevention: to work with young people before they become juvenile offenders so as to help them establish family relationships, build self esteem, choose a career, and get an education which will help them pursue that career, as well as to promote social justice regardless of race, class, gender, or age.
IX. THE GLOBAL CRIMINAL ECONOMY
    A. Global Crime–the networking of powerful criminal organizations and their associates in shared activities around the world-have expanded rapidly in the era of global communications and rapid transportation networks.
    B. Global networking and forming strategic alliances allow participants to escape police control and live beyond the laws of any one nation.
    C. Reducing global crime will require a global response, including the cooperation of law enforcement agencies around the world.

ANALYZING AND UNDERSTANDING THE BOXES

After reading the chapter and studying the outline, re-read the boxes and write down key points and possible questions for class discussion.

Sociology and Everyday Life: How Much Do You Know About Peer Cliques, Youth Gangs, and Deviance?

Key Points:

Discussion Questions:

1.

2.

3.

You Can Make a Difference: Seeing the Writing on the Wall—and Doing Something About It!

Key Points:

Discussion Questions:

1.

2.

3.

Sociology and Social Policy: Juvenile Offenders and "Equal Justice Under Law"

Key Points:

Discussion Questions:

1.

2.

3.

Sociology in Global Perspective: The Global Reach of Russian Organized Crime

Key Points:

Discussion Questions:

1.

2.

3.

PRACTICE TESTS

MULTIPLE CHOICE QUESTIONS

Select the response that best answers the question or completes the statement.

1. The text defines "deviance" as any
    a. aberrant behavior.
    b. behavior, belief, or condition that violates social norms.
    c. serious violation of consistent moral codes.
    d. perverted act.

2. _____ refer(s) to systematic practices developed by social groups to encourage conformity and to discourage deviance.
    a. Law
    b. Folkways
    c. Mores
    d. Social control

3. According to functionalists such as Emile Durkheim, deviance serves all of the following functions, **except**:
    a. deviance helps us to identify social dynamite and social junk in a society
    b. deviance clarifies rules
    c. deviance promotes social change
    d. deviance unites a group

4. According to _____ theory, people are sometimes exposed to cultural goals that they are unable to obtain because they do not have access to culturally-approved means of achieving those goals.
    a. containment
    b. status inaccessibility
    c. strain
    d. conflict

5. All of the following are included in Robert Merton's modes of adaptation to cultural goals and approved ways of achieving them, **except**:
    a. retribution.
    b. ritualism.
    c. retreatism.
    d. rebellion.

6. According to _____ theory, a teenager living in a poverty ridden area of a central city is unlikely to become wealthy through a Harvard education, but some of his desires may be met through behaviors such as theft, drug dealing, and robbery.
    a. deviance management
    b. control
    c. illegitimate opportunity structures
    d. critical

7. Which theory would suggest that people must be taught that the risks of engaging in criminal behavior far outweigh any benefits gained from their actions?
   a. postmodern.
   b. critical.
   c. social bonding.
   d. rational choice.

8. _____ theories suggest that the probability of delinquency increases when a person's social bonds are weak and when peers promote antisocial values and violent behavior.
   a. Deviance management
   b. Social bond theory
   c. llegitimate opportunity structures
   d. Critical

9. According to Edwin Lemert's typology, _____ deviance is exemplified by a person under the legal drinking age who orders an alcoholic beverage at a local bar but is not "caught" and labeled a deviant.
   a. primary
   b. secondary
   c. tertiary
   d. adolescent

10. The _____ approach argues that criminal law protects the interests of the affluent and powerful.
   a. functionalist
   b. liberal feminist
   c. interactionist
   d. conflict

11. _____ feminism explains women's deviance and crime as a rational response to gender discrimination experienced in work, marriage, and interpersonal relationships.
   a. Radical
   b. Communist
   c. Liberal
   d. Marxist

12. Marxist/Socialist feminists argue that women's deviance and crime occurs because:
   a. women are exploited by other women.
   b. women are exploited by capitalism and patriarchy.
   c. women experience gender discrimination in work, marriage, and interpersonal relationships.
   d. women are consumers and tend to purchase more than they can afford.

13. All of the following theorists are correctly matched with their theory, **except**:
    a. Howard Becker – labeling theory
    b. Robert Merton – strain theory
    c. Travis Hirschi – differential association
    d. Meda Chesney Lind – feminist approach

14. The _____ is the major source of information on crimes reported in the United States.
    a. Presidential Crime Reports
    b. Crime Statistical Lab Reports
    c. Civil Code Reports
    d. Uniform Crime Reports

15. Punishment for a _____ typically ranges from more than a year's imprisonment to death.
    a. felony
    b. misdemeanor
    c. parole violation
    d. morals crimes

16. Much _____ crime is a violation of positions of trust at the expense of the general public.
    a. occupational or corporate
    b. street
    c. organized
    d. conventional

17. Drug trafficking, prostitution, loan sharking, and money laundering are examples of _____ crime.
    a. occupational/white collar
    b. street
    c. organized
    d. conventional

18. All of the following are examples of political crime, **except**:
    a. unethical or illegal use of government authority for the purpose of material gain.
    b. money laundering.
    c. engaging in graft through bribery, kickbacks, or "insider" deals.
    d. dubious use of public funds and public property.

19. According to the text, rates of arrest
    a. are about the same for males and females at every age group and for most offenses.
    b. are slightly higher for females than males in the younger age levels and for violent crimes.
    c. are about the same for males and females for prostitution due to more stringent enforcement of criminal laws pertaining to male customers.
    d. remain higher for males than females at every age and for nearly all offenses.

20. All of the following are functions of punishment, **except**:
    a. deterrence
    b. retribution
    c. rehabilitation
    d. elimination of social dynamite and social junk

21. Topically, sociologists and criminologists define a _____ as a group of people, usually young, who band together for purposes generally considered to be deviant or criminal by the larger society.
    a. gang
    b. peer formation
    c. criminal conspiracy
    d. juvenile coalition

22. For many years, capital punishment (the death penalty) has been used in the United States as an appropriate and justifiable response to very serious crime. In 2006, _____ inmates were executed and more than 3,300 people awaited executions, having received the death penalty under federal law or the law of one of the 38 states that have the death penalty.
    a. 42
    b. 74
    c. 53
    d. 200

23. _____ imposes a penalty on the offender and is based on the premise that the punishment should fit the crime. The greater the degree of social harm, the more the offender should be punished.
    a. Punishment
    b. Rehabilitation
    c. Restoration
    d. Retribution

24. African Americans are overrepresented in arrest data. In 2005, African Americans made up about 12 percent of the U.S. population but accounted for _____ percent of all arrests.
    a. 28
    b. 15
    c. 50
    d. 60

25. All of the following crimes are considered to be felonies, except:
    a. homicide
    b. aggravated assault
    c. larceny
    d. rape

26. From the conflict perspective, members of stigmatized groups (such as welfare recipients, the homeless, and persons with disabilities) who are costly to society but relatively harmless are referred to as:
    a. social junk
    b. social throwaways
    c. city trash
    d. social dynamite

27. According to research conducted by sociologists Richard Cloward and Llyod Ohlin, _____ gangs emerge in communities that do not provide either legitimate or illegitimate opportunities.
    a. retreatist
    b. criminal
    c. conflict
    d. street

28. A minister who is opposed to war conducts a nonviolent protest at a local military installation, thus committing a trespassing violation. This person has engaged in:
    a. conformity
    b. innovation
    c. rebellion
    d. retribution

29. Sociologist _____ believed that deviance is rooted in societal factors such as rapid social change and lack of social integration among people. As social integration (bonding and community involvement) decreased, deviance and crime increased.
    a. Emile Durkheim
    b. Karl Marx
    c. Edwin Sutherland
    d. Robert Merton

30. _____ is the systematic study of crime and the criminal justice system, including the police, courts, and prison.
    a. criminology
    b. survey
    c. sociology
    d. ethnomethodology

31. _____ social control takes place through the socialization process. Individuals internalize societal norms and values that prescribe how people should behave and then follow those norms and values in their everyday lives.
    a. Internal
    b. Interior
    c. External
    d. Exterior

32. We are most familiar with _____ deviance, based on a person's intention or inadvertent actions. For example, a person may engage in intentional deviance by drinking too much or robbing a bank; or in inadvertent deviance by losing money in a Las Vegas casino or laughing at a funeral.
    a. intentional
    b. behavioral
    c. cultural
    d. social

33. Members of _____ seek to acquire a "rep" (reputation) by fighting over "turf" (territory) and adopting a value system of toughness, courage, and similar qualities.
    a. retreatist gang
    b. criminal gangs
    c. street gangs
    d. conflict gangs

34. Sociologist Travis Hirschi's _____ theory holds that the probability of deviant behavior increases when a person's ties to society are weakened or broken.
    a. illegitimate opportunity
    b. social control
    c. labeling
    d. social bond

35. Sociologist Walter Reckless suggested that many people do not resort to deviance because of _____ - such as supportive family and friends, reasonable social expectations, and supervision by others.
    a. outer containment
    b. informal containments
    c. inner containments
    d. social bonds

36. John is a drug addict who believes that using marijuana or other illegal drugs is no more deviant that drinking alcoholic beverages and therefore should not be stigmatized. John would be in the _____ of labeling.
    a. career stage
    b. primary stage
    c. delinquent stage
    d. tertiary stage

37. Prostitution might be explained as a reflection of society's double standard, whereby it is acceptable for a man to pay for sex but unacceptable for a woman to accept money for such services. This explanation reflects a _____ feminist approach.
    a. Marxist
    b. liberal
    c. conventional
    d. radical

38. A _____ is a serious crime for which punishment typically ranges from more than a year's imprisonment to death.
   a. felony
   b. homicide
   c. misdemeanor
   d. crime

39. Of all factors associated with crime, the age of the offender is one of the most significant. Arrest rates for index crimes are highest for people between the ages of:
   a. 10 and 18
   b. 13 and 25
   c. 36 and 45
   d. 26 and 35

40. The National crime Victimization Survey (NCVS) was developed by the Bureau of Justice Statistics as an alternate means of collecting crime statistics. The most recent (NCVS) indicates that _____ percent of violent crimes are not reported to the police.
   a. 30
   b. 40
   c. 50
   d. 80

TRUE/FALSE QUESTIONS

1. People may be regarded as deviant if they express a radical or unusual belief system.
   T   F

2. According to sociologists, deviance is relative.
   T   F

3. Deviance is found in all societies.
   T   F

4. A person who accepts culturally-approved goals, but rejects the approved means of achieving them is a representation of Mertons' ritualism.
   T   F

5. According to differential association theory, people learn the necessary techniques and motivation for deviant behavior from people with whom they associate.
   T   F

6. Moral entrepreneurs use their views of right and wrong to establish rules by which they expect other people to live.
   T   F

7. According the research of Chambliss, the group known as the "Saints" became negatively stigmatized and eventually dropped out of school.
   T   F

8. Members of stigmatized groups who are costly to society but relatively harmless are categorized as social dynamite.
   T   F

9. Karl Marx wrote extensively about deviance and crime.
   T   F

10. Recently, the radical feminist approach suggests that women's deviance and crime are a rational response to gender discrimination that women experience in families and the workplace.
   T   F

11. A felony is a serious crime with punishment from imprisonment to death.
   T   F

12. The UCR is published annually by the CIA.
   T   F

13. Moral crimes are often referred to as victimless crimes.
   T   F

14. According to the rational choice approach, terrorists are rational actors who constantly calculate the gains and losses of participation in violence and sometimes suicidal acts against others.
   T   F

15. A recent study by the justice department found that Native Americans are more likely to be victims of violent crimes than are members of any other racial ethic group.
   T   F

16. According to Box 7.1 "Sociology and Everyday Life", dealing with the issue of crime, many gang members continue their membership into adulthood.
   T   F

17. In Box 7.2"You Can Make a Difference," "Seeing the Writing on the Wall," one should confront or challenge a person who is tagging a wall or writing graffiti on a public space.
   T   F

18. According to Box 7.3, "Sociology and Social Policy, unlike adult offenders, juveniles are not always represented by legal counsel.
   T   F

19. In Box 7.4, "The Global Reach of Russian Organized Crime," many analysts believe that today organized crime in Russia is well under control of the Russian government.
T   F

20. In that same article, journalists Robert Friedman has documented how more that thirty Russian crime syndicates now operate in the United States.
T   F

FILL-IN-THE-BLANK QUESTIONS

1. _____ is the systematic study of crime and the criminal justice system.

2. According to _____, criminal behavior is learned within intimate personal groups.

3. According to the research of Chambliss, the _____ were negatively labeled whereas the _____ were positively labeled, even though they were both involved in criminal behavior.

4. The _____ views the cause of women's crime as originating in patriarchy.

5. _____ crime comprises illegal activities committed by people in the course of their employment or financial affairs.

6. _____ crimes are illegal acts committed by corporate employees on behalf of the corporation.

7. According to the research of _____ and _____ there are three basic gang types: criminal, conflict, and retreatist, emerge on the basis of what types of structures of illegitimate opportunities are available.

8. According to sociologist _____, people with economic and political power define as criminal any behavior that threatens their own interest.

9. _____ theory holds that the probability of deviant behaviors increases when a person's ties to society are weakened or broken.

10. _____ theory refers to the systematic practices that social groups develop in order to encourage conformity to norms, rules, and laws and to discourage deviance.

11. A process in which the prosecution negotiates a reduced sentence for the accused in exchange for a guilty plea is know as _____.

12. The _____ suggests that both deviant behavior and conventional behavior are learned through the same social processes.

13. The _____ is based on the assumption that women are exploited by both capitalism and patriarchy.

14. The Supreme Court decision, *Ring v. Arizona*, ruled that _____, not judges, must decide whether a convicted murderer should receive the death penalty.

15. The practice of _____ is the use of ethnic or racial background as a means of identifying criminal suspects and remains a highly charged issue in the criminal justice system.

SHORT ANSWER/ESSAY QUESTIONS

1. What is meant by deviant behavior? What are some factors that help determine whether an act will be viewed as deviant?
2. What are the key assumptions of the symbolic interactionist perspective in deviance?
3. What are the primary assumptions of conflict and feminist perspectives on deviance? What are some of the gaps in research that these two perspectives have attempted to fill?
4. What is meant by the global criminal economy? What are some examples? How can global crime be reduced?
5. How are age, race, class, and gender related to crime statistics?

STUDENT CLASS PROJECTS AND ACTIVITIES

1. The organized crime syndicate, the Mafia, has a long and interesting history. You are to research the origin, the purpose, and the growth of the Mafia in Italy and Sicily, noting some similarities and differences of each. In addition to tracing the history of both of these organizations, explain the growth of the Mafia in the United States. Explain some of the original neighborhood programs that the Mafia sponsored. Explain why you think it is difficult to eradicate this crime syndicate.
2. Research the topic of crime rates in a cross-cultural perspective, with specific emphasis on comparing the rates of the U.S. with those of other industrialized countries. Summarize your findings, provide some explanations for your findings, and provide a bibliography of your references. Submit your papers at the appropriate time, following the instructions you have been given in the writing of your paper.
3. As indicated in your textbook, global crime–the networking of powerful criminal organizations and their associates in shared activities around the world–is a relatively new phenomenon, although theses organizations have existed for many years in their country of origin. You are to: 1) select five global criminal organizations; 2) explain the origin, purpose, and growth of the organization; 3) describe some of their illegal activity; 4) examine and explain some key factors in the success of expansion of their criminal activity; 5) provide some possible solutions for reducing global crime; and 6) critique your solutions. What would it take for your solutions to work?
4. Research the identification and expansion of world terrorism. What nations of the world today support and/or sponsor world terrorism? Why? In your research, identify various diverse "cells" of terrorists. Who are their leaders? What theoretical approaches best explain the existence of terrorists? What are some possible solutions for eradicating world terrorism and crime? Provide bibliographic references in your writing up this project. Follow any additional instructions provided by your professor and submit your paper at the appropriate time.

INTERNET ACTIVITIES

1. "**Defining Deviancy Down**,"
   http://www.albany.edu/scj/jcjpc/vol2is5/deviancy.html, explains how Senator Moynihan's Misleading Phrase About Criminal Justice Is Rapidly Being Incorporated Into Popular Culture. Read how this occurs at the site above.
2. Videos available from **Court TV**, **http://www.courttv.com/video/**, on recent trials or specific topics in the criminal justice system typically are well researched, objectively presented, and of great interest to students. The Court TV website offers streaming video feeds that you could use if your class is connected to the Internet. Study specific court cases then follow the progress of each case from this web site. You can view video clips on your computer.
3. The Internet has become a highly useful source for obtaining statistical data and current information about crime in the United States. Search these sites and bring back to class important crime facts.
   http://www.ojp.usdoj.gov/bjs/pubalp2.htm,
   http://www.ojp.usdoj.gov/bjs/welcome.html,
   http://www.ojp.usdoj.gov/bjs/cvict.htm,
   http://www.ojp.usdoj.gov/bjs/drugs.htm
   http://www.ncjrs.gov

INFOTRAC COLLEGE EDITION EXERCISES

Visit the **InfoTrac College Edition** website at:
http://www.wadsworthmedia.com/webtutor/infotrac.htm. You will arrive at a screen that enables you to search topics.

1. Examine the **Juvenile Crime Bill of 1999** (draft). What kind of picture does it tend to paint about juvenile justice in America?
2. Search through the articles on **youth gangs**. Create your own questionnaire, in order to conduct a survey based on factual information you find during your search. Have students on your campus participate in your survey. Write up the results of your survey, following any specific directions provided by your instructor.
3. Search for information on **peer groups**. Type in the phrase *Age Groups* in the search box. Go to the periodical listings. Compose a list of research findings from these articles.
4. There are numerous subdivisions under **deviant behavior**. Work in small groups and select a category of interest and present your group's findings to the class. To enhance your class presentation, create a poster with facts that you find within these categories.
5. Using the periodical references on **deviant behavior**, bring to class those which focus on the theories of deviance from this chapter. (You are not allowed to use articles with the theory in the title!)
6. Look up the word **Terrorism** and compare the various pieces of legislation that have been enacted to address this expanding problem.

SOLUTIONS

MULTIPLE CHOICE QUESTIONS

1. B, p. 206	15. A, p. 223	29. A, p. 210
2. D, p. 208	16. A, p. 225	30. A, p. 209
3. A, p. 210	17. C, p. 225	31. A, p. 208
4. C, p. 210	18. B, p. 226	32. B, p. 206
5. A, p. 212	19. D, p. 228	33. D, p. 213
6. C, p. 212	20. D, p. 223	34. D, p. 219
7. D, p. 218	21. A, p. 205	35. C, p. 219
8. B, p. 219	22. C, p. 235	36. D, p. 220
9. A, p. 220	23. D, p. 233	37. D, p. 216
10. D, p. 214	24. A, p. 230	38. A, p. 223
11. C, p. 216	25. C, p. 223	39. B, p. 228
12. B, p. 216	26. A, p. 215	40. C, p. 226
13. C, p. 222	27. C, p. 213	
14. D, p. 223	28. C, p. 212	

TRUE/FALSE QUESTIONS

1. T. p, 206	8. F, p. 215	15. T, p. 230
2. T, p. 207	9. F, p. 215	16. T, p. 208
3. T, p. 210	10. F, p. 216	17. F, p. 213
4. F, p. 212	11. T, p. 223	18. T, p. 234
5. T, p, 217	12. F, p. 223	19. F, p. 238
6. T, p. 220	13. T, p. 224	20. T, p. 238
7. F, p. 220	14. T, p. 227	

FILL-IN-THE-BLANK QUESTIONS

1. Criminology, P. 209
2. differential association theory, p. 217
3. roughnecks, p. 220; saints, p. 220
4. radical feminist approach, p. 216
5. Occupational (white collar), p. 225
6. Corporate, p. 225
7. Cloward, p. 212; Ohlin, p. 212
8. Richard Quinney, p. 215
9. Social bond, p. 219
10. Social control, p. 208
11. Plea bargaining, p. 233
12. differential reinforcement theory, p. 218
13. Marxist (socialist), p. 116; feminist approach, p. 216
14. juries, p. 236
15. racial profiling, p. 232

8

CLASS AND STRATIFICATION IN THE UNITED STATES

BRIEF CHAPTER OUTLINE

CHAPTER SUMMARY

Social stratification is the hierarchical arrangement of large social groups based on their control over basic resources. A key characteristic of systems of stratification is the extent to which the structure is flexible. **Social mobility** is the movement of individuals or groups from one level in a stratification level to another. **Slavery**, a form of stratification in which people are owned by others, is a closed system. In a **caste system**, people's status is determined at birth based on their parents' position in society. The **class system**, which exists in the United States, is a type of stratification based on ownership of resources and on the type of work that people do. Karl Marx and Max Weber viewed class as a key determinant of social inequality and social change. According to Marx, capitalistic societies are comprised of two classes – the capitalists, who own the means of production, and the workers, who sell their labor to the owners. By contrast, Weber developed a multidimensional concept that focuses on the interplay of **wealth**, **prestige**, and **power**. Sociologists have developed several models of the class structure: one is the broadly based Weberian approach; the second is based on a Marxian approach. Throughout human history, people have argued about the distribution of scarce resources in society. Money, in the form of both income and wealth is unevenly distributed in the U. S. Among the most prosperous nations in today's world, the United States has the highest degree of inequality of income distribution. **Income**—the economic gain derived from wages, salaries, income transfers, or ownership of property—and **wealth**—the value of all of a person's or family's economic aspects—is unevenly distributed in the United States. The stratification of society into different social groups results in wide discrepancies in income, wealth, and access to available goods and services (including physical health, mental health, nutrition, housing, education, and safety). Sociologists distinguish between **absolute poverty**, which exists when people do not have the means to secure the basic necessities of life, and **relative poverty**, which exists when people may be able to afford basic necessities but are still unable to maintain an average standard of living. Poverty is highly concentrated according to age, gender and race/ethnicity. There are both economic and structural sources of poverty. Low wages are a key problem, as are unemployment and underemployment. The United States has attempted to solve the poverty problem, with the most enduring one referred to as social welfare. Functionalists' perspectives on the U.S. class structure view classes as broad groupings of people who share similar levels of privilege based on their roles in the occupational structure. According to the Davis-Moore thesis, positions that are most important within a society, requiring the most talent and training, must be highly rewarded. Conflict perspectives are based on the assumption that social stratification is created and maintained by one group in order to enhance and protect its own economic interests. Symbolic interactionists focus on the micro-level analysis; such as examining the social and psychological factors that influence the contributions of the wealthy to charitable and arts organizations. As the gap between rich and poor, employed and unemployed widens, social inequality will increase in the future if society does not address this problem. Given that the well-being of all people is linked, some analysts are urging a joint effort to regain the American Dream by attacking poverty.

LEARNING OBJECTIVES

After reading Chapter 8, you should be able to:

1. Define income and wealth and describe their relation to social class.

2. Describe Marx's perspective on class position and class relationships.

3. Describe current statistics about the poor in the U.S.

4. Compare Socioeconomic Status and social class.

5. Describe the ways that income is distributed in the U.S.

6. Compare functionalist and conflict approaches to measuring class.

7. Distinguish between absolute and relative poverty and describe the characteristics and lifestyle of those who live in poverty in the U.S.

8. Describe the contributions of the Symbolic Interactionist perspective to understanding social inequality.

9. Outline Weber's multidimensional approach to social stratification and explain how people are ranked on all three dimensions.

10. Discuss slavery and its relationship to global poverty.

11. Distinguish between functionalist and conflict explanations of social inequality.

12. Summarize the most important consequences of inequality in the U.S.

13. Describe the ways that SES has shaped the life choices of members of your own family.

14. Discuss the distribution of income and wealth in the U.S. and describe how this distribution affects life chances.

KEY TERMS

(defined at page number shown and in glossary)

KEY PEOPLE

(identified at page number shown)

Peter Blau and Otis Duncan, p. 254
Patricia Hill Collins, p. 255
Kingsley Davis and Wilbert Moore,
p. 273
Barbara Ehrenrich, p. 258
Dennis Gilbert and Joseph A. Kahl,
p. 256

Michael Harrington, p. 275
Karl Marx, p. 249
Diana Pearce, p. 271
Max Weber, p. 253
Eric Olin Wright, p. 259

CHAPTER OUTLINE

I. WHAT IS SOCIAL STRATIFICATION?
   A. **Social stratification** is the hierarchical arrangement of large social groups based on their control over basic resources.
   B. Max Weber's term **life chances** describes the extent to which persons within a particular layer of stratification have access to important scarce resources.
II. SYSTEMS OF STRATIFICATION
   A. Systems of stratification may be open or closed based on the availability of **social mobility** – the movement of individuals or groups from one level in a stratification system to another.
      1. **Intergenerational mobility** is the social movement experienced by family members from one generation to the next.
      2. **Intragenerational mobility** is the social movement of individuals within their own lifetime
   B. **Slavery**, a closed system, is an extreme form of stratification in which some people are owned by others.
      1. Some analysts suggest that, throughout recorded history, only five societies have been slave societies.
      2. Sociologist Collins suggests that the legacy of slavery is embedded in current patterns of prejudice and discrimination against African Americans.
      3. Engerman believes that slavery will not be totally abolished until debt bondage, child labor, contract labor and coerced work cease to exist throughout the world.
   C. A **caste system** is a system of social inequality in which people's status is permanently determined at birth based on their parents' ascribed characteristics.
      1. In India, caste is based in part on occupation; in South Africa it was based on racial classification.
      2. Cultural beliefs sustain caste systems.
      3. Caste systems grow weaker as societies industrialize; people start to focus on the type of skills needed for industrialization.

D. The class system is a type of stratification based on the ownership and control of resources and on the type of work people do.
   1. A class system is more open than a caste; boundaries between classes are less distinct.
   2. Status comes in part through achievement rather than entirely by ascription.
III. CLASSICAL PERSPECTIVES ON SOCIAL CLASS
   A. Karl Marx: Relationships to the Means of Production
      1. According to Marx, class position in capitalistic societies is determined by people's work situation, or relationship to the means of production.
         a. The **bourgeoisie** or **capitalist class** consists of those who privately own the means of production; the **proletariat**, or **working class**, must sell their labor power to the owners in order to earn enough money to survive.
         b. Class relationships involve inequality and exploitation; workers are exploited as capitalists expropriate a surplus value from their labor; continual exploitation results in workers' **alienation**, a feeling of powerlessness and estrangement from other people and from oneself.
      2. The **capitalist class** maintains its position by control of the society's superstructure – comprised of the government, schools, and other social institutions, which produce and disseminate ideas perpetuating the existing system; this exploitation of workers ultimately results in **class conflict**—the struggle between the capitalist class and working class.
   B. Max Weber: Wealth, Prestige, and Power
      1. Weber's multidimensional approach to stratification examines the interplay among wealth, prestige, and power as being necessary in determining a person's class position.
         a. Weber placed people who have a similar level of **wealth** – the value of all of a person or family's economic assets, including income, personal property, and *income* producing property – and income in the same class.
         b. **Prestige** is the respect or regard with which a person or status position is regarded by others, and those who share similar levels of social prestige belong to the same status group regardless of their level of wealth.
         c. **Power** – the ability of people or groups to carry out their own goals despite opposition from others – gives some people the ability to shape society in accordance with their own interests and to direct the actions of others.
      2. Wealth, prestige, and power are separate continuums on which people can be ranked from high to low; individuals may be high on one dimension while being low on another.
      3. **Socioeconomic status (SES)** – a combined measure that attempts to classify individuals, families, or households in terms of indicators such as income, occupation, and education – is used to determine class location.
IV. CONTEMPORARY SOCIOLOGICAL MODELS OF THE U.S. CLASS STRUCTURE
   A. The Weberian Model of the Class Structure
      1. The Upper (or Capitalist) Class is the wealthiest and most powerful class, comprised of people who own substantial income producing assets. About 1 percent of the population is included in this class.

2. The Upper-Middle Class is based on a combination of three factors: university degrees, authority and independence on the job, and high income. Examples of occupations for this class are highly educated professionals such as physicians, stockbrokers, or corporate managers. About 14 percent of the U. S. population is in this category.
3. The Middle Class has been traditionally characterized by a minimum of a high school diploma. Today, an entry-level requirement for employment in many middle-class occupations requires a two-year or four-year college degree. About 30 percent of the U. S. population is in this class even though most people in this country think of themselves as middle-class.
4. The Working Class is comprised of semiskilled machine operatives, clerks and salespeople in routine, mechanized jobs, and workers in **pink-collar occupations** – relatively low paying, nonmanual semiskilled positions primarily held by women. An estimated 30 percent of the U. S. population is in this class.
5. The Working Poor account for about 20 percent of the U. S. population and live from just above to just below the poverty line; they hold unskilled jobs, seasonal migrant employment in agriculture, lower paid factory jobs, and service jobs (e.g., such as counter help at restaurants).
6. The Underclass includes people who are poor, seldom employed, and are caught in long term deprivation that results from low levels of education and income and high rates of unemployment. About 3 to 5 percent of the U. S. population is in this category.
B. The Marxian Model of the U.S. Class Structure
1. These criteria can be used to determine the class placement for all workers in a capitalist society.
   a. Erik Olin Wright outlined four criteria for placement in the class structure: (a) ownership of the means of production; (b) purchase of the labor of others (employing others); (c) control of the labor of others (supervising others on the job); and (d) sale of one's own labor (being employed by someone else).
2. The Capitalist Class is composed of those who have inherited fortunes, own major corporations, or are top corporate executives who own extensive amounts of stock or control company investments.
3. The Managerial Class includes upper level managers – supervisors and professionals who typically do not participate in company wide decisions – and lower level managers who may be given some control over employment practices, such as the hiring and firing of some workers.
4. The Small-Business Class consists of small business owners, craftspeople, and some doctors and lawyers who may hire a small number of employees but largely do their own work.
5. The Working Class is made up of blue collar workers, including skilled workers (e.g., electricians, plumbers, and carpenters), unskilled blue collar workers (e.g., laundry and restaurant workers), and white collar workers who do not own the means of production, do not control the work of others, and are relatively powerless in the workplace. The working class constitutes about half of all employees in the U. S.

V. INEQUALITY IN THE UNITED STATES

A. Distribution of Income and Wealth
  1. Money is essential for acquiring goods and services. People without money cannot purchase things they need or desire. Income and wealth are very unevenly distributed in the United States. Among prosperous nations, the United States is the number one in inequality of income distribution.
  2. **Income Inequality** is the economic gain derived from wages, salaries, income transfers, and ownership of property. In the last two decades of the twentieth century, the gulf between the rich and the poor widened in the United States. Income distribution varies by race/ethnicity, class and gender.
  3. **Wealth Inequality.** Wealth includes property and other assets such as bank accounts, corporate stocks, bonds and insurances policies. An analysis of the Forbes 400 list of the wealthiest U.S. citizens found that 42 percent of the people on the list had inherited enough wealth to be on the list. Disparities in wealth are more pronounced across racial ethnic and gender lines.

B. Consequences of Inequality
  1. Physical and Mental Health and Nutrition
    a. As people's economic status increases, so does their health status; the poor have shorter life expectancies and are at greater risk for chronic illnesses and infectious diseases. About 46.6 million people in the United States are without health insurance coverage.
  2. Housing
    a. Homelessness is a major problem in the U.S.
    b. Lack of affordable housing and substandard housing are also central problems brought about by economic inequality.
    c. Good health and adequate nutrition are essential to good life chances; low income families are unable to provide adequate food for their children.
  3. Education
    a. Education and life chance are directly linked; while functionalists view education as an "elevator" for social mobility, conflict theorists stress that schools are agencies for reproducing the capitalist class system and perpetuating inequality in society.
    b. Poverty affects the ability of many young people to finish high school, much less enter college.
  4. Crime and Lack of Safety
    a. Both crime and lack of safety on the streets and at home are consequences of inequality.
    b. Poverty and violence are linked; street violence is often not random at all—but a response to profound social inequalities in the inner city.

VI. POVERTY IN THE UNITED STATES

A. Although some people living in poverty are unemployed, many hardworking people with full time jobs also live in poverty.
B. The official poverty line is based on what is considered to be the minimum amount of money required for living at a subsistence level.

C. Sociologists distinguish between **absolute poverty** – when people do not have the means to secure the most basic necessities of life – and **relative poverty** – when people may be able to afford basic necessities but still are unable to maintain an average standard of living.

D. Who Are the Poor?
1. Age: Children are more likely to be poor than older persons; older women are twice as likely to be poor as older men; older African Americans and Latino/as are much more likely to live below the poverty line than are non Latino/a whites.
2. Gender: About two thirds of all adults living in poverty are women; this problem is described as the **feminization of poverty** – the trend in which women are disproportionately represented among individuals living in poverty.
3. Race and Ethnicity: white Americans (non Latino/as) account for approximately two thirds of those below the official poverty line; however, a disproportionate percentage of the poverty population is made up of African Americans, Latino/as, and Native Americans with Native Americans among the most severely disadvantaged persons in the United States.

E. Economic and Structural Sources of Poverty
1. An economic source of poverty is the low wages paid for many jobs: half of all families living in poverty are headed by someone who is employed, and one third of those family heads work full time.
2. Poverty also is exacerbated by structural problems such as (a) *deindustrialization* – millions of U.S. workers have lost jobs as corporations have disinvested here and opened facilities in other countries where "cheap labor" exists – and (b) **job deskilling** – a reduction in the proficiency needed to perform a specific job that leads to a corresponding reduction in the wages paid for that job.

F. Solving the Poverty Problem
1. The United States has attempted to solve the poverty problem with social welfare programs; however, the primary beneficiaries are not poor.
2. Recent major suggestions for solving the poverty problem are: (a) changing past welfare programs, in the name of "welfare reform and establish state-level workfare programs and mandatory time limits on welfare benefits; and (b) providing better training and education in order that the poor might be able to "work their way out of poverty."

VII. SOCIOLOGICAL EXPLANATIONS OF SOCIAL INEQUALITY IN THE UNITED STATES
A. Functionalist Perspectives
1. According to the Davis-Moore thesis:
a. All societies have important tasks that must be accomplished and certain positions that must be filled.
b. Some positions are more important for the survival of society than others.
c. The most important positions must be filled by the most qualified people.
d. The positions that are the most important for society and require scarce talent, extensive training, or both, must be the most highly rewarded.

   e. The most highly rewarded positions should be those that are functionally unique (no other position can perform the same function), and those positions upon which others rely for expertise, direction, or financing.
  2. This thesis assumes that social stratification results in **meritocracy** – a hierarchy in which all positions are rewarded based on people's ability and credentials.
 B. Conflict Perspectives
  1. From a conflict perspective, inequality does not serve as a source of motivation for people; powerful individuals and groups use ideology to maintain their favored positions at the expense of others.
  2. Core values, laws, and informal social norms support inequality in the United States (e.g., legalized segregation and discrimination produce higher levels of economic inequality).credentials.
 C. Symbolic Interactionist Perspectives
  1. Symbolic interactionists focus on microlevel concerns such as the effects of wealth and poverty on people's lives.
  2. Some studies focus on the social and psychological factors that influence the rich to contribute to charitable and arts organizations.
  3. Other researchers have examined the social and psychological aspects of life in the middle class; few have examined rare insights into the social interactions between people from vastly different social classes.
VIII. U.S. STRATIFICATION IN THE FUTURE
 A. According to some social scientists, wealth will become more concentrated at the top of the U.S. class structure; as the rich have grown richer, more people have found themselves among the ranks of the poor.
 B. The gap between the earnings of workers and the income of managers and top executives in the U.S.A. has widened.
 C. Structural sources of upward mobility are shrinking while the rate of downward mobility has increased; the persistence of economic inequality is related to profound global economic changes.
 D. Some call for a united effort to regain the American Dream by attacking poverty.

ANALYZING AND UNDERSTANDING THE BOXES

After reading the chapter and studying the outline, re-read the boxes and write down key points and possible questions for class discussion.

Sociology and Everyday Life: How Much Do You Know About Wealth, Poverty, and the American Dream?

Key Points:

Discussion Questions:

1.

2.

3.

Framing Class in the Media: Taking the TV Express to Wealth and Upward Social Mobility

Key Points:

Discussion Questions:

1.

2.

3.

Sociology and Social Policy: Should Our Laws Guarantee People a Living Wage?

Key Points:

Discussion Questions:

1.

2.

3.

You Can Make a Difference: Feeding the Hungry

Key Points:

Discussion Questions:

1.

2.

3.

PRACTICE TESTS

MULTIPLE CHOICE QUESTIONS

Select the response that best answers the question or completes the statement.

1. Sociologists use the term _____ to refer to the hierarchical arrangements of large social groups based on their control over basic resources.
    a. social stratification
    b. social layering
    c. social distinction
    d. social accumulation

2. The extent to which individuals have access to important societal resources is known as:
    a. relative poverty
    b. absolute poverty
    c. social mobility
    d. life chances

3. A young woman's father is a carpenter; she graduates from college with a degree in accounting, becomes a CPA and has a starting salary that represents more money than her father ever made in one year. This illustrates _____ mobility.
    a. intragenerational
    b. intergenerational
    c. horizontal
    d. subjective

4. All of the following are true statements about slavery, **except**:
    a. Slavery is a closed system in which "slaves" are treated as property.
    b. Slaves were forcibly imported to the United States as a source of cheap labor.
    c. Slavery has ended throughout the world.
    d. Some people have been enslaved because of unpaid debts, criminal behavior, or war and conquest.

5. A _____ system is a system of social inequality in which people's status is permanently determined at birth based on their parents' ascribed characteristics.
   a. class
   b. slavery
   c. capitalist
   d. caste

6. A young woman who comes from an impoverished background works at two full-time jobs in order to save enough money to attend college. Ultimately, she earns a degree, attends law school, graduates with highest honors, and is hired by a firm at a starting salary of $75,000. This person has experienced _____ mobility.
   a. vertical, intragenerational
   b. vertical, intergenerational
   c. horizontal, class
   d. direct, vertical

7. The _____ system is a type of stratification based on the ownership and control of resources and on the type of work people do.
   a. class
   b. caste
   c. bureaucratic
   d. administrative

8. According to _____'s theory of class relations _____.
   a. Weber; the bourgeoisie consists of those who own the means of production.
   b. Marx; the proletariat consists of those who own the means of production.
   c. Marx; class relationships involve inequality and exploitation.
   d. Durkheim; wealth, prestige, and power are all important in determining a person's class position.

9. According to Karl Marx, _____ is a feeling of powerlessness and estrangement from other people and from oneself.
   a. alienation
   b. meritocracy
   c. class conflict
   d. classism

10. When workers overthrow the capitalists, according to Marx, they would eventually create a(n) _____ society.
    a. class
    b. caste
    c. egalitarian
    d. stratified

11. According to the _____ perspective, the explanation of social inequality in the U.S. is _____.
    a. symbolic interactionist; that the beliefs and actions of people reflect their class location in society.
    b. functionalist; that powerful groups use ideology to maintain their favored position at the expense of others.
    c. conflict; that some degree of social inequality is necessary for society to flow smoothly.
    d. absolute poverty; in comparison with the social status of others.

12. All of the following statements regarding Marx's analyses of class are correct, **except**:
    a. Class relationships involve inequality and exploitation.
    b. The exploitation of workers by the capitalist class ultimately will lead to the destruction of capitalism.
    c. The capitalist class maintains its position by control of the society's superstructure.
    d. Wealth, prestige, and power are separate continuums on which people can be ranked from high to low.

13. According to Max Weber, _____ is the respect or regard with which a person or status position is regarded by others.
    a. admiration
    b. power
    c. prestige
    d. rank

14. Entrepreneurs, as identified by Weber, are those who
    a. do not have to work.
    b. work for wages.
    c. are a privileged commercial class.
    d. live off their investments.

15. In 2005, ___ people in the United States were without health insurance coverage.
    a. 15 million
    b. 20 million
    c. 25 million
    d. 46.6 million

16. According to the Weberian model of the U.S. class structure, members of the _____ class have earned most of their money in their own lifetime as entrepreneurs, presidents of corporations, top level professionals, and so forth.
    a. upper-upper
    b. lower-upper
    c. upper-middle
    d. middle

162

17. The text points out that a combination of three factors qualifies people for the upper middle class. Which of the following is not one of these factors?
    a. university degrees
    b. authority and independence on the job
    c. inherited wealth
    d. high income

18. Over the past fifty years, Asian Americans, Latino/as, and African Americans have placed great emphasis on _____ as a means of attaining the American Dream.
    a. education
    b. affirmative action
    c. unemployment compensation
    d. vocational training

19. Women employed in pink-collar occupations are mainly classified in the:
    a. working class
    b. working poor
    c. middle class
    d. upper-middle class

20. In the Weberian Model of the U.S. class structure, medical technicians, nurses, lower-level managers, and semi-professionals make up the _____ class.
    a. upper
    b. middle
    c. working
    d. lower

21. The trend in which women disproportionately are represented among individuals living in poverty is referred to as:
    a. absolute poverty
    b. relative poverty
    c. situational poverty
    d. the feminization of poverty

22. All of the following are components of wealth, **except**:
    a. income
    b. a position on the local school board
    c. bank accounts
    d. insurance policies

23. In examining the unequal distribution of income and wealth in the United States, the text notes that
    a. the wealthiest 20 percent of households receive almost 50 percent of the total income "pie."
    b. the poorest 20 percent of households receive about 30 percent of the total income "pie."
    c. income inequalities between the rich and the poor narrowed greatly in 2007.
    d. Differences in median income of married couples and female-headed households narrowed by the year 2005.

24. In the last two decades of the twentieth century, the gulf between the rich and the poor
    a. greatly fluctuated.
    b. decreased dramatically.
    c. widened.
    d. stayed the same.

25. According to the functionalist explanation of social inequality,
    a. all societies have important tasks that must be accomplished and certain positions that must be filled.
    b. the most important positions must be filled by the most qualified people.
    c. the most highly rewarded positions should be those that are functionally unique and on which other positions rely.
    d. all of these choices

26. _____ refers to "the separation of the races."
    a. Slavery
    b. Racism
    c. Apartheid
    d. Racial conflict

27. According to sociologist Karl Marx, the _____ (capitalist class) consists of those who own the means of production – the land and capital necessary for factories and mines.
    a. poor
    b. bourgeoisie
    c. entrepreneurs
    d. rich

28. Prestige rankings have become the foundation for _____ research, which uses sophisticated statistical measurements to assess the influence of family background and education on people's occupational mobility and success.
    a. status attainment
    b. social
    c. class structure
    d. occupation position

29. According to the social class model developed by sociologists Dennis Gilbert and Joseph Kahl, members of the _____ have earned most of their money in their own lifetime as entrepreneurs, presidents of major corporations, sports or entertainment celebrities, or top-level professionals.
    a. upper-middle
    b. lower-upper
    c. upper-lower
    d. lower-upper

30. According to your text, all of the following are factors that have eroded the American Dream for the middle class, **except**:
    a. higher housing prices
    b. occupational insecurity
    c. negative media exposure
    d. cost of living squeeze

31. _____ attempt to determine what degree of control that workers have over the decision-making process and the extent to which they are able to plan and implement their own work.
    a. Conflict theorists
    b. Feminist theorists
    c. Functionalist theorists
    d. Postmodern theorists

32. According to Forbes magazine's 2004 list of the richest people in the world, Bill Gates (co-founder of Microsoft Corporation) was the wealthiest with a fortune of nearly $46.6 billion. According to sociologist Erik Wright's Marxian model, Bill Gates would be in the _____ class.
    a. capitalist
    b. entrepreneur
    c. working
    d. small-business

33. According to sociologist Dennis Gilbert, the top 5% of households alone received more than 20 percent of all income – an amount greater than that received by the bottom _____ of all households.
    a. 20%
    b. 60%
    c. 40%
    d. 80%

34. _____ theorists stress that schools are agencies for reproducing the capitalist class system and perpetuating inequality in society. Parents with limited income are not able to provide the same educational opportunities for their children as are families with greater financial resources.
    a. Conflict
    b. Davis-Moore
    c. Postmodern
    d. Functionalist

35. _____ is an extreme form of stratification in which some people are owned by others. It is a closed system in which people are treated as property and have little or no control over their lives.
    a. Slavery
    b. Caste system
    c. Class system
    d. Global

36. About two-third of all adults living in poverty are women. In 2004, single-parent families headed by women had a _____ percent poverty rate as compared with a 10 percent poverty rate for two-parent families.
    a. 45
    b. 35
    c. 25
    d. 15

37. All of the following are reasons why social inequality may increase in the United States in this century, **except**:
    a. the purchasing power of the dollar has stagnated or declined since the early 1970s
    b. federal tax laws in recent years have benefited corporation and wealthy families at the expense of middle- and lower-income families
    c. people living in poverty are demanding a larger share of the U.S. budget
    d. structural sources of upward mobility are shrinking while the rate of downward mobility has increased

38. Based on in-depth interviews and participant observation, sociologists Judith Rollins examined rituals of _____ that were often demanded by elite white women of their domestic workers, who were frequently women of color.
    a. deference
    b. inequality
    c. deviance
    d. disengagement

39. _____ focus on microlevel concerns and usually do not analyze larger structural factors that contribute to inequality and poverty.
    a. Feminists
    b. Symbolic interactionists
    c. Conflict
    d. Functionalists

40. All of the following statements regarding poverty and children are in the United States are TRUE, **except**:
    a. poor children are mainly from two parent households
    b. children in single-parent households headed by a woman are more likely to live in poverty
    c. many governmental programs established to alleviate childhood poverty have been cut back or eliminated
    d. most (55%) of white children, under 18, in female-headed households lived below the poverty line

TRUE/FALSE QUESTIONS

1. According to Max Weber, lifestyle describes the extent to which persons within a particular layer of stratification have access to important scarce resources.
   T   F

2. People no longer believe in the "American Dream."
   T   F

3. Both Karl Marx and Max Weber viewed class as an important determinant of social inequality.
   T   F

4. SES reflects the Weber multidimensional approach to determine social class.
   T   F

5. Most people in the U.S. identify themselves as the middle class.
   T   F

6. Of all the class categories, according to the Weberian Model, the one most shaped by formal education is the upper class.
   T   F

7. Accounting to the Weberian model of social class, an estimated thirty percent of the U.S. population is the working class.
   T   F

8. The working poor account for about 10 percent of the U.S. population.
   T   F

9. About 3 to 5 percent of the U.S. population is the underclass.
   T   F

10. People of color have owned small businesses in the United States throughout U.S. history.
    T   F

11. The Marxian model suggests that the working class consists of about 50 percent of all employees of the U.S.
    T   F

12. Relative poverty exists when people do not have the means to secure the most basic necessities of life.
    T   F

13. About two-thirds of all adults living in poverty are men.
    T   F

14. TANF, as a social welfare program, was replaced by AFDC.
    T   F

15. The Davis-Moore thesis assumes that social stratification results in meritocracy.
    T   F

16. According to Box 8.2, "Framing Class in the Media," the television program "Deal or No Deal" is "framed" to highlight the idea that getting rich in the United States is very difficult.
T   F

17. According to Box 8.3, "Should our Laws Guarantee People a Living Wage?," the high compensation packages received by many companies' chief executive officers is one of the major causes of social inequality in the United States.
T   F

18. According to Box 8.4, "You Can Make a Difference," about 3,000 homeless people in Washington D.C. are fed every day.
T   F

19. According to the photo essay, "What Keeps the American Dream Alive," millions of people around the world see the United States and believe in the American dream.
T   F

20. According to the photo essay, a large number of Americans each year win a very large lottery drawing.
T   F

FILL-IN-THE-BLANK QUESTIONS

1. _____ is the value of all of a person's or family's economic assets.

2. _____ is the ability of people or groups to achieve their goals despite opposition from others.

3. _____ is a combined measure that attempts to classify individuals, families, or households in terms of indicators such as income, occupation, and education to determine class.

4. Karl Marx identified _____ as the struggle between the capitalist class and the working class.

5. A reduction in the proficiency needed to perform a specific job that leads to a corresponding reduction in the wages for that job is known as _____.

6. Occupations that are relatively low-paying, non-manual, semi-skilled positions primarily held by women are called _____ occupations.

7. The poverty that exists when people do not have the means to secure the most basic necessities of life is known as _____ poverty.

8. _____ poverty exists when people may be able to afford basic necessities but are still unable to maintain an average standard of living.

9. The trend in which women are disproportionately represented among individuals living in poverty is known as _____.

10. A hierarchy in which all positions are rewarded based upon people's ability and credentials is termed _____.

11. The economic gain derived from wages, salaries, income, transfers, and ownership of property is identified as _____.

12. The movement of individuals or groups from one level in a stratification system to another is known as _____.

13. According to Marx, the working class, also known as the _____ are those who must sell their labor to the owners in order to survive.

14. According to the _____ thesis inequality is necessary for the smooth functioning of society.

15. The _____ is a type of stratification based on the ownership and control of resources and on the type of work people do.

SHORT ANSWER/ESSAY QUESTIONS

1. What is the "American Dream?" What assumption is it based upon? What are some of the facts about the "American Dream?"
2. What are some of the characteristics of the "new class society"? What are the two largest classes in the new class and why do they possess fundamentally different and opposed interests?
3. What are some of the consequences of inequality in the United States?
4. Who are the poor in the United States? Why does poverty persist in the United States? What are some economic and structural sources of poverty?
5. What do you think are your own social mobility chances? Will you be better or worse off than your parents? Your grandparents?

STUDENT CLASS PROJECTS AND ACTIVITIES

1. Using theories, concepts, facts, and discussions found in this chapter, as well as from other sources, write a proposal for a long-term plan that would reduce income inequality in our nation. Write your proposal as if you were going to present it to Congress. Document your facts and figures and present compelling evidence and information supporting your proposal. You must give sufficient goals and objectives of your proposal and provide proof that your plan is a feasible one. Type up your proposal and submit your paper at the appropriate time and in the style provided by your instructor.
2. Examine social stratification in either your own hometown, or the town/city in which your now reside. Conduct a driving route that will take you into a variety of neighborhoods that represent the six social classes as discussed on pages 280-283 of the text. You can probably secure census track data that will be particularly helpful for this project. Drive through the various neighborhoods recording the characteristics of the neighborhoods, the type of housing, probable lifestyles, the streets and the condition of the streets, the landscaping, any vehicles, and other

signs that would indicate social class. Look for symbols of wealth or poverty, such as fences, recreational facilities, statues, flagpoles, water fountains, trees, and noise level, or the absence of such. Record anything unique or special about each neighborhood. Look for children in the neighborhoods and describe the children and their activities. Look for signs of conspicuous consumption and/or conspicuous waste. Submit a five-page paper describing each neighborhood using the above information as specific guidelines for the descriptions. In the writing of your project, focus on what you learned and how you felt about conducting this project. Follow any other specific guidelines provided by your instructor (if there are any) and submit your paper in the appropriate format at the appropriate time.

3. Research the historical creation of the "War on Poverty" programs. What book, written by sociologist Michael Harrington, served as a "reader" to then President John F. Kennedy that documented the vast existence of U.S. poverty? What were some of that book's startling statistics? Which U.S. President declared war on poverty? When? Where? Why? How did the "name" of the programs—the War on Poverty—originate? What was the country's first response to this call to fight poverty?

4. What major programs were set up as a part of this "war?" Name and describe the programs. What eventually happened to the "War on Poverty?" What programs still exist? Write up and submit your paper, following any specific directions from your professor.

INTERNET ACTIVITIES

1. To read about welfare reform, go to the **Urban Institute**, **http://www.urban.org/**, a non-partisan economic and social policy research organization. Once there, look for the Focus on… box, and click on Welfare Reform to find current research on this topic.

2. The Bureau of Labor Statistics provides a wide range of economic data. Examine the **Economy at a Glance** section, **http://www.bls.gov/eag/**, which provides a regional and state-by-state breakdown of economic data. Then, read the section on Industry at a Glance, which provides the same kind of information on specific industries. Click on your state or the region in which you live to get current local information.

3. Research Harvard University's **Inequality and Social Policy Web Site Think Tank Links**: **http://www.ksg.harvard.edu/inequality/Focus/links.htm**. Explore some of the ideas that people are talking and "thinking" about in regards to inequality and social policy.

4. Use these websites to collect facts about poverty in the United States. Use critical thinking skills to construct reasoned judgments about "who are the poor?" **The U.S. Census Bureau Income Statistics http://www.census.gov/ftp/pub/hhes/www/income.html, The U.S. Census Bureau Poverty Statistics http://www.census.gov/ftp/pub/hhes/www/poverty.html, and The 2001 Health and Human Services 2001 Poverty Guidelines http://aspe.hhs.gov/poverty/01poverty.htm**

INFOTRAC COLLEGE EDITION EXERCISES

Visit the **InfoTrac College Edition** website at:
http://www.wadsworthmedia.com/webtutor/infotrac.htm. You will arrive at a screen that enables you to search topics.

1. Look up **Social Class**. Search for articles that address the plight of poor children. Bring a summary of the articles to class. Collectively construct a picture of the problems facing children in poverty.

2. Do a keyword search for the **American Dream**. What are the different ways that this term is used? What articles use the term to address social class issues? How often is this term used in a positive way? How has it become a negative symbol in our culture? Why do you think we use the word *dream* when we talk about our own social condition?

3. Look up **Karl Marx**. There are a number of references to his work *The Communist Manifesto*. You may want to actually read portions of this historic document.

4. Research five articles using InfoTrac. These articles must include all of the following terms: **Caste System**, **Class System**, **Slavery** and **Poverty**. State the journal, author, and title of each article. Once you complete this, write a brief synopsis on each of the 5 articles. Make certain that your paper meets your professor's requirements.

5. Discover the relationship between **social networks** and **occupational prestige**. An excellent source is the article, *Social Networks and Prestige Attainment: New Empirical Findings* (The American Journal of Economics and Sociology, Oct 1999).

SOLUTIONS

MULTIPLE CHOICE QUESTIONS

1. A, p. 244	15. D, p. 265	29. B, p. 256
2. D, p. 244	16. B, p. 256	30. C, p. 257
3. B, p. 244	17. C, p. 256	31. A, p. 274
4. C, p. 246	18. A, p. 256	32. A, p. 261
5. D, p. 248	19. A, p. 258	33. C, p. 263
6. B, p. 248	20. B, p. 257	34. A, p. 268
7. A, p. 248	21. D, p. 271	35. A, p. 245
8. C, p. 249	22. B, p. 263	36. B, p. 271
9. A, p. 252	23. A, p. 263	37. C, p. 275
10. C, p. 252	24. C, p. 263	38. A, p. 274
11. A, p. 275	25. D, p. 274	39. B, p. 274
12. D, p. 253	26. C, p. 248	40. D, p. 271
13. C, p. 253	27. B, p. 249	
14. C, p. 253	28. A, p. 254	

TRUE/FALSE QUESTIONS

1. F, p. 244	8. F, p. 258	15. T, p. 274
2. F, p. 246	9. T, p. 258	16. F, p. 250
3. T, p. 249	10. T, p. 261	17. T, p. 270
4. T, p. 254	11. T, p. 262	18. T, p. 276
5. T, p. 257	12. F, p. 269	19. T, p. 266
6. F, p. 256	13. F, p. 271	20. F, p. 266
7. T, p. 257	14. F, p. 273	

FILL-IN-THE-BLANK QUESTIONS

1. Wealth, p. 253
2. Power, p. 253
3. Socioeconomic status, p. 254
4. Class conflict, p. 252
5. job deskilling, p. 272
6. pink-collar, p. 258
7. absolute, p. 269
8. Relative, p. 269
9. feminization of poverty, p. 271
10. meritocracy, p. 274
11. income, p. 262
12. social mobility, p. 244
13. proletariat, p. 249
14. David-Moore, p. 273
15. class system, p. 248

9

GLOBAL STRATIFICATION

BRIEF CHAPTER OUTLINE

WEALTH AND POVERTY IN GLOBAL PERSPECTIVE
PROBLEMS IN STUDYING GLOBAL INEQUALITY
    The "Three Worlds" approach
    The Levels of Development Approach
CLASSIFICATION OF ECONOMIES BY INCOME
    Low-income economies
    Middle-income economies
    High-income economies
MEASURING GLOBAL WEALTH AND POVERTY
    Absolute, Relative, and Subjective Poverty
    The Gini Coefficient and Global Quality of Life Issues
GLOBAL POVERTY AND HUMAN DEVELOPMENT ISSUES
    Life expectancy
    Health
    Education and Literacy
    Persistent Gaps in Human Development
THEORIES OF GLOBAL INEQUALITY
    Development and Modernization Theory
    Dependency Theory
    World Systems Theory
    The New International Division of Labor Theory
GLOBAL INEQUALITY IN THE FUTURE

CHAPTER SUMMARY

Global stratification refers to the unequal distribution of wealth, power, and prestige on a global basis, resulting in people having vastly different lifestyles and life chances, both within and among the nations of the world. The social and economic gaps between the **developed nations** and the **developing nations** of the world are much more pronounced than they are in the United States. One approach to defining global stratification is through the "**Three Worlds**" **approach**. **First World** nations are the rich, industrialized countries having primarily capitalistic economies and democratic political systems. **Second World** nations are those countries having a moderate level of economic development and a moderate standard of living. **Third World** countries are the poorest countries with little or no industrialization, having lowest standards of living, shortest life expectancy, and highest mortality. Closely linked to the "three worlds" concept are the levels of development approach using terms such as "developed nations," "developing nations," "less-developed nations," and "underdevelopment." The World Bank classifies nations into three economic categories: low-income economies, middle-income economies, and high-income economies. Using the **gross domestic product** is now a means of measuring wealth and power on a global basis. Global poverty is sometimes defined in terms of *absolute*, *relative*, and *subjective* poverty. The World Bank uses the **Gini coefficient** as its measure of income inequality. Using the Human Development Index, the United Nations Development Program has established criteria for measuring the level of development in a country: life expectancy, education, and living standards. Because of **social exclusion**, a *fourth world* is developing; one where people are being systematically barred from access to positions that would enable their having an autonomous livelihood. Overall, the gap between the poorest nations and the middle-income nations has continued to widen. Social scientists use four primary theoretical perspectives to explain the persistence of global inequality: (1) development and modernization theory; (2) dependency theory; (3) world systems theory and (4) the new international division of labor theory. The future prospects of global inequality range from more to less optimistic predictions. Most analysts agree a nation of people can enjoy global prosperity only by ensuring that other people around the world have the opportunity to survive and thrive in their own surroundings.

LEARNING OBJECTIVES

After reading Chapter 9, you should be able to:

1. Define and describe global stratification.
2. Define and describe the "three worlds" approach used to classify nations of the world.
3. Explain the levels of development approach used for describing global stratification.
4. Explain how poverty is defined on a global basis.
5. Distinguish among absolute, relative, and subjective poverty.
6. Describe the future prospects of global inequality
7. Identify and explain the use of the Gini coefficient.
8. Explain the new international division of labor theory.
9. Describe some of the important ways that international aid has helped to fight global poverty and disease.
10. Classify and describe nations of the world by the three economic categories.
11. Compare and contrast the four major theories of global inequality.
12. Discuss global poverty and its effects upon human development.
13. Describe the contributions of the World Health Organization in addressing problems associated with global stratification.

KEY TERMS

KEY PEOPLE

CHAPTER OUTLINE

I. WEALTH AND POVERTY IN GLOBAL PERSPECTIVE
    A. *Global stratification* refers to the unequal distribution of wealth, power, and prestige on a global basis.
    B. Global stratification results in people having vastly different life styles and life chance both within and among the nations of the world.
    C. The world is divided into unequal segments characterized by *high-income countries*, *middle-income countries*, and *low-income countries*; however, economic inequality is not the only dimension of global stratification; social inequality may result from factors such as discrimination based on race, ethnicity, gender, or religion.
II. PROBLEMS IN STUDYING GLOBAL INEQUALITY
    A. One of the problems is determining what terminology should be used to refer to the distribution of resources in various nations.
    B. After World War II, the "three worlds" approach was utilized to distinguish among nations based upon their economic development and their standard of living.
        1. First World nations consist of the rich, industrialized nations that primarily have capitalistic economic systems and democratic political systems.
        2. Second World nations consist of the countries with at least moderate level of economic development and a moderate standard of living.

3. Third World nations consist of the poorest countries with little or no industrialization and the lowest standards of living, shortest life expectancies, and high mortality rate.
4. Recently, the term *Fourth World* is used to describe the "multiple black holes of **social exclusion**" of people in wide-range areas of the world, including those from sub-Saharan Africa to U.S. inner-city ghettos.
   C. The levels of development approach include concepts such as developed nations, developing nations, less-developed nations, and underdevelopment.
1. Leaders of the developed nations argue that economic development and growth is the primary way to solve poverty problems of the underdeveloped nations.
2. This viewpoint requires that people in the less-developed nations accept the beliefs and values of people in the developed nations.
3. If nations could increase the GNP, then social and economic inequality among their citizens could be reduced.
4. However, improving a country's GNP did not tend to reduce the poverty of the poorest people in that country and inequality increased even with greater economic development; some analysts blamed high rates of population growth in the underdeveloped nations.
III. CLASSIFICATION OF ECONOMIES BY INCOME
   A. The World Bank classifies nations into three economic categories.
   B. Low-income economies are nations that have a GNP per capita of $875 or less in 2005. About half of the world's population lives in the *fifty-four* low-income economies, with more women around the world tending to be more impoverished than men—a situation known as *the global feminization of poverty.*
   C. Middle-income economies are nations that have a GNP per capita of between $876 and $10,725 in 2005. About one-third of the world's population resides in the ninety-eight nations with middle-income economies. These nations typically have a higher standard of living and export diverse goods and services, ranging from manufactured goods to raw materials and fuels.
   D. High-income economies are found in fifty-six nations that have a GNI per capita of **more** than $10,725 in 2005 and continue to dominate the world economy, despite *capital flight* and *deindustrialization.*
IV. MEASURING GLOBAL WEALTH AND POVERTY
   A. In recent years, the United Nations and World Bank have begun to use the *gross domestic product* measurement—all the goods and services produced within a country's economy during a given year.
   B. Global poverty is sometimes defined in terms of *absolute*, *relative*, and *subjective poverty*.
   C. The Gini Coefficient and Global Quality of Life Issues
1. The World Bank uses as its measure of income inequality the *Gini coefficient,* which ranges from zero (when everyone has the same income) to 100 (when one person receives all the income).
2. Using this measure, the World Bank concluded that inequality increased in specific nations in 2005, such as Bulgaria, the Baltic countries, and the Slavic countries of the former Soviet Union.

V. GLOBAL POVERTY AND HUMAN DEVELOPMENT ISSUES
    A. Since the 1970s, the United Nations has focused on human development as a crucial factor fighting poverty.
    B. *Average life expectancy* has increased by about a third in the past three decades and is now more than 70 years in 87 countries; however, no country has reached a life expectancy for men of 80 years, but 19 countries have now reached an expectancy of 80 years or more for women.
    C. *Health* is defined by the World Health Organization as a stage of complete physical, mental, and social well-being and not merely the absence of disease or infirmity. Many people in low-income nations suffer from infectious and other diseases. New diseases have recently emerged in countries all over the world, while some middle-income countries are experiencing rapid growth in degenerative diseases such as cancer and coronary heart diseases.
    D. *Education* is fundamental to increasing literacy.
        1. What is literacy and why is it important for human   development?
        2. UNESCO defines a literate person as "someone who can, with understanding, both read and write a short, simple statement on their everyday life."
        3. The adult literacy rate in the low-income countries is about half of that of the high-income countries.
    E. *Persistent gaps* in human development paint an overall dismal picture for the world's poorest people.
        1. The gap between the poorest nations and the middle-income nations has continued to widen.
        2. Although more women have paid employment than in the past, more and more women are still finding themselves in poverty because of increases in single person and single-parent households headed by women and low-wage work employment.
VI. THEORIES OF GLOBAL INEQUALITY
    A. The most widely known theory is the **development and modernization theory**.
        1. This perspective links global inequality to different levels of economic development and suggests that low-income economies can move to middle- and high-income economies by achieving self-sustained economic growth.
        2. Along with intensive economic growth, the less developed nations can improve their standard of living with accompanying changes in people's beliefs, values, and attitudes toward work.
        3. Walt Roston suggested that all countries go through four stages of economic development: the *traditional stage*, *take-off stage*, *technological maturity*, and *high mass consumption*. In this fourth and final state the country reaches a high standard of living.
VII. **Dependency theory** states that global poverty can partially be attributed to the exploitation of the low-income countries by the high-income countries.
    A. Dependency theory has been more often applied to the newly industrializing countries (NICs) of Latin America.
    B. Scholars examining the NICs of East Asia found that dependency theory had little or no relevance to economic growth and development in that part of the world.

C. **World Systems theory** suggests that under capitalism, a global system is held together by economic ties.
   1. The capitalist world-economy is a global system divided into a hierarchy of three major types of nations: core, semiperipheral, and peripheral.
   2. **Core nations** are dominant capitalist centers characterized by high levels of industrialization and urbanization, such as the United States, Germany, and Japan.
   3. **Semiperipheral nations** are more developed than peripheral nations, but less developed than core nations, such as Taiwan, South Korea, Mexico, Brazil, India, Nigeria, and South Africa.
   4. **Peripheral nations** are dependent on core nations for capital, have little or no industrialization (other than brought in by core nations), and have uneven patterns of urbanization. Most low-income countries in Africa, South American and the Caribbean are peripheral Nations.
D. According to the new **international division of labor theory**, commodity production is being split into fragments that can be assigned to whichever part of the world can provide the most profitable combination of capital and labor.
   1. The global nature of these activities is referred to as *global commodity chains*, a complex pattern of international labor and production.
   2. *Producer-driven commodity chains* describe industries in which transnational corporations play a central part in controlling the production process.
   3. *Buyer-driven commodity chains* refers to industries wherein large retailers, brand-named merchandisers, and trading companies set up decentralized production network in various middle-and low-income counties.
VIII. GLOBAL INEQUALITY IN THE FUTURE
A. Social scientists describe an optimistic or a pessimistic scenario for the future depending upon the theoretical framework they apply in studying global inequality.
   1. Some analysts highlight the human rights issues embedded in global inequality; others focus primarily on an economic framework.
   2. In the future, continued population growth, urbanization, and environmental degradation threaten even the meager living conditions of those residing in high-income countries; the quality of life diminishes as natural resources are depleted.
   3. A more optimistic scenario suggests that, with modern technology and worldwide economic growth, it might be possible to reduce absolute poverty and to increase people's opportunities, ensuring that people all over the world have the opportunity to survive and thrive in their own surroundings.

ANALYZING AND UNDERSTANDING THE BOXES

After reading the chapter and studying the outline, re-read the boxes and write down key points and possible questions for class discussion.

Sociology and Everyday Life: How Much Do You Know About Global Wealth and Poverty?

Key Points:

Discussion Questions:

1.

2.

3.

Sociology in Global Perspective: Poverty and the Curse of civil War in Africa

Key Points:

Discussion Questions:

1.

2.

3.

Sociology and Social Policy: Should We in the United States Do Something About Child Labor in Other Nations?

Key Points:

Discussion Questions:

1.

2.

3.

You Can Make a Difference: Global Networking to Reduce World Hunger and Poverty

Key Points:

Discussion Questions:

1.

2.

3.

PRACTICE TESTS

MULTIPLE CHOICE QUESTIONS

Select the response that best answers the question or completes the statement.

1. The unequal distribution of wealth, power, and prestige on a global basis is referred to as
    a. global stratification.
    b. global layering.
    c. global distinction.
    d. global accumulation.

2. The income gap between the richest and the poorest 20 percent of the world population
    a. continues to widen.
    b. is greater in urban than in rural areas.
    c. has significantly decreased in the last decade.
    d. is slowly beginning to decline.

3. According to Gunnar Myrdal, a precondition for eliminating poverty in a society is to have
    a. a good public education system for all its citizens.
    b. a public works program.
    c. a democratic political system.
    d. greater social equality among its citizens.

4. The _____ approach was introduced by social analysts to distinguish among nations on the basis of their levels of economic development and the standard of living of their citizens.
    a. levels of development
    b. three worlds
    c. classification of economies
    d. global distinction

5. China and Cuba are classified as _____ nations.
    a. "developed world"
    b. "first world"
    c. "second world"
    d. "third world

6. First world nations are best represented by the countries of
    a. Japan, the United States, China.
    b. New Zealand, the United States, Great Britain.
    c. Canada, China, Korea.
    d. Russia, Australia, Japan.

7. When individuals and/or groups are systematically barred from access to positions that would enable them to have an autonomous livelihood in keeping with the social standards and values of a given social context, the process of _____ occurs.
    a. social isolation
    b. social exclusion
    c. social decay
    d. social stigma

8. The _____ Plan, named after a U. S. Secretary of State provided massive sums of money in direct aid and loans to rebuild countries destroyed during World War II.
    a. Albright
    b. Powell
    c. Marshall
    d. Rice

9. The term _____ refers to all the goods and services produced in a country in a given year.
    a. WHO
    b. GDP
    c. NIC
    d. GNI

10. Low-income economies are primarily found in _____ nations, where half of the world's population resides.
    a. Asian and South American
    b. Asian and African
    c. Eastern European and African
    d. South American and Asian

11. The situation wherein women around the world tend to be more impoverished than men is referred to as
    a. global feminization stratification.
    b. global feminization.
    c. global feminization of poverty.
    d. global feminization of labor.

12. Lower-middle income economies include the nations of
    a. Bolivia, Colombia, El Salvador.
    b. South Korea, the Ukraine, Mexico.
    c. Mexico, Guatemala, Russia.
    d. Mozambique, Bosnia, Turkey.

13. High-income economies are found in _____ nations.
    a. 56
    b. 15
    c. 38
    d. 25

14. The movement of jobs and economic resources from one nation to another is defined as:
    a. capital flight
    b. deindustrialization
    c. transnational flight
    d. economic mobility

15. The closing of plants and factories because of their obsolescence or employment of cheaper workers in other nations is known as:
    a. capital flight
    b. capital destabilization
    c. deindustrialization
    d. reindustrialization

16. The World Bank uses the term _____ to indicate all of the goods and services produced within a country's economy during a given year:
    a. GDP
    b. ENP
    c. GPA
    d. GPO

17. _____ poverty is a condition in which people do not have the means to secure the most basic necessities of life.
    a. subjective poverty
    b. relative poverty
    c. absolute poverty
    d. standard poverty

18. The World Bank uses the term _____ as its measure of income inequality.
    a. GNP
    b. Gini coefficient
    c. relative poverty
    d. Mrydal coefficient

19. According to the _____ theory, the capitalist world-economy is a global system divided into a hierarchy of three major types of nations.
    a. dependency
    b. development and modernization
    c. new international division of labor
    d. world systems

20. According to the _____ theory, commodity production is split into fragments that can be assigned to the part of the world that can provide the most profitable combination of capital and labor.
    a. world systems
    b. new international division of labor
    c. dependency
    d. development and modernization

21. The idea of _____ has become the primary means used in attempts to reduce social and economic inequalities and alleviate the worst effects of poverty in the less industrialized nations of the world.
    a. social equity
    b. development
    c. evolution
    d. degeneration

22. One of the primary problems encountered by social scientists studying _____ and social and economic inequality is what terminology should be used to refer to the distribution of resources in various nations.
    a. global stratification
    b. universal stratification
    c. planetary stratification
    d. central stratification

23. After _____, the terms "First World," "Second World," and "Third World" were introduced by social scientists.
    a. World War I
    b. World War II
    c. the Korean War
    d. the Vietnam War

24. The _____ was introduced by social analysts to distinguish among nations on the basis of their levels of economic development and the standard of living of their citizens.
    a. levels of development approach
    b. three worlds approach
    c. world systems approach
    d. global distinction approach

25. People living in _____ countries, the poorest countries, have little or no industrialization and the lowest standards of living, shortest life expectancies, and highest rates of mortality.
    a. first world
    b. second world
    c. third world
    d. fourth world

26. According to sociologist Manuel Castells, the _____ has disintegrated with the advent of the Information Age, a period in which industrial capitalism is being superseded by global informational capitalism.
    a. first world
    b. second world
    c. third world
    d. fourth world

27. In the United States, Native Americans are designated indigenous people who are descended from a country's aboriginal population and will be considered the:
    a. fourth world
    b. third world
    c. second world
    d. first world

28. Sociologist Manuel Castells stated that as informational capitalism grows around the world, the lack of regular work as a source of income is increasingly a key mechanism in:
    a. social isolation
    b. social inclusion
    c. social exclusion
    d. social exporting

29. According to Figure 9.1, the income gap between the richest and poorest people in the world_____ between 1960 and 2000.
    a. was less pronounced
    b. was most pronounced in the United States
    c. continued to decline
    d. continued to grow

30. The term _____ refers to material well-being that can be measured by the quality of goods and services that may be purchased by the per capita national income.
    a. standard of living
    b. standard of development
    c. standard of the economy
    d. standard of building

31. Among those most affected by poverty in low-income economies are
    a. women and children.
    b. adult men and women.
    c. aged men and women.
    d. aged women.

32. By the year 2000, the wealthiest 20 percent of the world population had almost _____ times the income of the poorest 20 percent.
    a. thirty
    b. fifty
    c. eighty
    d. ninety

33. The relationship between poverty and civil war is not limited to the Democratic Republic of Congo (DRC). According to reports by the ____, the poorest nations in the world are disproportionately likely to be at war, especially those nations which 25 percent of the national income comes from the export of natural resources.
    a. World News Today
    b. World Health Organization
    c. World Bank
    d. North American Free Trade Organization

34. According to the _____ theory, the traditional caste system becomes obsolete as industrialization progresses.
    a. world systems theory
    b. new international division of labor theory
    c. dependency theory
    d. development and modernization theory

35. A complex pattern of international labor and production processes that results in a finished commodity ready for sale in the marketplace is referred to as _____.
    a. global buyer-driven commodity
    b. global labor-intensive chains
    c. global commodity chains
    d. global production chains

36. Sociologist Immanuel Wallerstein stated that most low-income countries in Africa and South America are _____ nations that are dependent on other nations for capital, have little or no industrialization, and have uneven patterns of urbanization.
    a. peripheral
    b. core
    c. semiperipheral
    d. tertiary

37. What country has experienced a 270 percent increase in per capita income over the past twenty years?
    a. Singapore
    b. China
    c. Thailand
    d. South Korea

38. Health is defined in the Constitution of the World Health Organization as:
    a. a state of complete physical, mental, and social well-being
    b. absence of disease or infirmity
    c. a state of optimum health
    d. a state of mental happiness

39. The United Nations Educational, Scientific and Cultural Organization (UNESCO) defines a _____ person as "someone who can, with understanding, both read and write a short, simple statement on their everyday life".
    a. functional illiterate
    b. functional literate
    c. literate
    d. illiterate

40. Athletic footwear companies such as Nike and Reebok and clothing companies like The Gap and Liz Claiborne are example of _____ chains.
    a. individually-driven commodity
    b. producer-driven commodity
    c. international-driven commodity
    d. buyer-driven commodity

TRUE/FALSE QUESTIONS

1. Income disparities between rich and poor countries is greater than income disparities *within* countries.
   T   F

2. Gunner Mrydal believed that the poorest people living in extremely poor nations experience much greater economic hardship than do the poor who live in more affluent countries of the world.
   T   F

3. The Fourth World is a term *originally* designated for indigenous people who are descended from a country's aboriginal population.
   T   F

4. Castells uses the term "Fourth World" to refer to the new developing countries in the continent of Africa.
   T   F

5. In utilizing the "three worlds" approach for studying global inequality, North Korea is an example of a Third World Nation.

6. Ideas regarding underdevelopment were popularized by U.S. President Harry S. Truman.
   T   F

7. According to the development approach in studying global inequality, an increase in the standard of living of a century meant that a nation was moving toward economic development.
   T   F

8. High rates of population growth taking place in the underdeveloped nations have led to an increase in greater economic development of those countries.
   T   F

9. About half of the world's population lives in the 54 low-income economies, according to the World Bank classification.

10. According to the classification of economies by income, many countries referred to as "middle income" have very few of their population living in poverty.
    T   F

11. Subjective poverty is defined as a condition in which people do not have the means to secure the most basic necessities.
    T   F

12. Since the 1970s, the United Nations has more actively focused on human development, as measured by the HDI as a crucial factor for fighting poverty.
    T   F

13. The numbers of people dying worldwide from hunger-related diseases is the equivalent of 200 jumbo jet crashes per day with no survivors.
    T   F

14. About 17 million people die each year from diarrhea, malaria, tuberculosis, and other infectious and parasitic illnesses.
    T   F

15. According to UNESCO, women in low-income countries comprise about two-thirds of those who are illiterate.
    T   F

16. According to Box 9.1, the world's ten richest people are U.S. citizens.
    T   F

17. According to "Sociology and Everyday Life", Box 9.1, the richest fifth of the world's population receives about 50 percent of the total world income.
T   F

18. In analyzing poverty and the cause of civil war in Africa (Box 9.2), if it were not for civil war, the Democratic Republic of Congo could be a wealthy nation.
T   F

19. According to Box 9.3, the boycotts and public pressure to reduce or eliminate exploitative and dangerous child labor always produce good results.
T   F

20. According to Box 9.4, CARE assists the world's poor in their efforts to achieve social and economic well-being.
T   F

FILL-IN-THE-BLANK QUESTIONS

1. The unequal distribution of wealth, power, and prestige on a global basis is known as _____.

2. According to Manuel Castells, _____ is the process by which certain individuals and groups are systematically barred from access to positions that would enable them to have an autonomous livelihood.

3. According to Swedish economist _____, social inequality and economic inequality are the main causes of the poverty of a nation.

4. The concepts of underdevelopment and underdevelopment nations emerged out of the _____ following World War II.

5. The _____ is defined as all the goods and services produced within a country's economy during a given year.

6. When people may be able to afford basic necessities, but are still unable to maintain an average standard of living, measured by comparing one's person's income with the incomes of others is known as _____.

7. According to _____ theory, global poverty is partially caused by the exploitation of the low-income countries by the high-income countries.

8. Sociologist _____ is closely associated with World Systems Theory.

9. According to World Systems Theory _____ nations are dominant capitalist centers characterized by high levels of industrialization and urbanization.

10. According to World Systems Theory _____ nations are more developed than peripheral nations, but less developed than core nations.

11. According to the "lived experience" introductory section of this chapter, without the benefits of school meals furnished by the _____, Olympic medal winner and world marathon record holder Paul Tergat doubts if he would have become a successful long-distance runner.

12. Manuel Castells uses the term _____ to describe the "multiple black holes of social exclusion" existing throughout the planet.

13. Ideas regarding underdevelopment were popularized by President _____ in his inaugural address.

14. One major cause of shorter life expectancy in low-income countries is the high rate of _____.

15. According to the modernization theory of Walt W. Roston, societies in the _____ stage are slow to change because the people hold a fatalistic value system, do not subscribe to the work ethic, and save very little money.

SHORT ANSWER/ESSAY QUESTIONS

1. What do you think are the future prospects for greater equality across and within nations" Cite at least one theoretical framework from this chapter that provides an optimistic scenario and one that provides a pessimistic scenario for the future. Which do you agree with more? Why?
2. What is the global stratification and what effects does it have upon economic inequality?
3. How would the new international division of labor theory explain global stratification? What is meant by global commodity chains, producer-driven commodity chains and buyer-driven commodity chains?
4. What is literacy and why is it important for human development? Why is it especially important for women?
5. Why is the majority of the world's population growing richer while the poorest one-fifth of the world's population remains poor? How are global poverty and human development related? What steps could be taken to reduce global poverty?

STUDENT CLASS PROJECTS AND ACTIVITIES

1. Select one present independent country (or republic) that used to be a republic that comprised the former Union of Soviet Republics (before September, 1991): (1) Armenia; (2) Azerbaijan; (3) Belorussia; (4) ; (5) Georgia; (6) Kasakhstan; (7) Kirghizia; (8) Latvia; (9) Lithuania; (10) Moldavia; (11) Russian S.F.S.R.; (12) Tadshikistan; (13) Turkmenistan; (14) Ukraine; and (15) Uzbekistan. Or select one independent country that used to be part of a whole (such as Bosnia or Croatia or the former country of Yugoslavia). You are to prepare a research report of a minimum of five typewritten pages about their respective republic, or country. Your reports must include: (1) the population; (2) ethnic breakdown of various population groups; (3) date that the country first became a part of the whole and the date that the country became independent from the whole; (4) brief historical background of the country; (5) size (in square miles); (6) major goods, service, and raw products produced; (7) various religious practices; (8) language that is spoken; (9) typical lifestyle of the people; (10) the social and economic status of the people either within their own group or in comparison to an outside group, (11) political sentiments of the

people, i.e., is there a strong sense of nationalism; (12) any other information that enhances the understanding of this country or republic. Submit your paper at the appropriate time, following the instructions of your instructor.

2. Drawing upon World Systems Theory most closely associated with the work of Immanuel Wallerstein, select three countries that are members of ASEAN (Association of South East Asian Nations) who best represent the three major types of nations in the hierarchical position of nations in the global economy. Prepare a research report on the economic conditions of each of the three you selected, including information such as (1) the status of the major infrastructures of each country, such as their transportation systems, their power plants, their telecommunications systems; (2) the status of their financial capital and (3) the status of their human capital. i.e., the human skills needed to handle. Provide substantial explanations of why and how you selected the countries representing each specific category. Include an analysis of the resources and obstacles characterizing each specific country. Are there any global cities (or command posts) in any of the countries that you selected? Provide any other information that supports you selections.

3. In the later part of the twentieth century, countries in various parts of the globe began to form economic, political and/or social alliances. Research past and present alliances, noting the purpose of the alliance, the status of that alliance today, countries that are part of the alliance, and possible or probable future alliances. Some possibilities of the past, present, and future are (1) Atlantic Rim; (2) Asian Pacific (Pacific Rim); (3) the Group of Seven; (4) the Group of 77; (5) NAFTA; (6) Hemispheric Free Trade Area (HFTA); (7) Asia Pacific Economic Cooperation (APEC); (8) the EU (European Union); (9) North Atlantic Free Trade Area; (10) Mediterranean Rim; (11) NATO; and (12) OPEC. Research those alliances that were created during the years of this twenty-first century. Write up your paper in the specific form requested by your professor and submit it on the appropriate date.

4. Research the lives of the world's ten richest people, according to the latest Forbes magazine poll. (1) Provide a minimum one-paragraph biographical sketch on each of the "top ten." (2) Explain how each of them obtained their wealth. (3) What countries do they represent? (4) Who did they replace from the previous year's poll? (5) What are some similarities and differences among the top ten? Submit your paper at the appropriate time, following the instruction of your professor.

5. The world faces no shortage of problems—or of suggested ideas to solve them, such as the world population problems, energy problems, diseases, water and wealth, world poverty, as well as the sheer survival of animal, plant, and human species. According to the Special Issue of **Scientific American**, September, 2005, the world of nations is facing a unique turning point in the next twenty years, creating a "Crossroads for Planet Earth." This special edition puts forth a comprehensive plan for a bright future for planet earth beyond **2050**, responding to these problems. Research each topic in this special edition, or of a similar resource. Identify each specific world problem mentioned above, and provide a probable and possible solution to each. Write and submit your paper/project following specific instructions of your professor.

INTERNET ACTIVITIES

1. Visit the United Nations **Economic and Social Development** Web Site: **http://www.un.org/esa/**. This site has links to information about crime, trade, drugs, population, etc. from a global perspective.
2. Another organization to visit is the **World Bank Group**: **http://www.worldbank.org/**. To read about poverty, click on Poverty Reduction Strategy Papers. What are some of the proposed ways to reduce or end poverty?
3. **UNESCO** stands for the United Nations Educational, Scientific, and Cultural Organization: **http://www.unesco.org/**. What is the mission of UNESCO? To read a quick summary of important facts, click on Did You Know? This list of quick facts addresses literacy, primary education and childhood development, who finances education. How will reducing global illiteracy help the United States, and other countries?
4. A site devoted to international issues of women's health is the **International Women's Health Coalition**: http://www.iwhc.org/. Read about their population policies and efforts to increase AIDS awareness for women around the world.
5. From this site, **http://www.globalissues.org/TradeRelated/Poverty.asp**, you can take a surfing trip to gain facts about stratification, poverty and conflict. To generate a deeper level of interest bring facts and resources to class. A typed up "fact sheet" can be turned in for credit.
6. Visit this site, **http://www.heise.de/tp/english/special/eco/6099/1.html**, to read an essay entitled "Global Poverty in the Late 20th Century" by Michel Chossudovsky, Professor of Economics at the University of Ottawa. Write three critical thinking questions based on their reading of the essay.

INFOTRAC COLLEGE EDITION EXERCISES

Visit the **InfoTrac College Edition** website at: **http://www.wadsworthmedia.com/webtutor/infotrac.htm**. You will arrive at a screen that enables you to search topics.

1. Examine **Human Rights Violations** by searching for articles that address the human rights violations in Serbia, Russia, China, Saudi Arabia, and Sierra Leone to name a few.
2. Type in under the subject **Gini coefficient** and see how this measure of income inequality is used in actual research articles.
3. There are a number of articles under the subject **Women and Poverty**. Find articles related to global poverty. What cultural values are related in some of these articles?
4. There are a number of subdivisions under the subject search for **Developing Countries**. Under the subdivision Environmental Aspects students can learn about the interrelationship between the social structure and the physical environment.

SOLUTIONS

MULTIPLE CHOICE QUESTIONS

1. A, p. 282
2. A, p. 282
3. D, p. 283
4. B, p. 285
5. C, p. 285
6. B, p. 285
7. B, p. 286
8. C, p. 287
9. D, p. 287
10. B, p. 288
11. C, p. 290
12. A, p. 290
13. A, p. 291
14. A, p. 291
15. C, p. 291
16. A, p. 292
17. C, p. 292
18. B, p. 292
19. D, p. 302
20. B, p. 303
21. B, p. 285
22. A, p. 285
23. B, p. 285
24. B, p. 285
25. C, p. 285
26. B, p. 286
27. A, p. 286
28. C, p. 286
29. D, p. 282
30. A, p. 287
31. A, p. 288
32. C, p. 283
33. C, p. 289
34. D, p. 297
35. C, p. 304
36. A, p. 302
37. B, p. 291
38. A, p. 295
39. C, p. 295
40. D, p. 304

TRUE/FALSE QUESITONS

1. F, p. 283
2. T, p. 283
3. T, p. 286
4. F, p. 286
5. F, p. 285
6. T, p. 287
7. T, p. 287
8. F, p. 287
9. T, p. 288
10. F, p. 290
11. F, p. 292
12. T, p. 293
13. F, p. 295
14. T, p. 295
15. T, p. 295
16. F, p. 284
17. F, p. 284
18. T, p. 289
19. F, p. 296
20. T, p. 306

FILL-IN-THE-BLANK QUESTIONS

1. global stratification, p. 282
2. social exclusion, p. 286
3. Gunnar Myrdal, p. 283
4. Marshall Plan, p. 287
5. gross domestic product, p. 292
6. relative poverty, p. 292
7. dependency, p. 300
8. Immanuel Wallerstein, p. 301
9. core, p. 302
10. semiperipheral, p. 302
11. United Nations, p. 281
12. fourth world, p. 286
13. Harry Truman, p. 287
14. infant mortality, p. 293
15. traditional, p. 298

10

RACE AND ETHNICITY

BRIEF CHAPTER OUTLINE

CHAPTER SUMMARY

Issues of race and ethnicity permeate all levels of interaction in the United States. A **race** is a category of people who have been singled out as inferior or superior, often on the basis of physical characteristics such as skin color, hair texture, and eye shape. By contrast, an **ethnic group** is a collection of people distinguished, by others or by themselves, primarily on the basis of cultural or nationality characteristics. Race and ethnicity are ingrained in our consciousness and often form the basis of hierarchical ranking and determine who gets what resources. A **dominant group** is one that is advantaged and has superior resources and rights in a society while a **subordinate group** is one whose members, because of physical or cultural characteristics, are disadvantaged and subjected to unequal treatment by the dominant group and who regard themselves as objects of collective discrimination. **Prejudice** is a negative attitude based on faulty generalizations about the members of selected racial and ethnic groups and is rooted in **stereotypes** and *ethnocentrism*. **Racism** is a set of attitudes, beliefs, and practices that is used to justify the superior treatment of one racial or ethnic group and the inferior treatment of another racial or ethnic group. **Discrimination** – actions or practices of dominant group members that have a harmful impact on members of a subordinate group – may be either **individual** or **institutional discrimination** – involving day to day practices of organizations and institutions that have a harmful impact on members of subordinate groups. According to the interactionist contact hypothesis, increased contact between people from divergent groups should lead to favorable attitudes and behavior when a specific set of criteria are met. Two functionalist perspectives – **assimilation** and **ethnic pluralism** – focus on how members of subordinate groups become a part of the mainstream. Conflict theories analyze economic stratification and access to power in race and ethnic relations: caste and class perspectives, **internal colonialism**, **split labor market theory**, gendered racism, and racial formation theory. Critical race theory derives its foundation from the U.S. civil rights tradition and the writings of several civil rights leaders. The unique experiences of Native Americans, White Anglo Saxon Protestants/British Americans, African Americans, White Ethnics, Asian Americans, Latino/as (Hispanic Americans), and Middle Easterners are discussed, and the increasing racial ethnic diversity of the United States is examined. Globally, many racial and ethnic groups seek self-determination resulting in ethnic wars in some areas. In the United States, racial and ethnic diversity is increasing; several possibilities of this increase remain. It is predicted that by 2056, the roots of the average U.S. resident will not be white Europe. Some analysts argue that to eliminate racism, schools and work places must equalize opportunities for all people.

LEARNING OBJECTIVES

After reading Chapter 10, you should be able to:

1. Define prejudice and stereotypes and outline the major theories of prejudice.

2. Distinguish between assimilation and ethnic pluralism.

3. Define race and ethnic group and explain their social significance.

4. Describe symbolic interactionist perspectives on racial and ethnic relations.

5. Explain how the experiences of Native Americans have been different from those of other racial and ethnic groups in the United States.

6. Describe how the African American experience in the United States has been unique when compared with other groups.

7. Compare and contrast the experiences of racial and ethnic subordinate groups in the United States.

8. Explain the sociological usage of dominant group and subordinate group terminology.

9. Describe Robert Merton's typology of the relationship between prejudice and discrimination and be able to give examples of each.

10. Discuss discrimination and distinguish between individual and institutional discrimination.

11. Explain the key assumptions of conflict perspectives on racial and ethnic

12. Trace the intergroup relationships of racial and ethnic groups in the United States.

13. Explain why both assimilation and ethnic pluralism are functionalist perspectives on racial and ethnic relations.

14. Describe the ways that language can be used to perpetuate racial and ethnic stereotypes.

15. Discuss racial and ethnic struggles from a global perspective.

KEY TERMS

(defined at page number shown and in glossary)

KEY PEOPLE

(identified at page number shown)

CHAPTER OUTLINE

I.   RACE AND ETHNICITY
   A. A **race** is a category of people who have been singled out as inferior or superior, often on the basis of physical characteristics such as skin color, hair texture, and eye shape.
   B. An **ethnic group** is a collection of people distinguished, by others or by themselves, primarily on the basis of cultural or nationality characteristics.
   C. Social significance of race and ethnicity: Race and ethnicity are bases of hierarchical ranking in society; the dominant group holds power over other (subordinate) ethnic groups.
   D. Racial classifications in the U.S. census mirror how the meaning of race has continued to change over the past century in the U.S.
   E. A **dominant group** is one that is advantaged and has superior resources and rights in a society; a **subordinate group** is one whose members, because of physical or cultural characteristics, are disadvantaged and subjected to unequal treatment by the dominant group and who regard themselves as objects of collective discrimination.

II. PREJUDICE
   A. **Prejudice** is a negative attitude based on faulty generalizations about members of selected racial and ethnic groups. Prejudice is often based on stereotypes.
   B. **Stereotypes** are overgeneralizations about the appearance, behavior, or other characteristics of all members of a category. *Ethnocentrism* is maintained and perpetuated by stereotypes.
   C. **Racism** is a set of attitudes, beliefs, and practices that is used to justify the superior treatment of one racial or ethnic group and the inferior treatment of another.
   D. The frustration-aggression hypothesis states that people who are frustrated in their efforts to achieve a highly desired goal will respond with a pattern of aggression toward a **scapegoat** – a person or group that is incapable of offering resistance to the hostility or aggression of others.
   E. Theories of prejudice include the frustration aggression hypothesis, social learning theory, and the theory of the **authoritarian personality**, which is characterized by excessive conformity, submissiveness to authority, intolerance, insecurity, a high level of superstition, and rigid, stereotypic thinking.
   F. To measure prejudice, some sociologists use **social distance**, which is the extent to which people are willing to interact and establish relationships with members of racial and ethnic groups other than their own.

III. DISCRIMINATION
   A. **Discrimination** is defined as actions or practices of dominant group members that have a harmful impact on members of a subordinate group.
   B. Robert Merton identified four combinations of attitudes and responses:
      1. Unprejudiced nondiscriminators – persons who are not personally prejudiced and do not discriminate against others;

2. Unprejudiced discriminators – persons who may have no personal prejudice but still engage in discriminatory behavior because of peer group prejudice or economic, political, or social interests;
3. Prejudiced nondiscriminators – persons who hold personal prejudices but do not discriminate due to peer pressure, legal demands, or a desire for profits;
4. Prejudiced discriminators – persons who hold personal prejudices and actively discriminate against others.

C. Discriminatory actions vary in severity from the use of derogatory labels to violence against individuals and groups.
1. **Genocide** is the deliberate, systematic killing of an entire people or nation.
2. More recently, the term "ethnic cleansing" has been used to define a policy of "cleansing" geographic areas (such as in Yugoslavia) by forcing persons of other races or religions to flee – or die.

D. Discrimination also varies in how it is carried out.
1. **Individual discrimination** consists of one-on-one acts by members of the dominant group that harm members of the subordinate group or their property.
2. **Institutional discrimination** is the day-to-day practices of organizations and institutions that have a harmful impact on members of subordinate groups.

IV. SOCIOLOGICAL PERSPECTIVES ON RACE AND ETHNIC RELATIONS
A. Symbolic Interactionist Perspectives
1. The *contact hypothesis* suggests that contact between people from divergent groups should lead to favorable attitudes and behavior when a specific set of criteria is met.
2. However, scholars have found that increasing contact may have little or no effect on existing prejudices.

B. Functionalist Perspectives
1. **Assimilation** is a process by which members of subordinate racial and ethnic groups become absorbed into the dominant culture.
2. **Ethnic pluralism** is the coexistence of a variety of distinct racial and ethnic groups within one society.
3. **Segregation** is the spatial and social separation of categories of people by races, ethnicity, class, gender, and/or religion.

C. Conflict Perspectives
1. The **caste perspective** views racial and ethnic inequality as a permanent feature of U.S. society.
2. **Class perspectives** emphasize the role of the capitalist class in racial exploitation.
3. **Internal colonialism** occurs when members of a racial or ethnic group are conquered, or colonized, and forcibly placed under the economic and political control of the dominant group.
4. **Split labor market** refers to the division of the economy into two areas of employment, a primary sector composed of higher paid (usually dominant group) workers in more secure jobs, and a secondary sector comprised of lower paid (often subordinate group) workers in jobs with little security and frequently hazardous working conditions.

5. *Gendered racism* refers to the interactive effect of racism and sexism in the exploitation of women of color.

6. The *theory of racial formation* states that actions of the government substantially define racial and ethnic relations in the United States.

D. An Alternative Perspective: Critical Race Theory

1. One premise is that racism is so ingrained in the U.S. society that it appears to be ordinary and natural to many people.

2. *Interest convergence* is a crucial factor in bringing about social change.

3. Formal equality under the law does not necessarily equate to actual equality in society.

4. Ironies and contradictions exist in civil rights law, which are actually self-serving.

V. RACIAL AND ETHNIC GROUPS IN THE UNITED STATES

A. Native Americans

1. Historically, Native Americans experienced the following kinds of treatment in the United States:

a. genocide

b. forced migration

c. forced assimilation

2. Today, about two million Native Americans live in the United States (primarily in the southwest), and about one-third live on reservations.

3. Native Americans are the most disadvantaged racial or ethnic group in the United States in terms of income, employment, housing, nutrition, and health (especially among individuals living on reservations).

B. White Anglo-Saxon Protestants (British Americans)

1. Although many English settlers initially were indentured servants or sent here as prisoners, they quickly emerged as the dominant group, creating a core culture to which all other groups were expected to adapt.

2. Like other racial and ethnic groups, British Americans are not all alike; social class and gender affect their life chances and opportunities.

C. African Americans

1. Slavery developed with the establishment of the plantation system, which was heavily dependent on cheap and dependable manual labor.

2. Slavery was rationalized by stereotyping African Americans as inferior and childlike; however, some slaves and whites engaged in active resistance that eventually led to the abolition of slavery.

3. *Lynching* – a killing carried out by a group of vigilantes seeking revenge for an actual or imagined crime by the victim – was used by whites to intimidate African Americans into staying "in their place."

4. During World Wars I and II, African Americans were a vital source of labor in war production industries; however, racial discrimination continued both on and off the job.

a. After African Americans began to demand sweeping societal changes in the 1950s, racial segregation slowly was outlawed by the courts and the federal government.

b. Civil rights legislation attempted to do away with discrimination in education, housing, employment, and health care.

5. Today, African Americans make up about 13 percent of the U.S. population; many have made significant gains in education, employment, and income in the past three decades; however, other African Americans have not fared so well; for example, the African American unemployment rate remains twice as high as that of whites, and young people in central city areas face a bleak future.

D. The term *white ethnic Americans* is used to identify immigrants who came from European countries other than England. Ireland, Poland, Italy, Greece, Germany, Yugoslavia, Russia and other former Soviet republics are examples.

E. The term *Asian Americans* refers to many diverse groups with roots in Asia.

1. Chinese Americans
   a. The initial wave of Chinese immigration occurred between 1850 and 1880 when Chinese men came to the United States seeking gold in California and jobs constructing the transcontinental railroads.
   b. Chinese Americans were subjected to extreme prejudice and stereotyping; the Chinese Exclusion Act of 1882 was passed because white workers feared for their jobs.
   c. In the 1960s, the second and largest wave of Chinese immigration came from Hong Kong and Taiwan.
   d. Today, one third of all Chinese Americans were born in the United States; as a group, they have enjoyed considerable upward mobility, but many Chinese Americans live in poverty in Chinatowns.

2. Japanese Americans
   a. The earliest Japanese immigrants primarily were men who worked on sugar plantations in the Hawaiian Islands in the 1860s. The immigration of Japanese men was curbed in 1908, however, Japanese women were permitted to enter the U.S. for several more years because of the shortage of women.
   b. Internment: During World War II, when the United States was at war with Japan, nearly 120,000 Japanese Americans were placed in internment camps because they were seen as a security threat; many Japanese Americans lost all that they owned during the internment.
   c. In spite of the extreme hardship faced as a result of the loss of their businesses and homes during World War II, many Japanese Americans have been very successful.

3. Korean Americans
   a. The first wave of Korean immigrants were male workers who arrived in Hawaii between 1903 and 1910; the second wave came to the mainland following the Korean War in 1964 (e.g., the wives of servicemen, and Korean children who had lost their parents in the war); and the third wave arrived after the Immigration Act of 1965 permitted well-educated professionals to migrate to the U.S.
   b. Korean Americans have helped each other open small businesses by pooling money through the *kye* – an association that grants members money on a rotating basis to gain access to more capital.

4. Filipino Americans
    a. Most of the first Filipino immigrants were men who were employed in agriculture; following the Immigration Act of 1965, Filipino physicians, nurses, technical workers, and other professionals moved in large numbers to the U.S. mainland.
    b. Unlike other Asian Americans, most Filipinos have not had the start-up capital necessary to open their own businesses, and workers generally have been employed in the low-wage sector of the dual-labor market.
5. Indochinese Americans
    a. Most Indochinese Americans (including people from Vietnam, Cambodia, Thailand, and Laos) have come to the U.S. in the past two decades.
    b. Vietnamese refugees who had the resources to flee at the beginning of the Vietnam War were the first to arrive. Next came Cambodians and lowland Laotians, referred to as "boat people" by the media.
    c. Today, most Indochinese Americans are foreign born; about half live in western states, especially California. Even though most Indochinese immigrants spoke no English when they arrived in this country, some of their children have done very well in school and have been stereotyped as "brains."
F. Latino/as (Hispanic Americans)
    1. Mexican Americans or Chicano/as have experienced disproportionate poverty as a result of internal colonialism.
        a. More recently, Mexican Americans have been seen as cheap labor at the same time that they have been stereotyped as lazy.
        b. When anti-immigration sentiments are running high, Mexican Americans often are the objects of discrimination.
        c. Today, the families of many Mexican Americans have lived in the United States for four or five generations and have made significant contributions in many areas.
    2. When Puerto Rico became a possession of the United States in 1917, Puerto Ricans acquired U.S. citizenship and the right to move freely to and from the mainland While living conditions have improved substantially for some, others have continued to live in poverty in Spanish Harlem and other barrios.
    3. Cuban Americans have fared somewhat better than other Latinos. Early waves of Cuban immigrants were affluent business and professional people; the second wave of Cuban Americans in the 1970s fared worse—many had been released from prisons and mental hospitals in Cuba. More recent arrivals have developed their own ethnic and economic enclaves in Miami, and have become mainstream professionals and entrepreneurs.
G. Middle Eastern Americans
    1. Since 1970, many immigrants have arrived in the United States from Middle Eastern countries such as Egypt, Syria, Lebanon, Saudi Arabia, Kuwait, Yemen, Iran, and Iraq.
    2. While some are from working-class families, those coming from Lebanon, Syria, Iran, and Kuwait primarily came from middle and upper income family backgrounds.

3. In the United States, Islamic schools and centers often bring together people from a diversity of Middle Eastern countries such as Egypt and Pakistan. Islam is one of the fastest growing religious in the United States.

4. Middle Eastern Americans speak a variety of languages, have diverse religious backgrounds, and are often treated as categories rather than as individual human beings.

VI. GLOBAL RACIAL AND ETHNIC INEQUALITY IN THE FUTURE
   A. Worldwide Racial and Ethnic Struggles
      1. The cost of self-determination – the right to choose one's own way of life – often result in the loss of life and property in ethnic warfare such as in Yugoslavia, Spain, Britain, Romania, Russia, Moldova, Georgia, the Middle East, Africa, Asia, and Latin America.
      2. Today, the struggle between the Israeli government and Palestinian factions has heightened tensions around the world.
   B. Growing Racial and Ethnic Diversity in the United States
      1. Racial and ethnic diversity is increasing in the United States; by 2056, the roots of the average U.S. resident will be Africa, Asia, Hispanic countries, the Pacific Islands, and the Middle East – not white Europe.
      2. Interethnic tensions may ensure people may continue to employ *sincere fictions* – personal beliefs that are a reflection of larger societal mythologies, such as "I am not a racist" – even when these are inaccurate perceptions.
      3. Some analysts believe that there is reason for cautious optimism – throughout U.S. history subordinate racial and ethnic groups have struggled to gain the freedom and rights which were previously withheld from them, and movements comprised of both whites and people of color will continue to oppose racism in everyday life, to aim at healing divisions among racial groups, and to teach children about racial tolerance.

ANALYZING AND UNDERSTANDING THE BOXES

After reading the chapter and studying the outline, re-read the boxes and write down key points and possible questions for class discussion.

Sociology and Everyday Life: How Much Do You Know About Race, Ethnicity, and Sports?

Key Points:

Discussion Questions:

1.

2.

3.

Sociology in Global Perspective: Racism and Anti-Racism in European Football

Key Points:

Discussion Questions:

1.

2.

3.

Framing Race in the Media: Do Multiracial Scenes in Television Ads Reflect Reality?

Key Points:

Discussion Questions:

1.

2.

3.

You Can Make a Difference: Working for Racial Harmony

Key Points:

Discussion Questions:

1.

2.

3.

PRACTICE TESTS

MULTIPLE CHOICE QUESTIONS

Select the response that best answers the question or completes the statement.

1. _____ is a category of people who have been singled out as inferior or superior, often on the basis of real or alleged physical characteristics such as skin color, hair texture, eye shape, or other subjectively selected attributes.
   a. Ethnic group
   b. Race
   c. National group
   d. Sub-cultural group

2. All of the following are characteristics of ethnic groups, **except**:
   a. unique cultural traits
   b. a feeling of ethnocentrism
   c. territoriality
   d. the same skin color

3. According to the text, the terms "majority group" and "minority group" are
   a. accurate because "majority groups" always are larger in number than "minority groups."
   b. misleading because people who share ascribed racial or ethnic characteristics automatically constitute a group.
   c. less accurate terms than "dominant group" and "subordinate group," which more accurately reflect the importance of power in the relationships.
   d. no longer used because they are "politically incorrect."

4. _____ is a negative attitude based on faulty generalizations about members of selected racial and ethnic groups.
   a. Prejudice
   b. Discrimination
   c. Stereotyping
   d. Genocide

5. Prejudice can be _____ (bias in favor of a particular racial group) or _____ (bias against a particular racial group).
    a. on; off
    b. dominant; subordinate
    c. inward; outward
    d. positive; negative

6. According to the text, the use of Native American names, images, and mascots by sports teams is an example of
    a. prejudice.
    b. discrimination.
    c. stereotyping.
    d. genocide.

7. _____ is often hidden from sight and more difficult to prove.
    a. Subtle racism
    b. Open racism
    c. Closed racism
    d. Obvious racism

8. Racism tends to intensify in times of:
    a. low employment
    b. economic uncertainty
    c. low rates of immigration
    d. economic equality

9. All of the following are examples of overt racism, **except**:
    a. a police detective who repeatedly refers to African Americans in derogatory terms while discussing police department procedures with the author of a screenplay
    b. stereotyping
    c. segregation of students by race/ethnicity in public schools
    d. hate crimes perpetrated against victims because of their race/ethnicity, sexual orientation, or religion

10. The _____ hypothesis states that people who are disappointed in their efforts to achieve a highly desired goal will respond with a pattern of assertiveness toward others.
    a. social learning
    b. authoritarian personality
    c. frustration-aggression
    d. social convergence

11. When members of subordinate racial and ethnic groups become substitutes for the actual source of frustration within a community, they become:
    a. authoritarian personalities
    b. ethnocentric
    c. racists
    d. scapegoats

12. According to symbolic interactionists, prejudice results from:
    a. social learning
    b. anger and frustration
    c. the authoritarian personality
    d. discrimination

13. Psychologist Theodore W. Adorno and his colleagues concluded that highly prejudiced individuals tend to have a(n) _____ personality.
    a. frustrated-aggressive
    b. authoritarian
    c. racist
    d. discriminatory

14. Sociologist Emory Bogardus developed a scale to measure
    a. racial ethnic prejudice in society.
    b. stereotyping.
    c. social distance.
    d. income differentials by race/ethnicity.

15. According to sociologist Robert Merton's typology, an umpire who is prejudiced against African Americans and deliberately makes official calls against them when they are at bat is an example of a(n)
    a. unprejudiced nondiscriminator.
    b. unprejudiced discriminator.
    c. prejudiced nondiscriminator.
    d. prejudiced discriminator.

16. Institutional discrimination consists of
    a. one on one acts by members of the dominant group that harm members of the subordinate group or their property.
    b. day to day practices of organizations and institutions that have a harmful impact on members of subordinate groups.
    c. the division of the economy into two areas of employment, a primary sector or upper tier, and a secondary sector or lower tier.
    d. the deliberate, systematic killing of an entire people or nation.

17. A prejudiced judge giving harsher sentences to all African American defendants, and is not supported by the judicial system in that action, is an example of _____ discrimination.
    a. indirect institutionalized
    b. direct institutionalized
    c. small group
    d. isolate

18. _____ is the deliberate, systematic killing of an entire people or nation.
    a. Institutional discrimination
    b. Isolate discrimination
    c. Genocide
    d. Global prejudice

19. According to the contact hypothesis, contact between people from divergent groups should lead to favorable attitudes and behaviors when certain factors are present. Members of each group must have all of the following, **except**:
    a. equal status
    b. pursuit of the same goals
    c. being competitive with one another to achieve their goals
    d. positive feedback from interactions with each other

20. _____ is a process by which members of subordinate racial and ethnic groups become absorbed into the dominant culture.
    a. Assimilation
    b. Ethnic pluralism
    c. Accommodation
    d. Internal colonialism

21. _____ occurs when members of an ethnic group adopt dominate group traits, such as language, dress, values, religion, and food preferences.
    a. Acculturation
    b. Integration
    c. Amalgamation
    d. Miscegenation

22. _____ occurs when members of subordinate racial or ethnic groups gain acceptance in everyday social interaction with members of the dominant group.
    a. Amalgamation
    b. Miscegenation
    c. Integration
    d. Acculturation

23. _____ is the coexistence of a variety of distinct racial and ethnic groups within one society.
    a. Assimilation
    b. Ethnic pluralism
    c. Accommodation
    d. Internal colonialism

24. Switzerland has been described as a model of _____; over 6 million people with French, German, and Italian cultural heritages peacefully coexist there.
    a. miscegenation
    b. equalitarian
    c. integration
    d. inequalitarian pluralism

25. Based on early theories of race relations by W. E. B. Du Bois, sociologist Oliver Cox suggested that African Americans were enslaved because
    a. of prejudice based on skin color.
    b. they were the cheapest and best workers that owners could find for heavy labor in the mines and on plantations.
    c. they allowed themselves to be dominated by wealthy people.
    d. they agreed to work for passage to the United States, believing that they would be released after a period of indentured servitude.

26. According to sociologist William Julius Wilson, _____ is the most important factor to understanding to state of African Americans in the United States.
    a. class
    b. race
    c. cultural factors
    d. gender

27. According to _____, sports reflect the interests of the wealthy and powerful; athletes are exploited in order for coaches, managers, and owners to gain high levels of profit and prestige.
    a. interactionists
    b. sports theorists
    c. functionalists
    d. conflict theorists

28. The division of the economy into a primary sector, composed of higher paid workers in more secure jobs, and a secondary sector, composed of lower-paid workers in jobs with little security and hazardous working conditions is referred to as:
    a. the split labor market.
    b. racial formation.
    c. gendered racism.
    d. stacking.

29. The theory of _____ refers to the interactive effect of racism and sexism on the exploitation of women of color.
    a. gendered racism
    b. sexual racism
    c. colonial racism
    d. historical racial sexism

30. According to _____ theory, interest convergence is a critical factor in changing race and ethnic inequality.
    a. functionalist
    b. symbolic interactionist
    c. critical race
    d. ethnic study theory

31. The experiences of Native Americans in the United States have been characterized by:
    a. slavery and segregation
    b. restrictive immigration laws
    c. genocide, forced migration, and forced assimilation
    d. internment

32. The "Trail of Tears" was
    a. systematic discrimination against Latino/as.
    b. specific Jim Crow laws in 1950s.
    c. segregationist policies against Asian Americans.
    d. the forced migration of Native Americans.

33. President _____ was the first to establish a committee with the goal of ending discrimination in employment in government and its contractors.
    a. Truman
    b. Kennedy
    c. Johnson
    d. Nixon

34. Nearly 120,000 _____ were placed in internment camps, experiencing one of the worst forms of discrimination sanctioned by U.S. laws.
    a. Chinese Americans
    b. Korean Americans
    c. Filipino Americans
    d. Japanese Americans

35. Since World War II, many _____ have been very successful; their median income is more than 30 percent above the national average.
    a. Chinese Americans
    b. Filipino Americans
    c. Japanese Americans
    d. Korean Americans

36. The label _____ was first used by the U.S. government to designate people of Latin American and Spanish descent living in the United States.
    a. Mexican American
    b. Hispanic
    c. Chicano
    d. Latino/Latina

37. _____ live primarily in the southeast, especially Florida. Many were affluent professionals and businesspeople who fled their homeland after Fidel Castro's 1959 Marxist revolution.
    a. Puerto Rican Americans
    b. Mexican Americans
    c. Cuban Americans
    d. Central Americans

38. Middle Eastern women often experience discrimination for wearing a _____, or a "head-to-toe covering," showing only one's face.
    a. hijab
    b. burkah
    c. sari
    d. scarf

39. Following the September 11, 2001 terrorist attacks, the United States passed the _____ – a law giving the federal government greater authority to engage in searches and surveillance with less judicial review than previously.
    a. U.S. Homeland Security Act
    b. U.S. Patriot Act
    c. U.S. Immigration Act of 2001
    d. U.S. Protection Act

40. Throughout the world, many racial and ethnic groups seek _____, the right to choose their own way of life.
    a. self-determination
    b. freedom
    c. liberty
    d. civil disobedience

TRUE/FALSE QUESTIONS

1. Sociologists suggest that race is a socially constructed reality, not a biological one.
   T   F

2. Racial classifications in the U.S. census have remained unchanged for the past century.
   T   F

3. Sociologists emphasize that race is a socially constructed reality.
   T   F

4. Analysts have found that whites that accept racial stereotypes have no greater desire for social distance from people of color than do whites that reject negative stereotypes.
   T   F

5. The extermination of 6 million European Jews by Nazi Germany is an example of genocide.
   T   F

6. Discrimination is an individual activity; it is never based on norms of an organization or community.
   T   F

7. Equalitarian pluralism is an ideal typology that has never actually been identified in any society.
   T   F

8. De facto segregation refers to racial separation and inequality enforced by custom.
   T   F

9. The internal colonialism model helps explain the continued exploitation of immigrant groups such as the Chinese, Filipinos, and Vietnamese.
   T   F

10. The term "gendered racism" refers to the interactive effect of racism and sexism in the exploitation of women of color.
   T   F

11. White Anglo-Saxon Protestants (WASPS) have been the most privileged group in the United States.
   T   F

12. The term "white ethnics" was coined to identify immigrants who came from European countries other than England, such as Ireland, Poland, Germany, and Italy.
   T   F

13. White ethnics have not experienced discrimination in the United States.
   T   F

14. Americans are the fastest growing ethnic minority group in the United States.
   T   F

15. It is predicted that by 2056, the roots of the average U.S. resident will be in Africa, Asia, Hispanic countries, the Pacific Islands, and the Middle East.
   T   F

16. According to Box 10.1, all racial and ethnic groups in the United States have viewed sports as a means of social mobility.
   T   F

17. According to Box 10.2, "Sociology in Global Perspective," racism is an issue in European soccer just as it is an issue in U. S. sports.
   T   F

18. According to that same Box 10.2, soccer clubs must make regular public announcements at matches condemning racist chanting.
   T   F

19. According to Box 10.3, "Framing Race in the Media," multiracial scenes in television have no redeeming social values.
   T   F

20. According to Box 10.4, "You Can Make a Difference: Working for Racial Harmony," objectives could be reached over time.
   T   F

FILL-IN-THE-BLANK QUESTIONS

1. In the nineteenth century, biologists divided the world's population into three racial categories: _____, _____, and _____.

2. The term _____ refers to individuals who share some common cultural or national characteristics.

3. Sociologists emphasize that race is a _____, not a biological one.

4. _____ can either be a positive bias in favoring a group or can be a negative bias against a group.

5. _____ is an attitude that is used to justify the superior or inferior treatment of another racial or ethnic group.

6. Highly prejudiced individuals tend to have a _____ which is characterized by excessive conformity, intolerance, and rigid stereotypes.

7. _____ involves actions or practices of dominant group members that have a harmful impact on members of a subordinate group.

8. The day-to-day practices of organizations and institutions that have a harmful impact on members of subordinate groups is termed _____.

9. Segregation may be enforced by custom, _____, or by law which is an example of _____ segregation.

10. The process by which members of subordinate racial and ethnic groups become absorbed into the dominant culture is known as_____.

11. Dr. Martin Luther King, Jr. and the civil rights movement used the term _____ to refer to nonviolent action seeking to change a policy or law by refusing to comply with it.

12. _____ programs were instituted in public and private sector organizations in order to bring about greater opportunities for previously excluded racial and ethnic groups.

13. The _____ Exclusion Act of 1882 brought this group's immigration into the United State to a halt.

14. _____ were made wards of the government and were subjected to forced assimilation.

15. The term _____ is preferred by people who have their origins to Haiti, Puerto Rico, or Jamaica.

SHORT ANSWER/ESSAY QUESTIONS

1. How do racial groups, ethnic groups, minority and majority groups differ? What are some groups that may be one, but not any of the other?
2. What are some social factors that have prevented African Americans from achieving full equality in the United States?
3. What is the fastest growing ethnic minority group in the United States today? Why?
4. Why is education a crucial issue for Latino/as? Which Latino/a group has fared the best? Why?
5. Why are race and ethnicity so socially significant? In your response, provide both historic and contemporary explanations.

STUDENT CLASS PROJECTS AND ACTIVITIES

1. Investigate pluralism on your campus by conducting a brief campus study. Record your information in a written paper. Ask the following questions and include any others you may want to ask: (1) What is the number and percentage of the different ethnic or racial groups enrolled on the campus? (2) What is the number and percentage of "foreign" students? (3) What percentage of various ethnic groups holds important campus offices? Identify the office and the race or ethnicity of the office holder; (4) What percentage of racial or ethnic groups are in the campus sororities and fraternities and other organizations? (5) What percentage of racial or ethnic groups comprises both the male and female athletes who receive college or university scholarships? (6) Based on the research, evaluate your campus. How pluralistic is it? This information may be available from the admissions or registrar's office, student activities or athletic office. Write up your results following any additional instructions from your professor and submit your paper on the correct due date.
2. Conduct an informal survey of 20 people who appear to be an ethnic or racial minority in our country. Ask these specific questions and record the responses:
(1) Ask the respondent's name, general address, and specific race or ethnicity.
(2) Do you ever experience racism or bigotry or discrimination in some form in our society? (3) If so, what was it? (4) Can you remember the first time you were hurt in some way? (5) Can you remember a more recent time? (6) How did you respond? (7) How would you like to have responded? Provide any other questions that you would like to ask and then provide a summary, conclusion and evaluation of this project. Submit your papers at the appropriate time following any other specific instructions given by your instructor.
3. Examine your own ethnicity or racial heritage, if known. If your ethnicity or race is mixed, then select the one that you believe to be the most dominant heritage in your life. If your ethnicity or race is unclear, or not known, then examine your own socialization into an ethnic or racial group. Note that many people, because of adoption or other factors, may not know their original ethnic heritage, but may be able to identify their own racial heritage. Prepare a minimum five-page report on your experience as a member of your particular group. Gather as much background

information as possible in exploring your own ethnic or racial heritage. If known, trace the ethnic history of your own family, such as the approximate date of your ancestors' migration to the United States, the obstacles they faced, their cultural practices and experiences from childhood, usual activities, and family experiences. Include activities with extended kin members, type of education received, if any, and church and neighborhood experiences. Record any other event or experience that influenced your identity and membership in the group. Include any other information you desire, but do provide a conclusion and evaluation of this project. Submit your paper at the appropriate time following any other instructions provided by your professor.

INTERNET ACTIVITIES

1. Take a tour through cyberspace searching for general information about race and ethnicity. Search through these sites to collect relevant facts about racial and ethnic groups. These are good introductory sites for writing papers and collecting data: **http://eserver.org/race, http://www.trinity.edu/mkearl/race.html**
2. This site, **http://www.raleightavern.org/steele.html**, contains an essay by Tom Lovell on the work of Shelby Steele. Specific reference is made to Steele's term "race holding." What is the meaning of this term? What is the social impact of this term?
3. These are sites that are more specific to various racial and ethnic groups in the U.S. Students wishing to go more in depth with a study of specific racial and ethnic groups should start here. **African-American (http://www.aawc.com/aawc0.html), Latino (http://www.lif.org/), Asian-American (http://goldsea.com/Air/Issues/issues.html), and, Native American (http://www.hanksville.org/NAresources/).**

INFOTRAC COLLEGE EDITION EXERCISES

Visit the **InfoTrac College Edition** website at: **http://www.wadsworthmedia.com/webtutor/infotrac.htm**. You will arrive at a screen that enables you to search topics.
1. Scan the InfoTrac for articles dealing with **racism**. What are some common themes that emerge? (Qualitative research)
2. **Ethnic cleansing** remains a powerful tool for one group to eliminate another. There are two categories listed in InfoTrac: genocide and forced migration. Look for articles that describe where ethnic cleansing is taking place, and with what consequences.
3. Select one of the **ethnic minority groups** addressed in this chapter and research for information on that group as found in the archives of InfoTrac. Answer the following questions: What are some of the issues related to each group? How do research articles and newspaper articles about this group differ in their approach for addressing some of these issues? What do some of the articles suggest about the future of this groups integration into the dominant U.S.A. culture?
4. Search for these theoretical concepts in articles contained in the archives of Infotrac: Scapegoat, Contact Hypothesis, Psychological Assimilation, Internal Colonialism, Split Labor Market, and Critical Race Theory.

SOLUTIONS

MULTIPLE CHOICE QUESTIONS

1. B, p. 312	15. D, p. 320	29. A, p. 327
2. D, p. 313	16. B, p. 320	30. C, p. 327
3. C, p. 316	17. D, p. 320	31. C, p. 329
4. A, p. 316	18. C, p. 320	32. D, p. 330
5. D, p. 316	19. C, p. 321	33. B, p. 333
6. C, p. 316	20. A, p. 322	34. D, p. 337
7. A, p. 317	21. A, p. 322	35. C, p. 337
8. B, p. 317	22. C, p. 322	36. B, p. 339
9. B, p. 317	23. B, p. 323	37. C, p. 340
10. C, p. 317	24. B, p. 323	38. A, p. 341
11. D, p. 319	25. B, p. 325	39. B, p. 341
12. A, p. 319	26. B, p. 325	40. A, p. 342
13. B, p. 319	27. D, p. 326	
14. C, p. 319	28. A, p. 326	

TRUE/FALSE QUESTIONS

1. T, p. 312	8. T, p. 323	15. T, p. 343
2. F, p. 312	9. F, p. 326	16. F, p. 314
3. T, p. 316	10. T, p. 327	17. T, p. 318
4. F, p. 319	11. T, p. 331	18. T, p. 318
5. T, p. 320	12. T, p. 335	19. F, p. 325
6. F, p. 320	13. F, p. 335	20. T, p. 342
7. F, p. 323	14. T, p. 336	

FILL-IN-THE-BLANK QUESTIONS

1. Negroid, p. 312; Caucasian, p. 312
2. ethnicity, p. 313
3. socially constructed reality, p. 312
4. Prejudice, p. 316
5. Racism, p. 317
6. authoritarian personality, p. 319
7. Discrimination, p. 320
8. institutional discrimination, p. 320
9. de facto, p. 323; de jure, p. 323
10. assimilation, p. 322
11. civil disobedience, p. 333
12. Affirmative action, p. 333
13. Chinese, p. 336
14. Native Americans, p. 330
15. black, p. 332

11

SEX AND GENDER

BRIEF CHAPTER OUTLINE

CHAPTER SUMMARY

It is important to distinguish between **sex**, which refers to the biological and anatomical differences between females and males, and **gender**, which refers to the culturally and socially constructed differences between females and males found in the meanings, beliefs, and practices associated with "femininity" and "masculinity." Gender is socially significant because it leads to differential treatment of men and women; sexism (like racism) often is used to justify discriminatory treatment. **Sex** is not always clear-cut; a hormone imbalance before birth produces a **hermaphrodite**; while a **transsexual** identifies with the opposite gender. The closest approximation of a third sex in Western societies is a **transvestite**. **Sexual orientation** refers to an individual's preference for emotional-sexual relationships with members of the opposite sex, or both. **Gender** is a social construction with important consequences in everyday life. **Sexism** is linked to **patriarchy**, a hierarchical system in which cultural, political, and economic structures are male dominated. In analyzing gender stratification in pre-industrial societies, gender relationships vary. In most hunting and gathering societies, fairly equitable relationships exist because neither sex has the ability to provide all of the food necessary for survival. In horticultural societies, hoe cultivation is compatible with child care, and a fair degree of gender equality exists because neither sex controls the food supply while in pastoral societies, women have relatively low status. In agrarian societies, male dominance is very apparent; tasks require more labor and physical strength, and women are seen as too weak or too tied to child-rearing activities to perform these activities. In industrialized societies, a gap exists between unpaid work performed by women at home and paid work performed by men and women. In postindustrial societies, technology supports a service and information based economy; most adult women are in the labor force and have double responsibilities - those from the labor force and those from the family. The key agents of gender socialization are parents, peers, teachers and schools, sports, and the mass media, all of which tend to reinforce stereotypes of appropriate gender behavior. **Gender bias** and **sexual harassment** remain major problems in American society. Gender inequality results from the economic, political, and educational discrimination of women. In most workplaces, jobs are either gender-segregated or the majority of employees are of the same gender. Gender segregated occupations lead to a disparity, or **pay gap**, between women's and men's earnings. Even when women are employed in the same job as men, on average they do not receive the same pay. **Comparable worth** is the belief that wages should reflect the worth of a job, not the gender or race of the worker. Many women have a "second shift" because of their dual responsibilities for paid and unpaid work. According to functional analysts, husbands perform instrumental tasks of economic support and decision making, and wives assume expressive tasks of providing affection and emotional support for the family. Conflict analysts suggest that the gendered division of labor within families and the workplace results from male control and dominance over women and resources. **Feminism** is the belief that women and men are equal and that they should be valued equally and have equal rights. Although feminist perspectives vary in their analyses of women's subordination, they all advocate social change to eradicate gender inequality.

LEARNING OBJECTIVES

After reading Chapter 11, you should be able to:

1. Differentiate between hermaphrodites, transsexuals, and transvestites.

2. Differentiate between sex and gender roles.

3. Define gender identity and body consciousness.

4. Explain why definitions of sex are not always clear-cut

5. Distinguish between sex and gender and explain their sociological significance.

6. Outline the key assumptions of liberal, radical, socialist, and black (African American) feminism.

7. Use examples to explain the feminist perspective on gender equality.

8. Explain some of the important ways that people "do" gender in their daily interactions.

9. Define sexism and explain how it is related to discrimination and patriarchy.

10. Trace gender stratification from early hunting and gathering societies until today.

11. Describe the process of gender socialization and identify specific ways in which parents, peers, teachers, sports, and mass media contribute to the process.

12. Describe the relationship between gender roles, gender identity, and body consciousness.

13. Describe gender bias and explain how schools operate as a gendered institution.

14. Trace changes in labor force participation by women and note how these changes have contributed to the "second shift."

15. Describe functionalist and neoclassical economic perspectives on gender stratification and contrast them with conflict perspectives.

16. Discuss the gendered division of paid work and explain its relationship to the issue of pay equity or comparable worth.

17. Describe the most important ways that the mass media contributes to gender socialization.

KEY TERMS

(defined at page number shown and in glossary)

body consciousness, p. 352
comparable worth, p. 368
feminism, p. 375
gender, p. 352
gender bias, p. 362
gender identity, p. 352
gender role, p. 352
hermaphrodite, p. 349
homophobia, p. 352
matriarchy, p. 354

patriarchy, p. 354
primary sex characteristics, p. 348
secondary sex characteristics, p. 348
sex, p. 348
sexism, p. 354
sexual harassment, p. 362
sexual orientation, p. 350
transsexual, p. 349
transvestite p. 350
wage gap, p. 368

KEY PEOPLE

(identified at page number shown)

Ben Agger, p. 375
Susan Bordo, p. 353
Patricia Hill Collins, p. 359
Virginia Cyrus, p. 354
Paula England and Melissa Herbert, p. 370
Argie Hochschild, p. 371

Dorothy Holland and Margaret A. Eisenhart, p. 361
Elizabeth Higginbotham, p. 366
Judith Lorber, p. 352
Michael Messner, p. 363
Talcott Parsons, p. 371
Becky W. Thompson, p. 353

CHAPTER OUTLINE

I. SEX: THE BIOLOGICAL DIMENSION
    A. **Sex** refers to the biological and anatomical differences between females and males.
        1. **Primary sex characteristics** are the genitalia used in the reproductive process; **secondary sex characteristics** are the physical traits (other than reproductive organs) that identify an individual's sex.
    B. Hermaphrodites/transsexuals are examples of when sex is not always clear-cut.
        1. Hermaphrodites tend to have some combination of male and female genitalia.
        2. Transsexuals often feel they are the opposite sex from their sex organs.
    C. **Sexual orientation** is a preference for emotional sexual relationships with members of the opposite sex (heterosexuality), the same sex (homosexuality), or both (bisexuality).
II. GENDER: THE CULTURAL DIMENSION
    A. **Gender** refers to the culturally and socially constructed differences between females and males found in the meanings, beliefs, and practices associated with "femininity" and "masculinity."
        1. A microlevel analysis of gender focuses on how individuals learn **gender roles** – the attitudes, behavior, and activities that are socially defined as appropriate for each sex and are learned through the socialization process – and **gender identity** – a person's perception of the self as female or male. **Body consciousness**, how a person feels about his or her body, is a part of gender identity.
        2. A macrolevel analysis of gender examines structural features, external to the individual, which perpetuate gender inequality, including gendered institutions that are reinforced by a gendered belief system, based on ideas regarding masculine and feminine attributes that are held to be valid in a society.
    B. The Social Significance of Gender
        1. Gender is a social construction with important consequences in everyday life.
        2. Gender stereotypes hold that men and women are inherently different in attitudes, behavior, and aspirations.

C. Sexism
   1. **Sexism** – the subordination of one sex, usually female, based on the assumed superiority of the other sex – is interwoven with **patriarchy** – a hierarchical system of social organization in which cultural, political, and economic structures are controlled by men.
   2. **Matriarchy** is a hierarchical system of social organization in which cultural, political, and economic structures are controlled by women; few societies have been organized in this manner.
III. GENDER STRATIFICATION IN HISTORICAL AND CONTEMPORARY PERSPECTIVE
   A. Preindustrial Societies
      1. The earliest known division of labor between women and men is in *hunting and gathering societies*; men hunt while women gather food. A relatively equitable relationship exists because both sexes are necessary for survival.
      2. In *horticultural societies*, women make an important contribution to food production because hoe cultivation is compatible with child care; a fairly high degree of gender equality exists because neither sex controls the food supply; however, in pastoral societies, herding primarily is done by men; women contribute relatively little to subsistence production and thus have relatively low status.
      3. Gender inequality increases in *agrarian* societies as men become more involved in food production. Women are secluded, subordinated and sometimes mutilated.
   B. Industrial Societies
      1. In *industrial societies* – those in which factory or mechanized production has replaced agriculture as the major form of economic activity – the status of women tends to decline further;
      2. Gendered division of labor increases the economic and political subordination of women.
   C. Postindustrial Societies
      1. In *postindustrial societies,* technology supports a service and information - based economy.
      2. Most adult women are in the work force and have double responsibilities - those from the labor force and those from the family; today, more households are headed by women with no male present.
IV. GENDER AND SOCIALIZATION
   A. Parents and Gender Socialization
      1. From birth, parents act toward children on the basis of gender labels; children's clothing and toys reflect their parents' gender expectations.
      2. Boys are encouraged to engage in gender-appropriate behavior; they are not to show an interest in "girls'" activities.
   B. Peers and Gender Socialization
      1. Peers help children learn prevailing gender role stereotypes, as well as gender appropriate and inappropriate behavior.
      2. During adolescence, peers often are stronger and more effective agents of gender socialization than are adults.

3. Among college students, peers play an important role in career choices and the establishment of long term, intimate relationships.
  C. Teachers, Schools and Gender Socialization
    1. From kindergarten through college, schools operate as gendered institutions; teachers provide important messages about gender through both the formal content of classroom assignments and informal interactions with students.
    2. Teachers may unintentionally demonstrate **gender bias** – the showing of favoritism toward one gender over the other – toward male students.
  D. Sports and Gender Socialization
    1. The type of game played differs with the child's sex: from elementary school through high school, boys play football and other competitive sports while girls are cheerleaders, members of the drill team, and homecoming queens.
    2. For many males, sports participation and spectatorship is a training ground for masculinity; for females, sports still is tied to the male gender role, thus making it very difficult for girls and women to receive the full benefits of participating in such activities.
  E. Mass Media and Gender Socialization
    1. Gender stereotyping is found in media, ranging from children's cartoons to adult shows.
    2. On television, more male than female roles are shown, and male characters typically are more aggressive, constructive, and direct, while females are deferential toward others or use manipulation to get their way.
    3. Advertising also plays an important role in gender socialization.
  F. Adult Gender Socialization
    1. Men and women are taught gender-appropriate conduct in schools and the workplace.
    2. Different gender socialization may occur as people reach their forties and enter "middle age."
V. CONTEMPORARY GENDER INEQUALITY
  A. Gendered Division of Paid Work
    1. *Gender-segregated work* refers to the concentration of women and men in different occupations, jobs, and places of work.
    2. *Labor market segmentation* – the division of jobs into categories with distinct working conditions – results in women having separate and unequal jobs in the secondary sector of the split or dual labor market that are lower paying, less prestigious, and have fewer opportunities for advancement.
  B. Pay Equity (Comparable Worth)
    1. Occupational segregation contributes to a **pay gap** – the disparity between women's and men's earnings.
    2. **Comparable worth** is the belief that wages ought to reflect the worth of a job, not the gender or race of the worker.
  C. Paid Work and Family Work
    1. Although both men and women profess that working couples should share household responsibilities, researchers find that family demands remain mostly women's responsibility, even among women who hold full time paid employment.

2. Many women have a "double day" or "second shift" because of their dual responsibilities for paid and unpaid work.

VI. PERSPECTIVES ON GENDER STRATIFICATION
   A. Functionalists and neoclassical economic perspectives on the family view the division of family labor as ensuring that important societal tasks will be fulfilled.
      1. According to the human capital model, individuals vary widely in the amount of education and job training they bring to the labor market.
      2. From this perspective, what individuals earn is the result of their own choices and labor market demand for certain kinds of workers at specific points in time.
      3. Other neoclassical economic models attribute the wage gap to such factors as: (1) the different amounts of energy men and women expend on their work; (2) the occupational choices women make (choosing female-dominated occupations so that they can spend more time with their families); and (3) the crowding of too many women into some occupations (suppressing wages because the supply of workers exceeds demand).
   B. According to conflict perspectives, the gendered division of labor within families and the workplace results from male control of and dominance over women and resources.
      1. Although men's ability to use physical power to control women diminishes in industrial societies, they still remain the head of household, control the property, and hold more power through their predominance in the most highly paid and prestigious occupations and the highest elected offices.
      2. Conflict theorists in the Marxist tradition assert that gender stratification results from private ownership of the means of production; some men not only gain control over property and the distribution of goods but also gain power over women.
   C. Feminist Perspectives
      1. **Feminism** refers to a belief that women and men are equal and that they should be valued equally and have equal rights.
      2. In **liberal feminism**, gender equality is equated with equality of opportunity.
      3. According to **radical feminists**, male domination causes all forms of human oppression, including racism and classism.
      4. **Socialist feminists** suggest that women's oppression results from their dual roles as paid and unpaid workers in a capitalist economy. In the workplace, women are exploited by capitalism; at home, they are exploited by patriarchy.
      5. **Multicultural feminism** is based on the belief that women of color experience a different world than other people because of multilayered oppression based on race/ethnicity, gender, and class.

VII. GENDER ISSUES IN THE FUTURE
   A. During the twentieth century, women made significant progress in the labor force; laws were passed to prohibit sexual discrimination in the workplace and school; affirmative action programs helped make women more visible in education, the government, the professional world, and enter the political arena as candidates instead of volunteers.

B. Many men have joined feminist movements not only to raise their consciousness about men's concerns, but also about the need to eliminate sexism and gender bias.

C. However, in the midst of these changes, many gender issues remain unsolved, such as (1) gender segregation and the wage gap in the labor force, (2) women's burden of the "second shift" at home, (3) eating disorders for both women and some men, (4) body image and obsession with physical fitness, (5) possessions, especially for women as possessions or as sex objects.

ANALYZING AND UNDERSTANDING THE BOXES

After reading the chapter and studying the outline, re-read the boxes and write down key points and possible questions for class discussion.

Sociology and Everyday Life: How Much Do You Know About Body Image and Gender?

Key Points:

Discussion Questions:

1.

2.

3.

Sociology in Global Perspective: The Rise of Islamic Feminism in the Middle East?

Key Points:

Discussion Questions:

1.

2.

3.

Framing Suicide in the Media: "You Can Never Be Too Beautiful" and Teen Plastic Surgery

Key Points:

Discussion Questions:

1.

2.

3.

You Can Make a Difference: Joining Organizations to Overcome Sexism and Gender Inequality

Key Points:

Discussion Questions:

1.

2.

3.

PRACTICE TESTS

MULTIPLE CHOICE QUESTIONS

Select the response that best answers the question or completes the statement.

    1. _____ is the process of treating people as if they were *things*, rather than human beings. This occurs when we judge people on the basis of their physical appearance rather than on the basis of their individual qualities or actions.
        a. Stereotyping
        b. Objectification
        c. Sexism
        d. Imaging

2. According to the research of Hesse-Biber,
    a. women are more dissatisfied with their body weight and shape.
    b. women diet more than men.
    c. women express more anxiety about their bodies than men.
    d. all of these choices

3. Sociologists use the term _____ to refer to the biological attributes of men and women; _____ is used to refer to the distinctive qualities of men and men that are culturally created.
    a. gender, sex
    b. sex, gender
    c. primary sex characteristics, secondary sex characteristics
    d. secondary sex characteristics, primary sex characteristics

4. A(n) _____ is a person in whom sexual differentiation is ambiguous or incomplete.
    a. hermaphrodite
    b. transsexual
    c. transvestite
    d. homosexual

5. A person's perception of the self as female or male is referred to as one's:
    a. self concept
    b. gender identity
    c. gender role
    d. gender belief system

6. According to historian Susan Bordo, the anorexic body and the muscled body
    a. illustrate that eating problems and bodybuilding are unrelated.
    b. are opposites.
    c. exist on a continuum.
    d. are not gendered experiences.

7. According to social psychologist Hilary Lips, an example of _____ directed against men is the mistaken idea that it is more harmful for female soldiers to be killed in combat than male soldiers.
    a. gender stereotypes
    b. matriarchy
    c. patriarchy
    d. sexim

8. All of the following statements are **correct** regarding sexism, **except**:
    a. sexism, like racism, is used to justify discriminatory treatment
    b. evidence of sexism is found in the under-evaluation of women's work
    c. women may experience discrimination in leisure activities
    d. men cannot be the victims of sexist assumptions

9. Sociologist Virginia Cyrus suggest that under _____, men are seen as "natural" heads of the households, presidential candidates, corporate executives, college presidents, while women are seen as men's subordinates, playing supportive roles as housewives, mothers, nurses, and secretaries.
    a. patriarchy
    b. matriarchy
    c. polyarchy
    d. hierarchy

10. _____ refers to the means by which a society gains the basic necessities of life, including food, shelter, and clothing.
    a. Technoeconomic base
    b. Pastoralism
    c. Subsistence
    d. Standard of living.

11. In most hunting and gathering societies,
    a. a relatively equitable relationship exists because neither sex has the ability to provide all of the food necessary for survival.
    b. women are not full economic partners with men.
    c. relationships between women and men tend to be patriarchal in nature.
    d. menstrual taboos place women in subordinate positions by monthly segregation into menstrual huts.

12. Gender inequality and male dominance became institutionalized in _____ societies.
    a. hunting and gathering
    b. horticultural and pastoral
    c. agrarian
    d. industrial

13. All of the following are characteristics of agrarian societies that may contribute to gender inequality, **except**:
    a. private property
    b. male control over distribution of the surplus and the kinship system
    c. the cult of domesticity
    d. emphasis on legitimate heirs to inherit the surplus

14. In the industrial United States, men were responsible for being "breadwinners"; women were seen as "homemakers'; in the _____, the home became a private, personal sphere in which women created a haven for the family.
    a. "matriarchal theory"
    b. "cult of true womanhood"
    c. "gender equality theory"
    d. "theory of male dominance"

15. In postindustrial societies, 24/7 means that
    a. a person works twenty-four days in a month and is then off seven.
    b. a person is available twenty-four hours a day, seven days a week.
    c. out of twenty-four people, seven are employed directly or indirectly in a technological job.
    d. out of twenty-four qualified women who apply for positions, only sever secure eventual employment.

16. All of the following statements regarding gender socialization are correct, **except**:
    a. virtually all gender roles have changed dramatically in recent years
    b. many parents prefer boys to girls because of stereotypical ideas about the relative importance of males and females to the future of the family and society
    c. parents tend to treat baby boys more roughly than baby girls
    d. parents' choices of toys for their children are not likely to change in the near future

17. In their investigation of college women, anthropologists Dorothy Holland and Margaret Eisenhart determined that
    a. peer groups on college campuses are not organized around gender relations.
    b. the peer system propelled women into a world of romance in which their attractiveness to men counted most.
    c. the women's peers immediately influenced their choices of majors and careers.
    d. peer pressure did not involve appearance norms.

18. Teachers who devote more time, effort, and attention to boys than to girls in schools is an example of
    a. gender socialization.
    b. gender identity.
    c. gender role differentiation.
    d. gender bias.

19. A comprehensive study of gender bias in schools suggested that girls' self esteem is undermined in school through all of the following experiences, **except**:
    a. a relative lack of attention from teachers
    b. sexual harassment from male peers
    c. an assumption that girls are better in visual spatial ability, as compared with verbal ability
    d. the stereotyping and invisibility of females in textbooks

20. In discussing gender socialization in college environments, the text points out that
    a. women college professors encourage a more participatory classroom environment which includes both women and men.
    b. college instructors often pay more attention to women than men in their classes.
    c. men are called on less frequently, receive less encouragement, and often are interrupted, ignored, or devalued.
    d. college instructors want to eradicate gender bias more so than high school teachers.

21. Which of the following statements is **true** regarding men and women in higher education?
    a. Men are more likely than women to earn a college degree.
    b. Men tend to have higher grade point averages in college.
    c. Women typically have higher scores on standardized admissions tests like the Scholastic Assessment Test.
    d. Men constitute the majority of majors in architecture, engineering, computer technology, and physical science.

22. According to the text's discussion of gender and athletics,
    a. women who engage in activities that are assumed to be "masculine" (such as bodybuilding) may either ignore their critics or attempt to redefine the activity or its result as "feminine" or womanly.
    b. some women believe they are more likely to win women's bodybuilding competitions if they "overbuild" their bodies.
    c. being active in sports such as gymnastics makes women less likely to be the victims of anorexia and bulimia.
    d. women bodybuilders have learned that they are less likely to win competitions if they look like a fashion model.

23. Which of the following statements regarding gender issues and television are **true**?
    a. Television programs are sex-typed and white-male oriented.
    b. In television today, an equal number of males and females are depicted in leading roles.
    c. Most of the characters in educational programs have female names and voices.
    d. Gender stereotyping no longer exists in the media.

24. Regardless of women's marital status, women have been viewed as _____ wage earners in a male-headed household.
    a. critical
    b. primary
    c. prestigious
    d. supplement

25. According to recent research by sociologist Elizabeth Higginbotham, African American professional women
    a. find themselves limited in employment in certain sectors of the labor market.
    b. are concentrated in public sector employment.
    c. frequently are employed as public school teachers, welfare workers, librarians, public defenders, and faculty members at public colleges.
    d. all of these choices

26. _____ is the division of jobs into categories with distinct working conditions, which results in women having separate and unequal jobs.
    a. Gender bias
    b. Sexual division
    c. Labor market segmentation
    d. Gender disparity

27. Occupational segregation contributes to the _____ – the disparity between women's and men's earnings.
    a. wage gap
    b. gendered employment
    c. comparable worth distinction
    d. gendered earnings

28. Overall, women make approximately _____ cents for every $1 earned by men.
    a. 79
    b. 81
    c. 85
    d. 95

29. The belief that wages ought to reflect the worth of a job, not the gender or race of the worker, is referred to as:
    a. the earnings ratio
    b. pay equity
    c. non comparable worth
    d. the pay gap

30. According to the functionalist/human capital model,
    a. what women earn is the result of their own choices and the needs of the labor market.
    b. the gendered division of labor in the workplace results from male control of and dominance over women and resources.
    c. women's oppression results from their dual roles as paid and unpaid workers in a capitalist economy.
    d. none of these choices

31. Critics of functionalist and neoclassical economic perspectives have pointed out that these perspectives
    a. exaggerate the problems inherent in traditional gender roles.
    b. fail to critically assess the structure of society that makes educational and occupational opportunities more available to some than others.
    c. overemphasize factors external to individuals that contribute to the oppression of white women and people of color.
    d. focus on differences between men and women without taking into account the commonalities they share.

32. According to the conflict perspective on gender stratification,
    a. male dominance over women accumulates steadily due to survival demands in hunting and gathering societies.
    b. the gendered division of labor within families and in the workplace results from male control of and dominance over women and resources.
    c. in agrarian societies, male sexual dominance is almost nonexistent.
    d. gender stratification can be explained using the human capital model.

33. A women's group fights for better child care options, a woman's right to choose an abortion, and the elimination of sex discrimination in the workplace, arguing that the roots of women's oppression lie in women's lack of equal civil rights and educational opportunities. The platform of this group reflects:
    a. black (African American) feminism
    b. socialist feminism
    c. radical feminism
    d. liberal feminism

34. According to _____, male domination causes all forms of human oppression, including racism and classism.
    a. liberal feminism
    b. radical feminism
    c. socialist feminism
    d. multicultural feminism

35. _____ believe that the only way to achieve gender equality is to eliminate capitalism and develop an economy that would bring equal pay and equal rights to women.
    a. Multicultural feminism
    b. Socialist feminism
    c. Radical feminism
    d. Liberal feminism

36. In _____, race, class, and gender all simultaneously oppress African American women.
    a. liberal feminism
    b. socialist feminism
    c. multicultural feminism
    d. radical feminism

37. When studying women with eating problems, Becky Thompson found that
    a. more than half the women of color had been victims of sexual abuse, racism, or anti-Semitism.
    b. eating problems are not associated with sexual orientation.
    c. African American women from the rural South had been socialized to believe that skin color and hair did not determine beauty to the extent that weight did.
    d. eating problems primarily affect white, middle-class women.

38. According to journalist Susan Faludi, men
    a. are not influenced by the feminist movement.
    b. feel pressure to succeed at work as the family "bread-winner."
    c. worry that the feminist movement will hurt their ability to achieve status.
    d. are becoming increasingly aware of their body image.

39. Sociologist William Godde suggests that some men have felt "under attack" by women's demands for equality because
    a. they are being paid less than women in some specific jobs.
    b. they do not see themselves as responsible for patriarchy.
    c. they have tried to eliminate inequality.
    d. they view society now as female-dominated.

40. What is a major issue of the lower wages paid to women in the work force?
    a. Women's lower wages in the labor force raise men's wages.
    b. The pay gap between men and women continues to widen.
    c. Women's lower wages in the labor force suppress men's wages as well.
    d. If women receive higher wages, they will begin to dominate world markets.

TRUE/FALSE QUESTIONS

1. Many men compare themselves unfavorably to bodybuilders and believe they need to gain weight or muscularity.
   T   F

2. Most "sex differences" actually are socially constructed "gender differences."
   T   F

3. Gendered belief systems generally do not change over time.
   T   F

4. Sexism is the subordination of one sex, usually female, based on the assumed superiority of the other sex.
   T   F

5. The earliest known division of labor between women and men is in agrarian societies.
   T   F

6. As societies industrialize, the status of women tends to decline further.
T   F

7. The cult of true womanhood increased white women's dependence on men and became a source of discrimination against women of color.
T   F

8. In postindustrial societies, gendered segregation in the workforce dramatically decreases.
T   F

9. Unlike parents in the United States, parents in many other countries do not prefer boys to girls.
T   F

10. Children's toys reflect their parents' gender expectations.
T   F

11. In studies of parental gender socialization, gender-linked chore assignments occur less frequently in African American families.
T   F

12. Female peer groups place more pressure on girls to do "feminine" things than male peer groups place on boys to do "masculine" things.
T   F

13. By young adulthood, men and women no longer receive gender-related messages from peers.
T   F

14. Research shows the females receive more praise from teachers for their contribution in the class then do males.
T   F

15. The degree of gender segregation in the professional labor market has declined since the 1970s, but racial-ethnic segregation has remained deeply embedded in the social structures.
T   F

16. In discussing body image and gender, recent studies show that up to 95 percent of men express dissatisfaction with some aspect of their bodies.
T   F

17. According to Box 11.2, Rima Khoreibi has written a book about an Islamic superhero woman to demonstrate that sexism is deeply rooted in Islam.
T   F

18. In "Framing Gender in the Media," the news item suggests that the American Society for Aesthetic Plastic Surgery encourages teenage plastic surgery without question.
T   F

19. In Box 11.4, "You Can Make a Difference," "Take Back the Night" is an event that promotes sexism.
T   F

20. In the "Photo Essay," it is suggested that more men are assuming greater responsibility at home, doing household tasks and child rearing.
T   F

FILL-IN-THE-BLANK QUESTIONS

1. The word _____ is used to refer to the distinctive qualities of men and women that are culturally created.

2. A person in whom sexual differentiation is ambiguous or incomplete is known as a(n) _____.

3. A(n) _____ is a male who lives as a woman or a female who lives as a man but does not alter the genitalia.

4. Extreme prejudice directed at gays, lesbians, bisexuals, and others who are perceived as not being heterosexual is termed _____.

5. A person's perceptions of self as female or male is one's _____.

6. The subordination of one sex, usually female, based on the assumed superiority of the other sex is known as _____.

7. Gender _____ consists of showing favoritism toward one gender over the other.

8. _____ work refers to the concentration of women and men in different occupations.

9. The beliefs that wages ought to reflect the worth of a job, not the gender or the race of the worker, is known as _____.

10. The disparity between women's and men's earnings is the _____.

11. The attitudes, behavior, and activities that our socially defined as appropriate for each sex and are learned through the socialization process is known as a(n) _____.

12. A hierarchial system of social organization in which cultural, political, and economic structures are controlled by women is defined as a(n) _____.

13. _____ is the unwanted sexual advances, requests for sexual favors, or other verbal or physical conduct of a sexual nature.

14. _____ refers to an individual's preference for emotional-sexual relationships with members of the opposite sex, same sex, or both.

15. _____ refer to biological males who behave, dress, work, and are treated in most respects as a woman.

SHORT ANSWER/ESSAY QUESTIONS

1. Why and in what ways do societies differentiate people because of gender?
2. What causes gender inequality in the United States?
3. What are some of the consequences of gender inequality?
4. How do gender relations in both the labor force and in families change in postindustrial societies?
5. How does the nature of work affect gender equality in societies and all over the world?

STUDENT CLASS PROJECTS AND ACTIVITIES

1. Conduct an informal survey of twenty students on your campus about their opinions of the women's movement and feminism. Some suggested guideline questions to ask: (1) What is your name, age, gender, and college major? (2) Do you support women's rights? Why? Why not? (3) What is your opinion of the women's movement in our country? (4) Do you consider yourself a feminist? Why? Why not? (5) Has the women's movement improved your life in any way? If so, how? (6) Is there still a need for a strong women's movement? Why? Why not? (7) Does the present women's movement reflect the view of most women? Explain your answer. (8) Is a woman's place at home with her children, if it is economically feasible? (9) Do women today have more freedom than their mothers did? (10) Do women today enjoy life more than their mothers did? (11) Include any other information they may want to obtain from their respondents. In the writing up of this project, provide the responses to each particular question and a summary of the responses. Provide a conclusion and a personal evaluation of this project and submit it to your instructor at the appropriate time.

2. Research the history of sexual harassment in the United States. You are to research: (1) how the term came to be defined; (2) how it became an issue in our society; (3) how it has affected women in the workplace; (4) some specific examples of how it has affected women in the workplace; (5) the prevalence of sexual harassment in the workplace; (6) some court cases involving charges of sexual harassment and their outcomes. Include any other information in this research project. Provide in your paper a bibliography, a summary of your findings, and a personal evaluation of the project. Submit your paper to your instructor at the appropriate time.

3. This is a project that works best for a coed campus but it could easily be altered to fit your campus: You are to reverse your roles regarding dating on your college campus. The women are to initiate the date invitation either in person or by telephone, invite the young men out, at their expense. Each woman is to drive to her date's residence and pick him up. She should also help him get in and out of the car, open all doors for him, assist with sitting and removing and replacing coats, accept the check and pay for the same. She is, in the meanwhile, to maintain a conversation about normal interests, preferably with reversed themes and roles, such as sports and the stereotyped 'male talk' on campus. The female should initiate all conversation, remain in control of the conversation, and interrupt when necessary

to maintain control, remain aloof, when necessary, take up more space when sitting, and give all outward appearance of being completely in charge. The women should walk her date to the door and should initiate any signs of affection, if desired. The assignment for the male is this: The males in the class should initiate any activity in order to get asked out by the females, either by females in the class or on campus. The males are to remain rather passive, after making the first move. They are to act dependent and end every statement with a question, such as "What do you think?" "Is this right?" "Is this okay with you?", etc. They are to walk in front of their dates, and ask for assistance with the doors, with the car, and with their coats or hats. They are not to pay for anything, and are to sit quietly and listen politely to their date's conversation. They can never interrupt or take charge of the conversation in any way whatsoever. The females determine all the activity. Following your date, record your specific activities, actions, and reactions and respond, in writing, to the following questions: (1) Were you comfortable? Explain. (2) Did the behaviors seem amusing? (3) What communication problems arose? (4) Did habitual patterns of behavior surface? (5) How did you feel about this activity? (6) What specifically did you learn? (7) Record any other information, especially an evaluation of the project, and submit your paper to your instructor at the appropriate time.

4. Research the topic of oppression, resistance, and the women of Afghanistan(text, page 355). Several sources are mentioned in the bibliography of your textbook. In the writing up of your paper, include the following information: (1) Describe the plight of women and girls before, during, and after the Taliban's takeover of the country, and after the forced exit of the Taliban. (2) Describe the situation today. Is the situation much better today? Why or why not? (3) What problems remain for the females in Afghanistan today? (4) Describe how the major social institutions, the family, education, political, economic, and religious institutions relate to females today. (5) What specific rights do females have today; what specific rights are females denied? (6) Select another country in the world today where females experience subordination such as under the Taliban. Respond to all of the above questions and statements. In the writing up of your paper, provide any other information pertinent to this topic. (7) Provide a bibliography of your references. (8) Write up and submit your paper, following your professor's specific requirements.

INTERNET ACTIVITIES

1. Surf these websites in order to compare similarities and differences in the ways that both sexes are defining the issues.
   a. **Men's Studies Websites:**
      i. http://www.mensstudies.com/
      ii. http://www.ncfm.org/
      iii. http://www.xyonline.net/links.shtml
   b. **Feminist Websites:**
      i. http://vos.ucsb.edu/browse.asp?id=2711
      ii. http://bailiwick.lib.uiowa.edu/wstudies/
      iii. http://www.feministstudies.org/

2. **This is a giant listing** of books related to men's issues: **http://www.ncfm.org/read.htm**. It offers a great starting point for students interested in researching gender related topics.
3. Try and learn as much as you can about **gender** BEFORE you have your first class on this topic. Go to this site, **http://www.trinity.edu/mkearl/gender.html,** and in small groups come back to class and present what you believe to be the most essential aspects of this study.

INFOTRAC COLLEGE EDITION EXERCISES

Visit the **InfoTrac College Edition** website at: **http://www.wadsworthmedia.com/webtutor/infotrac.htm**. You will arrive at a screen that enables you to search topics.

1. Compare research on **Transsexualism**, **Hermaphroditism**, and **Transvestitism**. Search for these terms in the media as well and see how they are portrayed differently.
2. Gender differences in undergraduates' body esteem: the mediating effect of objectified body consciousness and actual/ideal weight discrepancy Search InfoTrac and read this article. Bring to class three essay questions. You can trade essay questions in class and provide answers in an informal small group discussion.
3. Go to InfoTrac and conduct a general search for the term **Gender**. You will find a multitude of articles in numerous subfields. Come to class with an idea or issue that is new to you. This is a good way to start a class discussion or introduce the topic of **gender**.
4. Look up the keyword **Fatherhood**. Be sure to look at "Fatherhood's Top Ten Films," by Julia Duin in *Insight on the News*, February 15, 1999. What films represent fathers in a positive light?
5. There are over a hundred articles on **Patriarchy** listed on InfoTrac. Write a brief essay on this subject and use at least three sources from this list. The essay should introduce the concept and then demonstrate an important dimension.

SOLUTIONS

MULTIPLE CHOICE QUESTIONS

1. B, p. 347
2. D, pp. 347-348
3. B, p. 348
4. A, p. 349
5. B, p. 352
6. C, p. 353
7. D, p. 354
8. D, p. 354
9. A, p. 354
10. C, p. 354
11. A, p. 355
12. C, p. 356
13. C, p. 356
14. B, p. 356

15. B, p. 357
16. A, pp. 358-359
17. B, p. 361
18. D, p. 362
19. C, p. 362
20. A, p. 362
21. D, p. 366
22. A, p. 364
23. A, pp. 364-365
24. D, p. 366
25. D, p. 366
26. C, p. 366
27. A, p. 367
28. B, p. 368

29. D, p. 368
30. A, p. 374
31. B, p. 374
32. B, p. 374
33. D, p. 375
34. B, p. 375
35. B, p. 376
36. C, p. 376
37. A, p. 377
38. D, p. 378
39. B, p. 378
40. C, p. 378

TRUE/FALSE QUESTIONS

1. T, p. 348
2. T, p. 352
3. F, p. 353
4. T, p. 354
5. F, p. 355
6. T, p. 356
7. T, p. 357

8. F, p. 357
9. F, p. 358
10. T, p. 359
11. T, p. 359
12. F, p. 360
13. F, p. 361
14. F, p. 362

15. T, p. 366
16. T, p. 351
17. F, p. 355
18. F, p. 366
19. F, p. 376
20. T, p. 373

FILL-IN-THE-BLANK QUESTIONS

1. gender, p. 352
2. hermaphrodite, p. 349
3. transvestite, p. 350
4. homophobia, p. 352
5. gender identity, p. 352
6. sexism, p. 354
7. bias, p. 362
8. Gender-segregated, p. 365
9. comparable, worth, p. 368
10. wage gap, p. 368
11. gender role, p. 352
12. matriarchy, p. 354
13. Sexual harassment, p. 362
14. Sexual orientation, p. 350
15. Berdaches, p. 350

12

AGING AND INEQUALITY BASED ON AGE

BRIEF CHAPTER OUTLINE

CHAPTER SUMMARY

Aging is the physical, psychological, and social processes associated with growing older. Over the past two decades, the older U.S. population has been increasing, while the proportion of young people is decreasing. As a result of these changing population trends, research on aging has grown dramatically according to the discipline of **gerontology**, the study of aging and older people. People are assigned roles and positions based on the age structure and role structure in a particular society. In hunting and gathering societies, people typically have shorter life expectancies and all are expected to work. Young people are valuable assets and older people may be perceived as being less productive. In horticultural, pastoral, and agrarian societies, people begin to live longer and begin to accumulate a surplus. However, life is still very hard for most people. In industrial societies, living standards improve and advances in medicine contribute to greater longevity for more people. In postindustrial societies, a shift from a primarily young society to an older society brings about major changes in societal patterns and in the needs of the population. According to a case study, the graying of the Japanese population has taken only 25 years, whereas it took almost a century in North America and Europe. This "graying" of the Japanese population may be bringing about a gradual change in the social importance of the elderly in that nation. Age differentiation in the United States produces **age stratification**, the inequalities, differences, segregation, or conflict between age groups. In contemporary society, various strata of the life course present unique problems at each level. The strata range from infancy and childhood, adolescence, young adulthood, middle adulthood, and late adulthood. **Ageism** – prejudice and discrimination against people on the basis of age – is reinforced by stereotypes based on one sided and exaggerated images of older people. Age, gender, and inequality are intertwined. Older women who retire from the work force do not garner economic security for their retirement years at the same rate that many men do. Age, race/ethnicity, and economic inequality are also closely intertwined; the lower income status of older American ethnic minority group members can be traced to patterns of limited economic opportunities throughout their lives. Medicare, Medicaid, and civil service pensions are the major **entitlements** for persons aged 65 and older living below the poverty line. Older people in rural areas are more likely to have lower incomes and tend to be classified as poor. **Elder abuse** includes physical abuse, psychological abuse, financial exploitation, and medical abuse or neglect of people age 65 or older. In examining the living arrangements for older adults, relatives provide most of the care, although support services help older individuals cope with their daily problems. Either by choice or necessity, some older adults move to smaller housing units or apartments. The most restrictive environment for older persons is the nursing home setting, even in a best case scenario. Functionalist explanations of aging – such as **disengagement theory** – focus on how older persons adjust to their changing roles in society by detaching themselves from their social role and prepare for their eventual death. **Activity theory** - a symbolic interactionist perspective – states that people change in late middle age and find substitutes for previous statuses, roles, and activities. Conflict theorists link the loss of status and power experienced by many older persons to their lack of ability to produce and maintain wealth in a capitalist economy. The association of death with the aging process contributes to ageism in our society.

Three well-known frameworks explain how people cope with the process of dying: the *stage-based approach*, the *dying trajectory*, and the *task-based approach*. How a person dies is shaped by many social and cultural factors. A **hospice** is an organization that provides a home-like facility or home-based care for people who are terminally ill. Over time, hospice care has moved toward hospital standards. In examining aging in the future, advances in medical technology may lead to a more positive outlook on aging, however with the increase of the aged population and the decrease of the birth rate, concerns over entitlements and other governmental assistance for the aged continue. Classism, racism, sexism, and ageism all restrict individual's access to valued goods and services in society. Older people continue to resist ageism through organizations such as the Grey Panthers, AARP, and the Older Women's League.

LEARNING OBJECTIVES

After reading Chapter 12, you should be able to:

1. Define aging and explain why research on aging has recently increased.

2. Compare historical and contemporary perspectives on age.

3. Explain the typical American living arrangements for older persons.

4. List and describe support services for older persons.

5. Explain the social significance of age.

6. Explain what the "graying" of Japan means for women in that society.

7. Describe trends in aging and explain how life expectancy has changed in the United States during the twentieth century.

8. Discuss ageism and describe the negative stereotypes associated with older persons.

9. Describe and provide explanations for elder abuse.

10. Distinguish between chronological age and functional age and note the social significance of each.

11. Trace the process of aging through the life course, noting the social consequences of age at each stage.

12. Distinguish between functionalist, symbolic interactionist, and conflict perspectives on aging.

13. Describe the relationship of wealth, poverty and aging.

14. Describe how age, gender and poverty are intertwined.

15. Describe how age, race/ethnicity, and economic inequality are intertwined.

16. Evaluate the three major frameworks that explain how people cope with death and dying.

17. Describe the most significant age-related inequalities experienced in the workforce.

KEY TERMS

(defined at page number shown and in glossary)

activity theory, p. 406
age stratification, p. 390
ageism, p. 395
aging, p. 383
chronological age, p. 384
cohort, p. 385
disengagement theory, p. 406

elder abuse, p. 403
entitlements, p. 401
functional age, p. 384
gerontology, p. 386
hospice, p. 410
life expectancy, p. 385

KEY PEOPLE

(identified at page number shown)

Elaine C. Cumming and William E. Henry, p. 406
Anne Fausto Sterling, p. 393
Melissa A. Hardy and Lawrence E. Hazelrigg, p. 400
Elisabeth Kübler Ross, p. 409
William C. Levin, p. 397
Patricia Moore, p. 398
William J. Wilson, p. 392

CHAPTER OUTLINE

I. THE SOCIAL SIGNIFICANCE OF AGE
    A. **Aging** is the physical, psychological, and social processes associated with growing older.
    B. **Chronological age** is a person's age based on date of birth; functional age is observable physical attributes (physical appearance, mobility, etc.) that are used to assign people to age categories.
    C. There are significant differences in **life expectancy** – the average length of time a group of individuals of the same age will live – based on racial ethnic and sex differences.
    D. Social gerontology is the study of the social (non-physical) aspects of aging; gerontology is the study of aging and older people.
    E. While young persons historically were considered "little adults" and were expected to do adult work, today the skills necessary for many roles are more complex and the number of unskilled positions is more limited.
    F. **Age stratification** refers to the inequalities, differences, segregation, or conflict between age groups.

II. AGE IN GLOBAL PERSPECTIVE
  A. In hunting and gathering societies, people typically have shorter life expectancies and all people are expected to work.
     1. Young people are assets in the nomadic lifestyle involved in hunting wild game.
     2. Older people may be perceived as being less productive.
  B. In horticultural, pastoral, and agrarian societies, death rates begin to decline.
     1. More people reach older ages, but life is still very hard.
     2. Accumulating a surplus becomes possible and older individuals, especially men, are often the most privileged in society.
     3. In agrarian societies, food surplus, because of farming, makes it possible for more people to live to adulthood and old age; in these societies, most older people live with the other family members.
  C. In industrial societies, living standards improve and advances in medicine contribute to greater longevity for more people while in postindustrial societies, a shift from a primarily young society to an older society brings about major changes in societal patterns and in the needs of the population.
  D. The graying of the Japanese population has taken only twenty-five years, whereas it took almost a century in North America and Europe.
     1. The sociocultural change and population shift may bring about gradual change in the social importance of the elderly in Japan.
     2. With over 60 percent of Japanese women in the labor force, greater pressure is being placed on Japanese policy makers to consider how the government can play a larger role in the care of the aging population rather than that care placed completely on the family, especially women.
III. AGE AND THE LIFE COURSE IN CONTEMPORARY SOCIETY
  A. Infancy and childhood are typically thought of as carefree years; however, children are among the most powerless and vulnerable people in society.
     1. Early socialization plays a significant part in children's experiences and their quality of life.
     2. Two-thirds of all childhood deaths are caused by injury; far too many children lose their lives at an early age due to the abuse, neglect, or negligence of adults.
  B. Adolescence roughly spans the teenage years; today it is a period in which young people are expected to continue their education and perhaps hold a part-time job.
     1. A variety of reports have labeled contemporary U.S. teenagers as a "generation at risk" because of many social problems, including crime and violence, teenage pregnancy, suicide, drug abuse and peer pressure.
     2. One way to save the adolescent generation of today is to reduce poverty among them and their families.
  C. Young Adulthood follows adolescence and lasts to about age 39.
     1. During this time, people are expected to get married, have children, and get a job.
     2. People who are unable to earn income and pay into social security or other retirement plans in early adulthood become disadvantaged in later life.

D. Middle Adulthood–roughly the ages of 40 to 65–did not exist in the United States until recently.
   1. The process of senescence, primary aging, occurs at this time.
   2. Women undergo menopause (the cessation of the menstrual cycle); men undergo a climacteric in which the production of the "male" hormone levels produces nervousness and depression.
   3. On the positive side, for some people, middle adulthood represents the time period of having the highest levels of income and prestige.
E. Late adulthood is generally considered to begin at age 65–the "normal" retirement age.
   1. *Retirement* is the institutionalized separation of an individual from an occupational position, with continuation of income through a retirement pension based on prior years of service.
   2. Some gerontologists subdivide late adulthood into three categories: (1) the "young-old (ages 65-74); (2) the "old-old (ages 75-85); and (3) the "oldest-old" (over age 85).
   3. The rate of biological and psychological changes in older persons may be as important as their chronological age in determining how they are perceived by themselves. A loss of height, impaired ability to function, heart attacks, strokes, and cancer are not uncommon at this stage of life; these changes can cause stress.
   4. Late adulthood may turn into a time of despair or one of integrity.

IV. INEQUALITIES RELATED TO AGING
A. **Ageism** – prejudice and discrimination against people on the basis of age, particularly when they are older persons – is reinforced by stereotypes of older persons as cranky, sickly, and lacking in social value.
B. If we compare wealth (all economic resources of value) with income (available money or its purchasing power), we find that older people tend to have more wealth but less income than younger people; among older people, there is a wider range of assets and income than in other age categories.
C. Age, gender, and inequality are intertwined; although middle-aged and older women make up an increasing portion of the work force, they are paid much less than men their age, receive raises at a slower rate, and still work largely in gender-segregated jobs.
D. Ninety percent of all retired people in the U.S. draw Social Security benefits, which are a form of **entitlements** – benefits paid by the government. Supplemental Social Income, Medicaid, civil service pensions, and Medicare are among the primary entitlements for those over the age of 65. Medicare is a nationwide health care program for persons age 65 and over who participate in Social Security or "buy into" the program.
E. Age, race/ethnicity and economic inequality are closely intertwined.
   1. The lower income states of old American ethnic minority group members can be traced to patterns of limited economic opportunities throughout their lives.

F. Older people in rural areas are more likely to have lower incomes and tend to be classified as poor.
   1. The rural elderly have lower lifetime earnings, limited savings, and fewer opportunities for part-time work.
   2. The homes of the rural elderly have lower value and are in greater need of repair.
G. **Elder abuse** refers to physical abuse, psychological abuse, financial exploitation, and medical abuse or neglect of people age 65 or over.

V. LIVING ARRANGEMENTS FOR OLDER ADULTS
A. Support services, homemaker services and day care centers help older people cope with the problems of daily care.
B. Nursing homes are the most restrictive environments for older persons.
   1. Financing of long-term care is a problem for most older persons.
   2. Medicare excludes most long-term care; most private insurance policies also exclude long-term care.

VI. SOCIOLOGICAL PERSPECTIVES ON AGING
A. Functionalists examine how older persons adjust to their changing roles. According to **disengagement theory**, older persons make a normal and healthy adjustment to aging when they detach themselves from their social roles and prepare for their eventual death.
B. **Activity theory** - a symbolic interactionist perspective on aging – states that people tend to shift gears in later life and find substitutes for previous statuses, roles, and activities.
C. Conflict theorists note that, as people grow older, their power tends to diminish unless they are able to maintain their wealth. Consequently, those who have been disadvantaged in their younger years become even more so in late adulthood.

VII. DEATH AND DYING
A. In contemporary industrial societies, death is looked on as unnatural because it has been removed from everyday life: most deaths occur among older persons and in institutionalized settings.
B. Three well-known frameworks provide explanations of how people cope with the process of dying.
   1. The *stage-based approach*, popularized by Elizabeth Kübler-Ross, who proposed five stages in the dying process:
      a. denial ("Not me!")
      b. anger ("Why me?")
      c. bargaining ("Yes me, but....")
      d. depression and sense of loss
      e. acceptance
   2. The *dying trajectory*, which focuses on the perceived course of dying and the expected time of death.
   3. The *task-based approach* is based on the assumption that the dying person can and should go about daily activities and completing physical, social, or spiritual tasks.

C. A **hospice** is a homelike facility that provides supportive care for patients with terminal illnesses; over time, hospice care has moved toward hospital standards.
VIII.   AGING IN THE FUTURE
    A. By the year 2050, there will be approximately 80 million persons age 65 and over, as compared with 35 million in 2000. More people will survive to age 85 – or even 95 and over.
    B. A 1994 report warned that this increase (coupled with a decreasing birth rate) may result in entitlements consuming nearly all federal tax revenues by the year 2012, leaving the government with no money for anything else.
    C. New biomedical research discoveries may lead to more positive aging; however, many of the benefits and opportunities of living in a highly technological, affluent society are not available to all.

ANALYZING AND UNDERSTANDING THE BOXES

After reading the chapter and studying the outline, re-read the boxes and write down key points and possible questions for class discussion.

Sociology and Everyday Life: How Much Do You Know Aging and Age-Based Discrimination?

Key Points:

Discussion Questions:

1.

2.

3.

Sociology and Social Policy: Driving While Elderly: Policies Pertaining to Age and Driving

Key Points:

Discussion Questions:

1.

2.

3.

Framing Aging in the Media: "Just When You Thought You Were Too Old for Romance"

Key Points:

Discussion Questions:

1.

2.

3.

You Can Make a Difference: Getting Behind the Wheel to Help Older People: Meals on Wheels

Key Points:

Discussion Questions:

1.

2.

3.

PRACTICE TESTS

MULTIPLE CHOICE QUESTIONS

Select the response that best answers the question or completes the statement.

1. All of the following statements regarding aging are correct, **except**:
   a. Aging is the physical, psychological, and social processes associated with growing older.
   b. Aging is an inevitable consequence of living.
   c. Aging is a socially significant process.
   d. In the United States, aging is valued status.

2. _____ age refers to a person's age based on date of birth; by contrast, _____ age refers to observable individual attributes such as physical appearance, mobility, strength, coordination, and mental capacity that are used to assign people to age categories.
   a. Obvious, subjective
   b. Subjective, obvious
   c. Chronological, functional
   d. Functional, chronological

3. The median age of the U.S. population is
   a. increasing due to an increase in life expectancy combined with a decrease in birth rates.
   b. decreasing due to an increase in birth rates.
   c. roughly the same as it had been for most of the twentieth century due to stabilization of both birth and death rates.
   d. unaffected by the aging of the Baby Boomers.

4. Two hundred years ago, the age spectrum was divided into
   a. babyhood and adulthood.
   b. babyhood, a very short childhood, and adulthood.
   c. babyhood, childhood, adulthood, and old age.
   d. infancy, childhood, adolescence, adulthood, and old age.

5. In postindustrial societies, the largest employers are in the fields of _____ and _____ which may benefit older people.
   a. health, agriculture
   b. education, health
   c. education, textile production
   d. textile production, service industries

6. In the U.S., most of all childhood deaths today are caused by:
   a. HIV/AIDS
   b. injuries
   c. communicable diseases
   d. birth defects

7. According to the text, middle adulthood represents the time when many people
   a. are strapped for cash because their children are in college.
   b. may have grandchildren who give them another tie to the future.
   c. have a decline in income as they approach their retirement years.
   d. grow discontented with their spouse of many years.

8. Which of the following statements is **correct** regarding Alzheimer's disease?
   a. Most older people suffer from Alzheimer's disease.
   b. Most people with Alzheimer's disease have an extremely short life expectancy.
   c. About 55 percent of all organic mental disorders in the older population are caused by Alzheimer's disease.
   d. In 1995, researchers found a possible cure for Alzheimer's disease.

9. According to Erik Erikson, older people must resolve a tension of _____ versus _____ in their lives.
   a. integrity, despair
   b. activity, disengagement
   c. stability, change
   d. autonomy, entitlements

10. Ageism is
    a. prejudice and discrimination against people on the basis of age, particularly when they are younger persons.
    b. prejudice and discrimination against people on the basis of age, particularly when they are older persons.
    c. gradually being overcome by positive depictions of older persons in the media.
    d. experienced equally by both women and men.

11. When Patricia Moore, at age 27, disguised herself as an 85-year-old woman and went to a grocery store, she learned that
    a. people permitted her to go to the head of the checkout line because of her age.
    b. grocery store clerks treated her as a preferred customer.
    c. people tried to have as little to do with her as possible.
    d. other people's reactions to her changed when she played the role of an older person.

12. In a recent study, gerontologists Melissa H. Hardy and Lawrence E. Hazelrigg found that
    a. gender is more directly related to poverty in older persons than is race/ethnicity, educational background, or occupational status.
    b. race/ethnicity is more directly related to poverty in older persons than is gender, educational background, or occupational status.
    c. educational background is more directly related to poverty in older persons than is gender, race/ethnicity, or occupational status.
    d. occupational status is more directly related to poverty in older persons than is gender, race/ethnicity, or educational background.

13. All of the following statements are **correct** regarding elder abuse, **except**:
    a. abuse and neglect of older persons has received increasing public attention in recent years
    b. financial exploitation is a form of elder abuse
    c. more than 1.6 million older people in the United States are victims of elder abuse
    d. nursing home personnel are the most frequent abusers of older persons

14. Studies have shown that _____ were the most frequent abuses of older persons.
    a. nursing home personnel
    b. younger siblings
    c. daughters
    d. sons

15. Functionalist perspectives on aging focus on
    a. how older persons adjust to their changing roles in society.
    b. the connection between personal satisfaction in a person's later years and a high level of activity.
    c. reasons why aging is especially problematic in contemporary capitalist societies.
    d. how people are consigned to the sick role.

16. _____ theory states that people tend to shift gears in late-middle age and find substitutes for previous statuses, roles, and activities.
    a. Continuity
    b. Disengagement
    c. Activity
    d. Engagement

17. _____ theorists view aging as problematic in contemporary capitalistic societies.
    a. Functionalist
    b. Activity
    c. Conflict
    d. Symbolic Interactionists

18. The _____ framework focuses on the perceived course of dying and the expected time of death.
    a. dying trajectory
    b. stage-based approach
    c. task-based approach
    d. Kübler-Ross approach

19. All of the following about hospice care are true **except**:
    a. a hospice provides care for the terminally ill
    b. over time, hospice care has moved away from hospital standards
    c. hospice philosophy asserts that people should participate in their own care and control
    d. the approach is family-based

20. All of the following are organizations fighting ageism except:
    a. Gray Panthers
    b. Aged United
    c. Older Women's League
    d. American Association of Retired Persons

21. Almost _____ of the people living in the high-income countries today are expected to survive to age 65. By sharp contrast, in low-income countries such as Zambia, Uganda, and Rwanda, nearly _____ of the total population in each nation is not expected to survive to age 40.
    a. 75 percent, 35 percent
    b. 80 percent, 40 percent
    c. 85 percent, 45 percent
    d. 90 percent, 50 percent

22. When people say "Act your age," they are referring to _____ age – a person's age based on date of birth.
    a. functional
    b. normative
    c. chronological
    d. sequential

23. When feminist scholars state that functional age is subjective for men and women, they mean
    a. women and men age in similar fashion.
    b. women and men are evaluated differently.
    c. men and women are defined about evenly.
    d. society evaluates men and women using the same standards.

24. As they age, men are believed to become more _____ and women are thought to become more _____.
    a. distinguished, grandmotherly
    b. helpless, powerful
    c. active, passive
    d. mentally depressed, physically active

25. The age at which one-half the people in a population are younger and the other half are older is referred to as the _____ age of that population
    a. mean
    b. modal
    c. median
    d. standard

26. People born between 1946 and 1964 are commonly referred to as _____.
    a. World War II baby products
    b. Baby Boomers
    c. Cold War baby explosion
    d. population dynamics

27. Referred to as the _____, the aging of the U. S. population resulted from an increase in life expectancy combined with a decrease in birth rates.
    a. golden pond ERA
    b. graying of America
    c. demographic transition
    d. functional stage

28. In 1900, about 4 percent of the U.S. population was over age 65. In 2005, approximately _____ percent of the population was 65 or over.
    a. 9
    b. 13
    c. 20
    d. 24

29. According to Census Bureau projections, about _____ of the population will be at least age 65 by 2050.
    a. 20 percent
    b. 25 percent
    c. 33 percent
    d. 35 percent

30. One of the fastest growing segments of the U.S. population is made up of _____. This cohort is expected to almost double in size between 2000 and 2025.
    a. teenagers.
    b. those between ages 25-35.
    c. those between ages 45-55.
    d. those age 85 and over.

31. The age pyramid that depicts the current distribution of the U.S. population can best be described as
    a. a perfect pyramid.
    b. a pyramid that is larger at the base because of an increase in birth rates.
    c. a pyramid that bulges in the middle because of the Baby Boomers and is smaller at the base because of declining birth rates among post-Baby Boomers.
    d. a pyramid that bulges at the very top because of the extraordinary growth of the old-old population.

32. _____ is the study of the nonphysical aspect of aging.
    a. Human gerontology
    b. Urban gerontology
    c. Cultural gerontology
    d. Social gerontology.

33. If the present trend continues in Japan, by 2025 people age 65 and over will make up about _____ percent of the total population.
    a. 20
    b. 25
    c. 30
    d. 50

34. Statistics from the Children's Defense Fund point out the perils of infancy childhood; everyday in the United States, _____ infants are born into poverty.
    a. 1 out of every 2
    b. 1 out of every 3
    c. 1 out of every 4
    d. 1 out of every 5

35. As compared with all other racial-ethnic categories, _____ have the highest rate of motor vehicle death.
    a. Native American children
    b. Asian American children
    c. African American children
    d. Hispanic American children

36. The text points out that adolescence
    a. is a period in which the individual is still allowed to act "childish".
    b. is a time when young people are expected to continue their education and perhaps hold a part-time job.
    c. is a time when young men and women are not expected to hold a job while they go to school.
    d. is a period in which the individual is accorded adult status.

37. Young adulthood follows adolescence and lasts to about age_____.
    a. 25
    b. 30
    c. 39
    d. 45

38. Middle adulthood refers to people between the ages of:
    a. 30 and 50
    b. 40 and 55
    c. 40 and 65
    d. 55 and 70

39. The percentage of the population age 65 and above varies from state to state – from a low 6.6 percent in _____ to a high of 16.8 in _____.
    a. Wyoming, California.
    b. Utah, Hawaii.
    c. Vermont, Texas.
    d. Alaska, Florida.

40. If current trends continue, by the year 2050 there will be an estimated _____ million people age 65 and older.
    a. 40
    b. 50
    c. 70
    d. 80

TRUE/FALSE QUESTIONS

1. A group of people born within a specific period in time is known as a cohort.
   T   F

2. As men age, they are believed to become less powerful when compared to women.
   T   F

3. About 4 percent of the U.S. population was over age 65 in 1900; by 2000, that population increased by 20 percent.
   T   F

4. Gerontology is the study of the non-physical aspects of aging.
   T   F

5. In hunting and gathering societies, the elderly have great prestige because they accumulate a surplus of goods.
   T   F

6. The graying of Japan has taken longer than in North America and Europe.
   T   F

7. Adolescence is a twentieth century concept.
   T   F

8. In the U.S., young adulthood lasts to about age 39.
   T   F

9. Few of the wealthiest people in the United States are over 65 years of age.
   T   F

10. Age, race/ethnicity, gender, and economic inequality are interrelated.
    T   F

11. Nearly half of all women over age sixty-five are widowed and living alone on fixed incomes.
    T   F

12. According to analysts, Social Security keeps more white men out of poverty than it does white women and people of color.
    T   F

13. Older Native Americans are among the most disadvantaged of all categories.
    T   F

14. Only about five percent of frail, older persons are currently in nursing homes.
    T   F

15. Over time, the hospice movement has moved away from hospital standards to more home-based care.
    T   F

16. According to Box 12.1, the rate of elder abuse in the United States has been greatly exaggerated by the media.
    T   F

17. According to the article on driving while elderly, most states have policies regarding drivers over the age of 65.
    T   F

18. In Box 12.3, the article suggests that many films today showing what it means to be older are depicting a positive image of aging and romance.
    T   F

19. According to Box 12.4, some persons under the age of 60 benefit from the Meals on Wheels program.
    T   F

20. Meals on Wheels, according to Box 12.4, is largely a grassroots operation.
    T   F

FILL-IN-THE-BLANK QUESTIONS

1. A person's age based on date of birth is one's _____ age, while observable individual attributes is one's _____ age.

2. The average number of years that a group of people born in the same year could be expected to live is known as _____.

3. A group of people born within a specific period of time is known as a(n) _____.

4. The study of aging and older people is known as _____.

5. The inequalities differences, segregation, or conflict between age groups is known as _____.

6. Prejudice and discrimination against people on the basis of age is termed _____.

7. Certain benefit payments paid by the government are known as _____.

8. A national health care program for persons aged 65 and older who are covered by Social Security or who buy into the program is known as _____.

9. Psychiatrist _____ proposed a five stage theory of the process of dying.

10. A(n) _____ is an organization that provides home-based or a homelike facility care for terminally-ill people.

11. _____ states that people tend to shift gears in late middle age and find substitutes for previous statuses, roles and activities.

12. The first stage in the dying process, according to Kübler-Ross, is _____.

13. Ninety percent of all people in the United States over age 65 draw _____.

14. _____ states that older persons make a normal and healthy adjustment to aging when they detach themselves from their social roles and prepare for their eventual death.

15. The _____ approach is based upon the assumption that the dying person should go about daily activities.

SHORT ANSWER/ESSAY QUESTIONS

1. Is the United States graying? If so, what is the significance of this graying?
2. How are race, class, gender, and aging related?
3. What is the typical living arrangement for older adults in the United States?
4. What are some examples of ageism in the United States?
5. What are some examples of problems of the elderly, primarily problems of elderly women?

STUDENT CLASS PROJECTS AND ACTIVITIES

1. Conduct research on the history of the original Social Security legislation in our country. In your research, examine these issues and provide responses to the following questions: (1) What was the original purpose of the act, and what were the major provisions of the act? (2) Did most legislators support the act? Why? (3) What groups were opponents of the act? (4) What was the political position of the President toward the act? (5) What year did the legislation become law? (6) Why? (7) Why was the United States one of the last industrialized nations in the world to provide its elderly citizens with old-age insurance? (8) What western country was the first to enact old-age benefits? (9) List at least ten different countries that had some type of old-age insurance programs before the United States. (10) How was the program originally funded? What did the employer contribute? (11) How did workers become eligible for full benefits? (12) Evaluate the effectiveness of this legislation. Did it meet its goals? Why? Why not? (13) Personally critique this act. Do you agree with the premises of Social Security? Why? Why not? Submit your written paper to your instructor at the appropriate time, following any instructions given to you by your instructor.

2. You should conduct research of an old-age insurance program found in one other country for the purpose of comparing the current state of its program with the current "crisis" of Social Security in the United States. Some possible choices: Germany, Denmark, Sweden, United Kingdom, Austria, France, Ireland, Italy, Spain, Belgium, Greece, the Netherlands, New Zealand, Australia, and Japan. Examine the history of the legislation, and answer the following questions: (1) What was the purpose of the act, and what were the major provisions of the act? (2) What groups supported the act? Why? (3) What groups opposed the act? Why? (4) What was the political climate of the time; did the leader of the country support the act? Why? (5) When did the act become law? (6) How is the program financed? (7) How do people qualify for the program? (8) What is its current status? Are there any indications of a financial crisis for the program? (9) Briefly compare this country's program with that of the Social Security program in the United States. Are there any similarities? Some differences? (10) Provide any other information that will enhance the research, including a conclusion and a personal evaluation. Submit your paper, following your instructor's guidelines, at the appropriate time.

3. Choose at least two elderly people who are accessible to you (such as grandparents, neighbors, retired faculty, etc.) and interview them about their experiences in aging. If they do not object, you may use a tape recorder to record the interview. In the interview, ask them to describe some of the important events that occurred in their childhood, adolescence, adulthood and, most recently, in the past several years. For consistency in your interviews and written papers, answer each of these questions: (1) How do they describe themselves? (2) What has been some of the most important events in their lives? Remember to cover childhood, adolescence, and adulthood. (3) How do members of their family, friends, and other people treat them? Is this different from previous times? How? (4) Do they think of themselves as old? Why? Why not? (5) What is one thing that they like about their lives today? (6) What is one thing that they liked about their yesteryear life? (7) What

types of celebrations were most important in their lives? Describe them. (8) What was the most significant event in their life up to this point? (9) What is one thing they would change if they could? (10) What is one thing they would never change? (11) How do they think the majority of society views them today? How do they feel about this view? (12) Do they have any additional statements or thoughts they would like to make? Summarize the major points of the interviews and present this information in a written paper. Submit your paper to your instructor at the appropriate time.

4. The year 2005 was the midpoint of a decade that presented a unique picture of the world's population. Before 2000, "young people" always outnumbered "old people." From 2000 forward, "old people" began to and will continue to outnumber "young people." The graying of the world's population is not moving in unison around the globe. Search through references, such as the book by Robert Ashley and Amanda Barusch, 2004, *Social Forces and Aging: An Introduction to Social Gerontology* and the special magazine issue of <u>Scientific American</u>, September 2005, to research and respond to these questions: (1) What are at least six major impacts of the "graying" the world?" (2) What countries represent this "graying" process? (3) Select two of these representative countries and compare and contrast their plans to address their ever-growing aging population concerns and needs, such as: **A**. health care; **B**. specific "aging" diseases **C**. the dependency ratio and the **D**. sustainability of the elderly population. (4) List some countries that continue to have half of their population aged 23 years old or younger. Provide several explanations for this phenomenon and the resultant effects of the younger populations. (5) Finally, provide your own aging and population projections to the year 2050. (6) Submit your paper as per your professor's instructions.

INTERNET ACTIVITIES

1. To learn more about research being conducted on **aging**, go the site for the **National Institute on Aging**: **http://www.nia.nih.gov/**. Click on *Research Programs* (behavioral and social research) to read about current research on **aging**. Look for research on myths about **aging** and how drug costs affect the elderly.

2. A very powerful interest group is the **AARP** (The American Association of Retired Persons), **http://www.aarp.org/**. Click on AARP in your state and then select either a region or your state specifically to read about the services that are provided by the AARP to members. Next, select some other states for purposes of comparison. For example, Florida and Arizona have large older populations. Does the AARP have a stronger presence in those states, and does it offer more services?

3. The **Administration on Aging**, **http://www.aoa.dhhs.gov/**, provides extensive information on **aging** for professionals in the field of **aging**, service providers, and policy makers, as well as older Americans themselves. Find out more about this agency at the web site.
4. You can calculate your own life expectancy from the Living to 100 Life Expectancy Calculator web site: http://www.livingto100.com/.
5. To find health-related research, such as research on Alzheimer's, go to the **National Institute of Health** website, **http://www.nih.gov/**, and do a search for **aging** research.

INFOTRAC COLLEGE EDITION EXERCISES

Visit the **InfoTrac College Edition** website at: **http://www.wadsworthmedia.com/webtutor/infotrac.htm**. You will arrive at a screen that enables you to search topics.

1. Look for newspaper articles on **Age Discrimination**. Compare these with research articles on the same topic. Does it appear that the general public really understands this issue?
2. Examine research on **Elder Abuse** in the home and in nursing homes. Take a look at the *California Elder Abuse and Dependent Adult Civil Protection Act*. What factors seem to contribute to **elder abuse**?
3. Write an essay on the **Social Aspects of Aging**. There are a number of journal and news articles on this sub field.
4. Look for articles that explore **Aging** in other countries, especially Asian countries.

SOLUTIONS

MULTIPLE CHOICE QUESTIONS

1. D, p. 383
2. C, p. 384
3. A, p. 384
4. B, p. 386
5. B, p. 389
6. B, p. 391
7. B, p. 393
8. C, p. 394
9. A, p. 395
10. B, p. 395
11. C, p. 399
12. A, p. 400
13. D, p. 403
14. D, p. 403

15. A, p. 406
16. C, p. 406
17. C, p. 407
18. A, p. 409
19. B, p. 410
20. B, p. 411
21. D, p. 383
22. C, p. 384
23. B, p. 384
24. A, p. 384
25. C, p. 384
26. B, p. 384
27. B, p. 385
28. B, p. 385

29. A, p. 385
30. D, p. 385
31. C, p. 386
32. D, p. 386
33. B, p. 389
34. D, p. 390
35. A, p. 390
36. B, p. 391
37. C, p. 392
38. C, p. 392
39. D, p. 393
40. D, p. 410

TRUE/FALSE QUESTIONS

1. T, p. 385
2. F, p. 384
3. F, p. 385
4. T, p. 386
5. F, p. 388
6. F, p. 389
7. T, p. 391

8. T, p. 392
9. F, p. 400
10. T, p. 401
11. T, p. 400
12. T, p. 401
13. T, p. 402
14. T, p. 405

15. F, p. 410
16. F, p. 386
17. F, p. 396
18. T, p. 398
19. T, p. 408
20. T, p. 408

FILL-IN-THE-BLANK QUESTIONS

1. chronological, p. 384; functional, p. 384
2. life expectancy, p. 385
3. cohort, p. 385
4. gerontology, p. 386
5. age stratification, p. 390
6. ageism, p. 395
7. entitlements, p. 401

8. Medicare, p. 401
9. Elizabeth Kübler-Ross, p. 409
10. hospice, p. 410
11. activity theory, p. 406
12. denial, p. 409
13. Social Security, p. 401
14. Disengagement theory, p. 406
15. task-based, p. 409

13

THE ECONOMY AND WORK IN GLOBAL PERSPECTIVE

BRIEF CHAPTER OUTLINE

CHAPTER SUMMARY

While economists attempt to explain how the limited resources and efforts of a society are allocated among competing ends, sociologists focus on interconnections among the economy, other social institutions, and the social organization of work, at both the microlevel and the macrolevel of analysis. The **economy** is the social institution that ensures the maintenance of society through the production, distribution, and consumption of goods and services. Preindustrial economies are characterized by **primary sector production** in which workers extract raw materials and natural resources from the environment and use them without much processing. Industrial economies engage in **secondary sector production**, which is based on the processing of raw materials (from the primary sector) into finished goods. The economic base in postindustrial economies shifts to **tertiary sector production** – the provision of services rather than goods. As an ideal type, **capitalism** has four distinctive features: private ownership of the means of production, pursuit of personal profit, competition, and lack of government intervention. **Socialism** is characterized by public ownership of the means of production, the pursuit of collective goals, and centralized decision making. In a **mixed economy**, elements of a capitalist, market economy are combined with elements of a command, socialist economy. Mixed economies are often referred to as **democratic socialism** – an economic and political system that combines private ownership of some of the means of production, governmental distribution of some essential goods and services, and free elections. For many people, jobs and professions are key sources of identity. Functionalists, conflict theorists, and symbolic interactionists view the economy and the nature of work from a variety of perspectives. In examining how societies organize work, analysts find that as societies grow larger in population and more complex in their division of labor, different categories of work are identified. **Occupations** are categories of jobs that involve similar activities at different work sites. **Professions** are high status, knowledge-based occupations characterized by abstract, specialized knowledge, autonomy, authority over clients and subordinate occupational groups, and a degree of altruism. Those in managerial occupations typically are responsible for workers, physical plants, equipment, and the finances of a bureaucratic organization. **Marginal jobs** are those that do not comply with the following employment norms: job content should be legal, jobs should be covered by government regulations, jobs should be relatively permanent, and jobs should provide adequate hours and pay in order to make a living. **Contingent work** is part-time work, temporary work, and **subcontracted work** that offers advantages to employers but may be detrimental to workers. **Unemployment** remains a problem for many workers. There are three types of unemployment: cyclical, seasonal, and structural. The **unemployment rate** is the percentage of unemployed persons in the labor force actively seeking jobs. U.S. labor unions have been credited with improving their work environment and gaining some measure of control over their work-related activities. Many workers have joined unions in order to gain strength through collective actions. The number of persons in the U.S. having physical or mental disabilities is increasing. In 1990, the United States became the first nation to formally address the issue of equality for persons with a disability. Future trends of the U.S. economy and the global economy all indicate that the U.S. will remain a major player in the world economy and that transnational corporations will become even more significant. A global workplace is emerging in which telecommunications networks will link workers in distant locations; however, as nations become more dependent on each other they also become more competitive in the economic sphere; therefore changes in the global economy may require people of all nations to make changes in their past traditional ways of doing things.

LEARNING OBJECTIVES

After reading Chapter 13, you should be able to:

1. Define socialism and describe its major characteristics.

2. Trace the major historical changes that have occurred in economic systems and note the most prevalent form of production found in each.

3. Define contingent work and identify the role subcontracting often plays in contingent work.

4. Describe the four distinctive features of "ideal" capitalism and explain why pure capitalism does not exist.

5. Discuss the major characteristics of professions and describe the process of deprofessionalization.

6. Describe the purpose of the economy and distinguish between sociological perspectives on the economy and the study of economics.

7. Compare and contrast capitalism, socialism, and mixed economies.

8. Distinguish between the functionalist and conflict perspectives on the economy and work.

9. Distinguish between the primary and secondary labor markets and describe the types of jobs found in each.

10. Identify the occupational categories considered to be marginal jobs and explain why they are considered to be marginal.

11. Describe some of the means by which workers resist working conditions they consider to be oppressive.

12. Distinguish between the various types of unemployment.

13. Describe job satisfaction and alienation, and explain the impact of each on workers.

14. Trace the historical development of labor unions.

15. Compare scientific management (Taylorism) with mass production through automation (Fordism), noting the strengths and weaknesses of each.

16. Explain why the unemployment rate may not be a true reflection of unemployment in the United States.

KEY TERMS

(defined at page number shown and in glossary)

capitalism, p. 424
conglomerates, p. 426
contingent work, p. 439
corporations, p. 424
democratic socialism, p. 430
economy, p. 416
interlocking corporate directorates,
p. 427
labor union, p. 424
marginal jobs, p. 437
mixed economy, p. 430
occupations, p. 432

oligopoly, p. 426
primary labor market, p. 432
primary sector production, p. 417
professions, p. 433
secondary labor market, p. 432
secondary sector production, p. 420
shared monopoly, p. 426
socialism, p. 428
subcontracting, p. 439
tertiary sector production, p. 421
transnational corporations, p. 425
unemployment rate, p. 442

KEY PEOPLE

(identified at page number shown)

Esther Ngan Ling Chow, p. 444
Henry Ford, p. 436
Karen J. Hossfeld, p. 438
Karl Marx, p. 428
Ruth Milkman, p. 431

George Ritzer, p. 421
Denise Segura, p. 437
Adam Smith, p. 427
Frederick Winslow Taylor, p. 435
Thorstein Veblen, p. 420

CHAPTER OUTLINE

I. COMPARING THE SOCIOLOGY OF ECONOMIC LIFE WITH ECONOMICS
    A. Economists attempt to explain how the limited resources and efforts of a society
       are allocated among competing ends; sociologists focus on interconnections
       between the economy, other social institutions, and the social organization of
       work.
    B. At the macrolevel, sociologists may study the impact of transnational
       corporations on industrial and low-income nations; at the microlevel, they may
       study people's satisfaction with their jobs.
II. ECONOMIC SYSTEMS IN GLOBAL PERSPECTIVES
    A. The **economy** is the social institution that ensures the maintenance of society
       through the production, distribution, and consumption of goods and services.
    B. In preindustrial societies, most workers engage in **primary sector production** –
       the extraction of raw materials and natural resources from the environment.

C. Industrialization brings sweeping changes to the economy as new forms of energy and machine technology proliferate as the primary means of producing goods and most workers engage in **secondary sector production** – processing raw materials into finished goods.

D. A postindustrial economy is based on tertiary sector production – the provision of services (such as food service, transportation, communication, education, and entertainment) rather than goods.

III. CONTEMPORARY WORLD ECONOMIC SYSTEMS

A. **Capitalism** is an economic system characterized by private ownership of the means of production, from which personal profits can be derived through market competition and without government intervention.

B. "Ideal" capitalism has four distinctive features:
1. Private ownership of the means of production
2. Pursuit of personal profit
3. Competition
4. Lack of government intervention.

C. However, ideal capitalism does not exist in the U.S. for a number of reasons, including the presence of:
1. **Oligopolies** – where several companies control an entire industry.
2. **Shared monopolies** – where four or fewer companies supply 50 percent or more of a particular market.
3. Mergers and acquisitions across industries create **conglomerates** – combinations of businesses in different commercial areas, all owned by one holding company.
4. Competition is reduced by **interlocking corporate directorates** – members of the board of directors of one corporation who also sit on the board(s) of other corporations.
5. Government intervention occurs in the form of regulations after some individuals and companies in pursuit of profits have run roughshod over weaker competitors; however, much "government intervention" has been in the form of aid to business (tax credits, loan guarantees, etc.).

D. **Socialism** is an economic system characterized by public ownership of the means of production, the pursuit of collective goals, and centralized decision making.

E. "Ideal" socialism has three distinctive features:
1. Public ownership of the means of production
2. Pursuit of collective goals
3. Centralized decision making

F. A **mixed economy** combines elements of a market economy (capitalism) with elements of a command economy (socialism).

IV. PERSPECTIVES ON ECONOMY AND WORK IN THE UNITED STATES

A. From a functionalist perspective, the economy is a vital social institution because it is the means by which goods and services are produced and distributed.
1. When the economy runs smoothly, other parts of society function more effectively; however, if the system becomes unbalanced, maladjustment occurs.

2. The business cycle is the rise and fall of economic activity relative to long-term growth in the economy.
B. From a conflict perspective, business cycles are the result of capitalist greed; in order to maximize profits, capitalists suppress the wages of workers.
  1. As the prices of the products increase, the workers are not able to purchase them in the quantities that have been produced.
  2. Consequently, surpluses occur that cause capitalists to reduce production, close factories, lay off workers, and thus contribute to the growth of the reserve army of the unemployed which then helps to reduce the wages of the remaining workers. In some situations, workers are replaced with machines or nonunionized workers.
C. Symbolic interactionists examine factors that contribute to worker satisfaction.
  1. Work is an important source of self-identity for many people.
  2. Work can help people feel positive about themselves or it can cause them to feel alienated.
  3. Job satisfaction is highest when employees have some degree of control over their work, when they feel that they are part of the decision-making process, when they are not too closely supervised, and when they feel they play an important part in the outcome.

V. THE SOCIAL ORGANIZATION OF WORK
A. Sociologists who focus on microlevel analysis are interested in how the economic system and the social organization of work affect people's attitudes and behavior. Interactionists examine factors that contribute to job satisfaction or feelings of alienation.
B. Job Satisfaction and Alienation
  1. Job satisfaction refers to an attitude that people experience about their work, which results from (a) their job responsibilities, (b) the organizational structure in which they work, and (c) their individual needs and values.
  2. Self-actualization occurs when people feel a sense of accomplishment and fulfillment as a result of their work. Studies have found that worker satisfaction is highest when employees have some degree of control over their work and are not too closely supervised.
  3. Alienation occurs when workers' needs for identity and meaning are not met, and when work is done strictly for material gain, not a sense of personal satisfaction.
C. **Occupations** are categories of jobs that involve similar activities at different work sites.
  1. The **primary labor market** is comprised of high-paying jobs with good benefits that have some degree of security and the possibility for future advancement.
  2. The **secondary labor market** is comprised of low paying jobs with few benefits and very little job security or possibility for future advancement.

D. **Professions** are high-status, knowledge-based occupations that have five major characteristics:
1. Abstract, specialized knowledge
2. Autonomy
3. Self-regulation
4. Authority
5. Altruism
E. The term "manager" often is used to refer to executives, managers, and administrators who typically have responsibility for workers, physical plants, equipment, and the financial aspects of a bureaucratic organization.
1. Scientific Management (Taylorism) was developed by industrial engineer **Frederick Winslow Taylor** to increase productivity in factories by teaching workers to perform a task in a concise series of steps; paying workers only for the number of units they produced also contributed to the success of Taylorism.
2. Mass Production through Automation (Fordism) incorporated hierarchical authority structures and scientific management techniques into the manufacturing process. Assembly lines, machines, and robots became a means of technical control over the work process.
F. Lower Tier of the Service Sector and Marginal Jobs.
1. Positions in the lower tier of the service sector are part of the **secondary labor market**, characterized by low wages, little job security, few chances for advancement, higher unemployment, and very limited (if any) unemployment benefits.
2. Examples include janitors, waitresses, messengers, lower-level sales clerks, typists, file clerks, migrant laborers, and textile workers.
3. Many jobs in this sector are **marginal jobs** that differ from the employment norms of the society in which they are located. In the U.S., these norms are: (a) job content is legal; (b) the job is covered by government work regulations; (c) the job is relatively permanent; and (d) the job provides adequate pay with sufficient hours of work each week to make a living.
4. More than 11 million workers are employed in personal service industries, such as eating and drinking places, hotels, laundries, beauty shops, and household service, primarily maid service.
G. **Contingent work** is part time work, temporary work, and subcontracted work that offers advantages to employers but often is detrimental to the welfare of workers.
1. Employers benefit by hiring workers on a part-time or temporary basis; they are able to cut costs, maximize profits, and have workers available only when they need them.
2. **Subcontracting** – a form of economic organization in which a larger corporation contracts with other (usually smaller) firms to provide specialized components, product, or services – is another form of contingent work that cuts employers costs but often at the expense of workers.

VI. UNEMPLOYMENT
    A. There are three major types of unemployment: (1) cyclical unemployment occurs as a result of lower rates of production during recessions in the business cycle; (2) seasonal unemployment results from shifts in the demand for workers based on conditions such as weather, or seasonal demands such as holidays and summer vacations; and (3) structural unemployment which arises because the skills demanded by employers do not match the skills of the unemployed or because the unemployed do not live where the jobs are located.
    B. The **unemployment rate** is the percentage of unemployed persons in the labor force actively seeking jobs.
    C. Unemployment compensation provides unemployed workers with short-term income while they look for other jobs.

VII. WORKER RESISTANCE AND ACTIVISM
    A. Labor Unions
        1. U.S. labor unions have been credited with gaining an eight hour work day, a five day work week, health and retirement benefits, sick leave and unemployment insurance, and workplace health and safety standards for many employees through **collective bargaining** – negotiations between labor union leaders and employers on behalf of workers.
        2. Although the overall number of union members in the United States has increased since the 1960s, the proportion of all employees who are union members has declined.
    B. Absenteeism, Sabotage, and Resistance
        1. Absenteeism is one means by which workers resist working conditions they consider to be oppressive.
        2. Other workers use sabotage to bring about informal work stoppages, such as "throwing a monkey wrench in the gears" to halt the movement of the assembly line.
        3. While most workers do not sabotage machinery, a significant number do resist what they perceive to be oppression from supervisors and employers.

VIII. EMPLOYMENT OPPORTUNITIES FOR PERSONS WITH A DISABILITY
    A. The number of persons in the United States having physical or mental disabilities is increasing because:
        1. Advances in medical technology now make survival possible.
        2. As people live longer they are more likely to experience disabling diseases.
        3. Persons born with serious disabilities are more likely to survive infancy.
    B. In 1990, the United States passed the Americans with Disabilities Act.
        1. The ADA forbids discrimination on the basis of disability.
        2. Despite the law, about two-thirds of working-age persons with a disability in the U.S. are unemployed today.

IX. THE GLOBAL ECONOMY IN THE FUTURE
   A. The U.S. economy will experience dramatic changes in the next century as workers may find themselves fighting for a larger piece of an ever shrinking economic pie which includes trade deficit and a large national debt.
   B. Workers increasingly may be fragmented into two major divisions – those who work in the innovative, primary sector and those whose jobs are located in the growing secondary, marginal sector of the labor market.
   C. Most futurists predict that economic interdependence and competition will become even more significant in the global economy of the future, and multinational corporations will become even less aligned with the values of any one nation.

ANALYZING AND UNDERSTANDING THE BOXES
After reading the chapter and studying the outline, re-read the boxes and write down key points and possible questions for class discussion.

Sociology in Everyday Life: How Much Do You Know About the Economy and the World of Work?

Key Points:

Discussion Questions:

1.

2.

3.

Sociology in Global Perspective: McDonald's Golden Arches as a Lightning Rod

Key Points:

Discussion Questions:

1.

2.

3.

You Can Make a Difference: Creating Access to Information Technologies for Workers with Disabilities

Key Points:

Discussion Questions:

1.

2.

3.

Sociology and Social Policy: Does Globalization Change the Nature of Social Policy?

Key Points:

Discussion Questions:

1.

2.

3.

PRACTICE TESTS

MULTIPLE CHOICE QUESTIONS

Select the response that best answers the question or completes the statement.

1. The _____ is the social institution that ensures the maintenance of society through the production, distribution, and consumption of goods and services.
    a. economy
    b. government
    c. state
    d. political arena

2. The owner of a large corporation has amassed several million dollars that she utilizes in expanding her enterprise. This wealth is referred to as:
   a. services
   b. labor
   c. goods
   d. capital

3. Most of sub-Saharan Africa is highly dependent on _____ sector production.
   a. primary
   b. secondary
   c. tertiary
   d. quartiary

4. Industrial economies are characterized by _____ sector production.
   a. primary
   b. secondary
   c. tertiary
   d. quartiary

5. According to the Census Bureau 2007, the majority of U.S. jobs are in
   a. service occupations.
   b. sales and office occupations.
   c. management, professional, and related occupations.
   d. production, transportation, and material moving occupations.

6. According to the text, all of the following are features of capitalism, **except**:
   a. private ownership of the means of production
   b. pursuit of personal profit
   c. competition
   d. governmental intervention in the marketplace

7. Huge corporations developed during the period of:
   a. early capitalism
   b. early monopoly capitalism
   c. advanced monopoly capitalism
   d. post capitalism

8. If Sony, Philips, Time Warner, and only a few other companies control the entire music industry, this arrangement would be a(n):
   a. oligopoly
   b. monopoly
   c. conglomerate
   d. amalgamation

9. _____ is an economic system characterized by public ownership of the means of production, the pursuit of collective goals, and centralized decision making.
   a. Capitalism
   b. Communism
   c. Socialism
   d. Communitarianism

10. Democratic socialism is characterized by:
    a. extensive government action to provide support and services to its citizens.
    b. private ownership of the means of production, from which personal profits can be derived through market competition and without government intervention.
    c. public ownership of the means of production, the pursuit of collective goals, and centralized decision making.
    d. private ownership of some of the means of production, governmental distribution of some essential goods and services, and free elections.

11. According to the _____ perspective, when the economy runs smoothly, other parts of society function more effectively.
    a. conflict
    b. functionalist
    c. interactionist
    d. neo-Marxist

12. Job satisfaction refers to people's attitudes toward their work, based on:
    a. their job responsibilities
    b. the organizational structure in which they work
    c. their individual needs and values
    d. all of these choices

13. Studies have found that job satisfaction is highest when workers
    a. have some degree of control over their work.
    b. are removed from the decision making process.
    c. know that their supervisors closely watch all aspects of their work.
    d. do not feel the pressure of playing an important part in the outcome.

14. All of the following are characteristics of professions, **except**:
    a. broad based knowledge on a wide variety of topics
    b. authority
    c. concern for others, not just self-interest
    d. self regulation

15. Scientific management (Taylorism) is characterized by
    a. mass production.
    b. automation.
    c. time and motion studies.
    d. Fordism.

16. Fast-food restaurants like McDonald's and Burger King illustrate
    a. the industrial society.
    b. the piece rate system.
    c. Fordism.
    d. robotics.

17. According to the text, _____ jobs differ from employment norms of the society in which they are located.
    a. primary tier
    b. marginal
    c. peripheral
    d. criminal

18. Studies of immigrant women workers in the Silicon Valley have concluded that
    a. capitalist economic development in the past decade has been relatively free of racism as a method of labor division and control.
    b. these workers recently have joined unions and are now demanding higher wages.
    c. women currently make up nearly 100 percent of the high tech workforce in California.
    d. gender segregation is apparent on high tech assembly lines.

19. Which one of these is not a major type of unemployment?
    a. vertical
    b. seasonal
    c. cyclical
    d. structural

20. Despite the Americans with Disabilities Act, about _____ of working-age persons with a disability are unemployed today.
    a. one-fourth
    b. one-third
    c. two-thirds
    d. two-fifths

21. All of the following have been identified as characteristics central to the postindustrial economy, except:
    a. information displaces property as the central preoccupation in the economy
    b. workplace culture shifts toward factories and away from increased diversification of work settings, the workday, and the employee
    c. conventional boundaries between work and home (public life and private life) are breached
    d. workplace culture shifts away from factories and toward increased diversification of work settings, the workday, and the employee

22. _____ includes a wide range of activities, such as fast-food service, transportation, communication, education, real estate, advertising, sports, and entertainment.
    a. Primary sector production
    b. Secondary sector production
    c. Tertiary sector production
    d. Quartiary sector production

23. Under early monopoly capitalism (1890-1940), most ownership rapidly shifted to
    a. share holders.
    b. individuals.
    c. corporations.
    d. bureaucracies.

24. _____ are large-scale organizations that have legal powers, such as the ability to enter into contracts, and buy and sell property, separate from their individual owners.
    a. Corporations
    b. Bureaucracies
    c. Conglomerates
    d. Interlocking directorates

25. In 1937, _____ held their first sit-down strike. The workers occupied the facilities but refused to work, thus paralyzing production. The sit-down strike represented a new approach that came to dominate U.S. labor activism.
    a. American Motor workers
    b. Hewlett-Packard workers
    c. General Motors workers
    d. Ford Motor workers

26. _____ are non-union workers who try to take over the jobs of workers who are on strike.
    a. Scabs
    b. Bargainers
    c. Scums
    d. Free Riders

27. _____ provides unemployed workers with short-term income while they look for other jobs.
    a. Compensatory income
    b. Unemployment income
    c. Unemployment compensation
    d. Income dispersion

28. In which one of the following countries has labor union membership declined?
    a. Germany
    b. United States
    c. Netherlands
    d. Ireland

29. In 2005, the highest rates of union members in the United Sates were among _____.
    a. African American women
    b. white men
    c. Latinos/Latinas
    d. African American men

30. All of the following except _____ are means by which workers resist working conditions that they consider to be oppressive.
    a. absenteeism
    b. sabotage
    c. collective bargaining
    d. resistance

31. An estimated _____ persons in the United States have one or more physical or mental disabilities that differentially affect their opportunities for employment.
    a. 1.2 million
    b. 12 million
    c. 25 million
    d. 48 million

32. Information technology professionals and U.W. white-collar employees in many other fields have lost their jobs in recent years. _____ refers to the practice of U.S. companies moving certain operations outside of this country.
    a. Off-shoring
    b. On-shoring
    c. Out-shoring
    d. Far-shoring

33. _____ are tangible objects that are necessary (such as food, clothing, and shelter) or desired (such as DVDs and electric toothbrushes).
    a. Products
    b. Materials
    c. Productions
    d. Goods

34. _____ are intangible activities for which people are willing to pay, such as dry cleaning, a movie, or medical care.
    a. Services
    b. Goods
    c. Products
    d. Info-tracs

35. Steel workers who process metal ore and autoworkers who then convert the ore into automobiles, trucks, and busses are engaged in _____ production.
    a. secondary sector
    b. tertiary sector
    c. essential sector
    d. primary sector

36. The ostentatious display of symbols of wealth, such as owning numerous mansions and expensive works of art, and wearing extravagant jewelry and clothing are examples of:
    a. conspicuous consumers
    b. conspicuous consumption
    c. conspicuous leisure
    d. conspicuous waste

37. According to sociologist George Ritzer, the number of lower-paying, second-tier service sector positions actually has increased, often at the expense of workers, which he referred to as the _____ of society.
    a. Marxism
    b. McDonaldization
    c. Fordism
    d. Taylorism.

38. All of the following are distinctive features of "ideal" capitalism, except:
    a. private ownership of the means of production
    b. pursuit of personal profit
    c. a guarantee of government intervention
    d. competition

39. As workers grew tired of toiling for the benefit of capitalists instead of for themselves, some of them banded together to form the first national labor union, the _____, in 1869.
    a. American Federation of Labor
    b. Teamsters
    c. Knights of Labor
    d. Congress of Industrial Organizations

40. Competition is reduced over the long run by _____, where members of the board of directors of one corporation also sit on the board(s) of other corporations.
    a. oligopolies
    b. shared monopolies
    c. conglomerates
    d. interlocking corporate directorates

TRUE/FALSE QUESTIONS

1. The economy is the social institution responsible for the production, distribution, and consumption of goods and services.
   T   F

2. Preindustrial economies typically are based on the extraction of raw materials and natural resources from the environment.
   T   F

3. U.S. labor unions came into existence in the twentieth century when workers became tired of toiling for the benefit of capitalists.
   T   F

4. A conglomerate exists when four or fewer companies supply 50 percent or more of a particular market.
   T   F

5. According to Karl Marx, socialism and communism are virtually identical.
   T   F

6. According to a functionalist perspective, some problems in society are linked to peaks and troughs in the business cycle.
   T   F

7. Conflict theorists view capitalism as the solution to society's problems.
   T   F

8. The primary labor market consists of low-paying jobs with few benefits.
   T   F

9. Sociologists categorize most doctors, engineers, lawyers, professors, computer scientists, and certified public accounts as "professionals."
   T   F

10. Children whose parents are professionals are more likely to become professionals themselves.
    T   F

11. Scientific management increased worker satisfaction.
    T   F

12. Positions in the lower tier of the service sector are part of the secondary labor market, characterized by low wages, little job security, few chances for advancement, higher unemployment rates, and very limited unemployment benefits.
    T   F

13. Private household workers are among the most powerless because they cannot rely on any resources to protect them from abuse.
    T   F

14. Temporary workers are one of the slowest growing segments of the contingent work force.
    T   F

15. The United States is the first nation to formally address the issue of equality for persons with a disability.
    T   F

16. According to Box 13.1, workers' skills are usually upgraded when new technology is introduced in the workplace.
    T   F

17. According to Box 13.2, McDonald's represents a form of U.S. cultural imperialism.
    T   F

18. Most Americans accept the fact that blind persons can find a use for a computer and make a business out of it, according to Box 13.3.
    T   F

19. The internet has changed, for the better, the lives of blind people, according to Box 13.3.
   T   F

20. According to Box 13.4, the world is flat, which means that there is less level global playing field in business in the twenty-first century.
   T   F

FILL-IN-THE-BLANK QUESTIONS

1. The extraction of raw materials and natural resources is known as _____ production.

2. Postindustrial economics are characterized by _____ production.

3. A(n) _____ is a group of employees who join together to bargain with an employer over wages, benefits, and working conditions.

4. Large corporations that are headquartered in one country but sell and produce goods and services in many countries are known as _____.

5. A(n) _____ exists when several companies overwhelmingly control an entire industry.

6. Economist _____ wrote the classical book *An Inquiry in to the Nature and Causes of the Wealth of Nations*.

7. Karl Marx described _____ as a temporary stage en route to an ideal communist society.

8. _____ revolutionized management with a system called _____.

9. Fordism is named for _____ who used _____ technique in his manufacturing of automobiles.

10. In 1990, the United States became the first nation to formally address the issue of _____.

11. Frederick Taylor conducted _____ studies of workers in order to increase worker production.

12. _____ work is part-time, temporary, or subcontracted work.

13. In their classic books _____ and _____ Marx and Engels predicted the destruction of capitalism.

14. The _____ is the rise and fall of economic activity relative to long-term growth in the economy.

15. _____ are high-status, knowledge-based occupations.

SHORT ANSWER/ESSAY QUESTIONS

1. How far has the United States moved into postindustrialization? What are some of the fastest growing U.S. occupations?
2. What are the occupational categories considered to be marginal jobs in the United States? What are considered to be marginal? What is it meant by internationalization of marginal jobs?
3. What are the three major contemporary economic systems? How do they differ?
4. What are some similarities, and differences, in scientific management and mass production?
5. How do the functionalists, conflict theorists, and symbolic interactionists explain work in the United States?

STUDENT CLASS PROJECTS AND ACTIVITIES

1. This project calls for you to compare the U.S. budget deficit today to that for the years 1980-2006. During the years 1980-1992, the federal budget deficit shot up to between 3% and 7% of the gross domestic product. In the years of Presidents Kennedy, Johnson and Nixon, the deficits generally hovered at a relatively harmless 1% or 2% of the GDP, except for a very brief increase to 3% at the end of the Johnson Administration to help pay for the Vietnam War. You are to research the reasons that federal spending outstripped revenues at such as astonishing rate of increase during the Reagan-Bush years. Analyze the effectiveness of "supply-side" economics during these years (the idea that the U.S. could cut income taxes while simultaneously pay for massive increases in defense and certain highly popular domestic programs). Depict the federal budget deficit by presentation of graphs (probably a bar graph would serve best here) from the years 1980 to 2006. Then compare the federal budget deficit of the mid 1990s to today's deficit. Present at least four well thought out solutions for cutting the deficit. Critique some of the present policies enacted in order to cut the deficit. Some research sources for this project would be in the government documents section of your library, or from the Office of Management and Budget, U.S. Government. In the writing up of this paper, present a bibliography, as well as a well-written paper.
2. This project calls for you to conduct interviews. You are to interview one person who is employed in the primary labor market, one in the secondary labor market, and one in the tertiary labor market. You are to provide the names, company address, sex, race, and age of the workers. You are to ask questions about (1) the worker's family and educational background; (2) training requirements for the job; (3) how and why they entered this particular segment of the labor market; (4) their typical day of work; (5) employment benefits such as paid vacations, paid holidays, sick leave, health insurance, etc.; (6) job security and advancement possibilities; and (7) worker satisfaction. In the writing up of this project, include a summary of the responses to each specific question and an evaluation of this project. Submit your paper to your instructor at the appropriate time; following any other directions you have been given.

3. Trace the history (including the origin and the growth) of labor unions in the United States, for both male and female workers. Include in your research a statement of the present status of labor unions in the present American marketplace.
4. Trace the growth, development, and change of blue-collar and white-collar jobs in the American work force. Include in your research the present status of blue-collar versus white-collar jobs. Some possible sources are: the Economic Policy Institute (a think-tank in Washington D.C.) and in government documents such as the *Historical Statistics of the United States*. Include, when feasible, any charts and graphs, and provide a summary and conclusion of your research.

INTERNET ACTIVITIES

1. To read about one of the most powerful labor organizations in the United States, go to the **AFL-CIO web site**: **http://www.aflcio.org/**. Once there, click on unions affiliated with the AFL-CIO. As you can observe, a very large number of unions are associated with this organization.
2. To read some frequently asked questions about capitalism, go to the "Capitalism: Frequently Asked Questions" site:
   http://lilt-vetri.lilt.ilstu.edu/rrpope/rrpopepwd/articles/capitalism.html
   Look at questions about the relationship between capitalism and democracy, and is capitalism a just social system.
3. Check out **Forbes** listing of the richest people in the world:
   http://www.forbes.com/2005/03/09/bill05land.html. Do a web search for some of these people and see what businesses they are associated with.

INFOTRAC COLLEGE EDITION EXERCISES

Visit the **InfoTrac College Edition** website at:
http://www.wadsworthmedia.com/webtutor/infotrac.htm. You will arrive at a screen that enables you to search topics.

1. Search for articles on both **Seasonal** and **Structural Unemployment**. Bring to class a brief summary that compares and contrasts what we know about these two forms of unemployment.
2. **The Virtual Seminar in Global Political Economy** - The Virtual Seminar is a true international college of the Internet, with on-line classes for college credit exploring questions such as the Third World Debt crisis and international social movements. Search the InfoTrac for this topic and report your findings.
3. Look up articles on **Economic Interdependence**. Scan these articles and make a list of related issues to understand in order to get the full picture of this growing phenomenon.

SOLUTIONS

MULTIPLE CHOICE QUESTIONS

1. A, p. 416
2. D, p. 417
3. A, p. 417
4. B, p. 420
5. C, p. 421
6. D, p. 424
7. B, p. 424
8. A, p. 426
9. C, p. 428
10. D, p. 430
11. B, p. 430
12. D, p. 431
13. A, p. 431
14. A, p. 433

15. C, p. 435
16. C, p. 436
17. B, p. 437
18. D, p. 438
19. A, p. 441
20. C, p. 445
21. B, p. 421
22. C, p. 421
23. C, p. 424
24. A, p. 424
25. C, p. 425
26. A, p. 442
27. C, p. 442
28. B, p. 443

29. D, p. 433
30. C, p. 444
31. D, p. 444
32. A, p. 415
33. D, p. 415
34. A, p. 416
35. A, p. 420
36. B, p. 421
37. B, pp. 421-422
38. D, p. 424
39. C, p. 424
40. D, p. 427

TRUE/FALSE QUESTIONS

1. T, p. 416
2. T, p. 417
3. T, p. 424
4. F, p. 426
5. F, p. 428
6. T, p. 430
7. F, p. 431

8. F, p. 432
9. T, p. 433
10. T, p. 434
11. F, p. 436
12. T, p. 432
13. T, p. 438
14. F, p. 439

15. T, p. 445
16. F, p. 417
17. T, p. 423
18. F, p. 446
19. T, p. 446
20. T, p. 448

FILL-IN-THE-BLANK QUESTIONS

1. primary sector, P. 417
2. tertiary sector, p. 421
3. labor union, p. 424
4. transnational corporation, p. 425
5. oligarchy, p. 426
6. Adam Smith, p. 427
7. socialism, p. 428
8. Frederick Winslow Taylor, p. 435; scientific management, p. 435
9. Henry Ford, p. 436; scientific management, p. 436
10. disability, p. 445
11. time and motion, p. 435
12. Contingent work, p. 439
13. *Communism Manifesto*, p. 428; *Das Kapital,* p. 428
14. business cycle, p. 430
15. Professions, p. 433

14
POLITICS AND GOVERNMENT IN GLOBAL PERSPECTIVE

BRIEF CHAPTER OUTLINE

CHAPTER SUMMARY

The relationship between politics, power and authority is a strong one in all countries. **Politics** is the social institution through which power is acquired and exercised by some people or groups. **Power** – the ability of persons or groups to carry out their will even when opposed by others – is a social relationship involving both leaders and followers. Most leaders seek to legitimate their power through **authority** – power that people accept as legitimate rather than coercive. **Government** is the formal organization that regulates relationships among members of a society and those outside its borders. Some social analysts refer to the government as the **state** - the political entity that possesses a legitimate monopoly of the use of force within its territory to achieve its goals. According to Max Weber, there are three ideal types of authority: (1) **traditional**, (2) **charismatic**, and (3) **rational-legal** (**bureaucratic**). There are four main types of contemporary political systems: monarchy, authoritarianism, totalitarianism, and democracy. In a democracy the people hold the ruling power either directly or through elected representatives. There are two key perspectives on how power is distributed in the United States. According to the **pluralist model**, power is widely dispersed throughout many competing interest groups. According to the **elite model**, power is concentrated in a small group of elites and the masses are relatively powerless. The **power elite** are comprised of influential business leaders, key government leaders, and the military. **Political parties** are organizations whose purpose is to gain and hold legitimate control of government. The Democrats and the Republicans have dominated the U.S. political system since the mid-nineteenth century, although party loyalties have been declining in recent years. People learn political attitudes, values, and behaviors through **political socialization**. The vast governmental bureaucracy is a major source of power and carries out the actual functioning of the government. One way that special interest groups exert powerful influence on the bureaucracy is through the iron triangle of power - a three-way arrangement in which a private interest group of a congressional committee and bureaucratic agency make the final decision on a political issue. The military bureaucracy is so wide-ranging that it encompasses the **military-industrial complex** – the mutual interdependence of the military establishment and private military contractors. This complex is supported by **militarism** – a societal focus on military ideals and an aggressive preparedness for war. **Terrorism**, the use of calculated, unlawful physical force or threats of violence against a government, organization, or individual to gain some political, religious, economics or social objective, has recently created worldwide fear. Three types of terrorism-revolutionary terrorism, state-sponsored terrorism, and repressive terrorism-all extract a massive toll by producing rampant fear, widespread loss of human life, and extensive destruction of property. Recent acts of terrorism directed against the United States have been facts of life in some other countries for many years. **War**, an organized, armed conflict between nations or distinct political factions, has both direct and indirect effects. In examining politics and government in the future, one group of analysts view the U.S. as becoming less democratic. Other analysts focus more narrowly on specific questions of concern. Our response to these concerns will have a profound effect on people and governments all around the globe.

LEARNING OBJECTIVES

After reading Chapter 14, you should be able to:

1. Explain the relationship between politics, government, and the state, and note how political sociology differs from political science.

2. Distinguish between power and authority, and describe the three major types of authority.

3. State the major elements of the pluralist (functionalist) model of power and political systems.

4. State the major elements of elite (conflict) models of power and political systems.

5. Discuss the characteristics of the federal bureaucracy and explain what the term "permanent government" means.

6. Describe the military-industrial complex and explain why it is called an iron triangle.

7. Describe the purpose of political parties.

8. Compare and contrast governments characterized by monarchy, authoritarianism, totalitarianism, and democracy.

9. Describe how elite (conflict) models of power differ from pluralist (functionalist) models.

10. Explain the relationship between political socialization, political attitudes, and political participation.

11. Discuss militarism and explain why support for this ideology has been so strong in the United States.

12. Describe American cultural reactions to war and terrorism both here and abroad.

13. Analyze how well U.S. political parties measure up to the ideal-type characteristics.

14. Examine the significance of current world events on the immediate and long term future.

KEY TERMS

(defined at page number shown and in glossary)

authoritarianism, p. 461
authority, p. 456
charismatic authority, p. 457
democracy, p. 463
elite model, p. 465
government, p. 454

militarism, p. 476
military-industrial complex, p. 475
monarchy, p. 459
pluralist model, p. 463
political action committees, p. 464
political party, p. 468

KEY PEOPLE

(identified at page number shown)

CHAPTER OUTLINE

I. POLITICS, POWER, AND AUTHORITY
    A. **Politics** is the social institution through which power is acquired and exercised by some people and groups.
    B. In contemporary societies, the primary political system is the **government** – the formal organization that has the legal and political authority to regulate the relationships among members of a society and between the society and those outside its borders.
    C. Sociologists often refer to the government as the **state** – the political entity that possesses a legitimate monopoly over the use of force within its territory to achieve its goals.
    D. While political science primarily focuses on power and the distribution of power in different types of political systems, **political sociology** examines the nature and consequences of power within or between societies and focuses on the social circumstances of politics and the interrelationships between politics and social structures.
    E. Power and Authority
        1. **Power** is the ability of persons or groups to achieve their goals despite opposition from others.
        2. **Authority** is power that people accept as legitimate rather than coercive.
    F. Ideal Types of Authority
        1. According to Max Weber, there are three ideal types of authority:
            a. **Traditional authority** is power that is legitimized on the basis of long standing custom.
            b. **Charismatic authority** is power legitimized on the basis of a leader's exceptional personal qualities or the demonstration of extraordinary insight and accomplishment that inspire loyalty and obedience from followers.

    c. **Rational-legal authority** is power legitimized by law or written rules and regulations.

    2. These three types of authority demonstrate how different bases of legitimacy are tied to a society's economy.

II. POLITICAL SYSTEMS IN GLOBAL PERSPECTIVE

  A. Emergence of Political Systems

    1. Hunting and gathering societies do not have political institutions as such because they have very little division of labor or social inequality.

    2. Political institutions first emerged in agrarian societies as they acquire surpluses and develop greater social inequality.

    3. Nation-states – political organizations that have recognizable national boundaries within which their citizens possess specific legal rights and obligations – developed first in Spain, France, and England between the twelfth and fifteenth centuries.

  B. **Monarchy** is a political system in which power resides in one person or family and is passed from generation to generation through lines of inheritance.

  C. **Authoritarianism** is a political system controlled by rulers who deny popular participation in government.

  D. **Totalitarianism** is a political system in which the state seeks to regulate all aspects of people's public and private lives.

  E. **Democracy** is a political system in which the people hold the ruling power either directly or through elected representatives.

III. PERSPECTIVES ON POWER AND POLITICAL SYSTEMS

  A. Functionalist Perspectives: According to the **pluralist model**, power in political systems is widely dispersed throughout many competing **special interest groups** – political coalitions comprised of individuals or groups that share a specific interest they wish to protect or advance with the help of the political system.

    1. Functionalist Perspectives: According to the pluralist model, power in political systems is widely dispersed throughout many competing special interest groups – political coalitions comprised of individuals or groups that share a specific interest they wish to protect or advance with the help of the political system.

      a. Decisions are made on behalf of the people by leaders who engage in a process of bargaining, accommodation, and compromise.

      b. Competition among leadership groups (such as leaders in business, labor, education, law, medicine, consumer groups, and government) protects people by making the abuse of power by any one group more difficult.

      c. People can influence public policy by voting in elections, participating in existing special interest groups, or forming new ones to gain access to the political system.

      d. Power is widely dispersed in society; leadership groups that wield influence on some decisions are not the same groups, which may be influential in other decisions.

      e. Public policy is not always based on majority preference; it is the balance between competing interest groups.

2. Over the last two decades, **special interest groups** have become more involved in "single issue politics," such as abortion, gun control, gay and lesbian rights, or environmental concerns.
3. **Political action committees (PACs)** are organizations of special interest groups that fund campaigns to help elect (or defeat) candidates based on their stances on specific issues.

B. Conflict Perspectives: According to the **elite model**, power in political systems is concentrated in the hands of a small group of elites and the masses are relatively powerless.
1. Key elements of this model:
   a. Decisions are made by the elite, possessing greater wealth, education, status, and other resources than does the "masses" it governs.
   b. Consensus exists among the elite on the basic values and goals of society; however, consensus does not exist among most people in society on these important social concerns.
   c. Power is highly concentrated at the top of a pyramid-shaped social hierarchy; those at the top of the power structure come together to set policy for everyone.
   d. Public policy reflects the values and preferences of the elite, not the preferences of the people.
2. According to C. Wright Mills, the **power elite** is comprised of leaders at the top of business, the executive branch of the federal government, and the military (especially the "top brass" at the Pentagon). The elites have similar class backgrounds and interests.
   a. The corporate rich are the most powerful because of their unique ability to parlay the vast economic resources at their disposal into political power.
   b. At the middle level of the pyramid, Mills placed the legislative branch of government, interest groups, and local opinion leaders.
   c. The bottom (and widest layer) of the pyramid is occupied by the unorganized masses that are relatively powerless and vulnerable to economic and political exploitation.
3. **G. William Domhoff** referred to elites as the *ruling class* – a relatively fixed group of privileged people who wield sufficient power to constrain political processes and serve underlying capitalist interests.
   a. Individuals in the upper echelon are members of a business class based on the ownership and control of large corporations.
   b. The intertwining of the upper class and the corporate community produces cohesion at both the economic and social levels.
   c. Members of the ruling class also are linked though exclusive social clubs, expensive private schools, debutante parties, listings in the Social Register, and other upper class indicators.

d. The corporate rich and their families influence the political process in three ways:
   i. they influence the candidate selection process by helping to finance campaigns and providing favors to political candidates;
   ii. through participation in the special interest process, they are able to gain favors, tax-breaks, regulatory rulings, and other governmental supports;
   iii. they gain access to the policy-making process by holding prestigious positions on governmental advisory committees, presidential commissions, and other governmental appointments.
4. Class Conflict Perspectives
   a. Most contemporary elite models are based on the work of **Karl Marx**; however, there are divergent viewpoints about the role of the state within this perspective.
   b. While instrumental Marxists argue that the state acts invariably to perpetuate the capitalist class, structural Marxists contend that the state is not simply a passive instrument of the capitalist class.
5. Critique of Pluralist and Elite Models
   a. The Pluralist model emphasizes that many different groups compete for power and advantage in society.
   b. Domhoff argues U.S. society appears to be pluralist when actually it is elitist.
   c. Mills suggests that the economic, political, and military elite are interrelated and may be a relatively cohesive group.
IV. THE U.S. POLITCIAL SYSTEM
   A. Political Parties and Elections
      1. A **political party** is an organization whose purpose is to gain and hold legitimate control of government; they affect elections.
         a. Parties develop and articulate policy positions; educate voters about the issues and simplify the choices for them; recruit candidates who agree with those policies, and help those candidates win office; and, when elected, hold the candidates responsible for implementing the party's policy positions.
         b. Since the Civil War, two political parties – the Democratic and the Republican – have dominated the political system in the United States and confronted two broad types of concerns.
            i. Social issues are those relating to moral judgments or civil rights.
            ii. Economic issues involve the amount that should be spent on government programs and the extent to which these programs should encourage a redistribution of income and assets.
   B. Political Participation and Voter Apathy
      1. **Political socialization** is the process by which people learn political attitudes, values, and behavior. For young children, the family is the primary agent of political socialization.
      2. Socioeconomic status affects people's political attitudes, values, and beliefs.

3. Political participation occurs at four levels:
   a. voting
   b. attending and taking part in political meetings
   c. actively participating in political campaigns
   d. running for or holding political office.
4. At most, about 10 percent of the voting age population in this country participates at a level higher than simply voting, and less than 51 percent of the voting age population voted in the 2000 presidential election.

V. GOVERNMENTAL BUREAUCRACY
A. Characteristics of the Federal Bureaucracy
   1. The size and scope of government has grown in recent decades partially because of dramatic increases in technology and in demands from the public that the government "do something" about various problems facing society.
   2. Much of the actual functioning of the government is carried on by the permanent government in Washington that is made up of top tier, civil service bureaucrats who have built a major power base.
   3. The governmental bureaucracy has been able to perpetuate itself and expand because it has many employees with highly specialized knowledge and skills who cannot easily be replaced by those from the "outside."
B. The Iron Triangle and the Military Industrial Complex
   1. The *iron triangle* is a three-way arrangement in which a private-interest group (usually a business corporation), a congressional committee or subcommittee, and a bureaucratic agency make the final decision on a political issue that is to be decided by that agency.
   2. A classic example of the iron triangle is an arrangement referred to as the **military-industrial complex** – the mutual interdependence of the military establishment and private military contractors that started during World War II and has continued to the present.
      a. Between 1945 and early 1990s, the U.S. government spent an estimated $10.2 trillion for national defense (in constant 1987 dollars). In the 1990s, the U.S. spent more than $270 billion per year on the U.S. military budget.
      b. Several of the largest multinational corporations are among the largest defense contractors, and the defense divisions of these companies have a virtual monopoly over defense contracts. The military budgets for 2003 and 2004 were almost $400 billion per year.
      c. Some members of Congress have actively supported the military industrial complex because it provided a unique chance for them to provide economic assistance for their local, voting constituencies in the form of funding for defense related industries, military bases, and space centers in their home state. These activities are known as *pork* or *pork barrel projects*.

VI. THE MILITARY AND MILITARISM
A. **Militarism** is a societal focus on military ideals and an aggressive preparedness for war; militarism is supported by core U.S. values such as patriotism, courage, reverence, loyalty, obedience, and faith in authority. Sociologists have proposed several reasons for militarization:

1. The economic interests of capitalists, college and university faculty and administrators who are recipients of research grants, workers, labor union members, and others who depend on military spending.
   2. The role of the nation and its inclination toward coercion in response to perceived threats.
   3. The relationship between militarism and masculinity.
   B. Gender, Race and the Military
   1. The introduction of the all-volunteer force in the early 1970s and the end of the draft shifted the focus of the military from training "good citizen soldiers" to an image of the "economic person" – one who enlists in the military in the same way that a person might take a job in the private sector.
   2. Considerable pressure has been placed on the military to recruit women; in 2002, women made up approximately 15 percent of the total uniformed force.
   3. Militarization uses and affects women differently based on their ethnic or racial group; by 2001, African American women made up 47 percent of all enlisted (non-officer) women in the United States Army, a percentage four times their proportion compared with all United States women.
   4. People of color may enlist in the military, not because of their militaristic tendencies, but because of the more limited options available to them in society at large, especially if they come from lower-income family backgrounds.
VII. TERRORISM and WAR
   A. **Terrorism** is the use of calculated, unlawful physical force or treats of violence against a government, organization, or individual to gain some political, religious, economic, or social objective.
   B. Types of Terrorism
   1. *Political terrorism* uses intimidation, coercion, threats of harm, and other violence that attempts to bring about a significant change in or overthrow an existing government. There are three types of political terrorism.
      a. *Revolutionary terrorism* refers to acts of violence against civilians that are carried out by enemies of the government who want to bring about political change.
      b. *State-sponsored terrorism* occurs when a government provides financial resources for terrorists who conduct their activities in other nations.
      c. *Repressive terrorism* is conducted by a government against its own citizens for the purpose of protecting an existing political order.
   C. Terrorism in the United States
   1. The years 1995 and 2001 stand out in recent U.S. history as the two worst terrorist attacks that ever occurred in the continental United States.
   2. Bombings, biological and chemical attacks are now reality in the United States.
   3. Many people now rely on government officials to calm their fears.

D. War
   1. **War** – organized, armed conflict between nations or distinct political factions-includes both declared and undeclared wars.
   2. Casualties in conventional warfare pales when compared to biological or chemical warfare.
   3. Patriotism, even when temporary, causes a renewed interest in how the government is run, and affords an expression of loyalty to the exiting order, although some criticism remains.
E. Politics and Government in the Future
F. One group of analysts provides a distressed view about the future of the U.S.
G. Other views focus more narrowly on specific questions of concern seeking solutions.

ANALYZING AND UNDERSTANDING THE BOXES

After reading the chapter and studying the outline, re-read the boxes and write down key points and possible questions for class discussion.

Sociology and Everyday Life: How Much Do You Know About the Media?

Key Points:

Discussion Questions:

1.

2.

3.

Sociology in Global Perspective: The European Union: Transcending National Borders and Governments

Key Points:

Discussion Questions:

1.

2.

3.

Framing Politics in the Media: Hero Framing and the Selling of an Agenda

Key Points:

Discussion Questions:

1.

2.

3.

You Can Make a Difference: Keeping an Eye on the Media

Key Points:

Discussion Questions:

1.

2.

3.

PRACTICE TESTS

MULTIPLE CHOICE QUESTIONS

Select the response that best answers the question or completes the statement.

1. The social institution through which power is acquired and exercised by some people and groups is known as:
   a. government
   b. politics
   c. the state
   d. the military

2. According to the text, the state is
   a. the social institution through which power is acquired and exercised by some people and groups.
   b. the formal organization that has the legal and political authority to regulate the relationships among members of a society and between the society and those outside its borders.
   c. the political entity that possesses a legitimate monopoly over the use of force within its territory to achieve its goals.
   d. the political entity that seeks to regulate all aspects of people's public and private lives.

3. _____ is the power that people accept as legitimate rather than coercive.
   a. Influence
   b. Clout
   c. Authority
   d. Legitimation

4. Traditional authority is based on:
   a. a leader's exceptional personal qualities
   b. written rules and regulations of law
   c. documents such as the U.S. Constitution
   d. long-standing custom

5. Napoleon, Julius Caesar, Martin Luther King, Jr., Caesar Chavez, and Mother Teresa are examples of:
   a. charismatic authority
   b. traditional authority
   c. rational-legal authority
   d. nontraditional authority

6. All of the following statements regarding racialized patriarchy are correct, **except**:
   a. Zillah R. Eisenstein coined the term "racialized patriarchy"
   b. Racialized patriarchy is closely intertwined with rational-legal authority
   c. Racialized patriarchy is the continual interplay of race and gender
   d. Racialized patriarchy remains a reality in both preindustrial and industrialized nations

7. A unit of political organizations that has recognizable national boundaries where citizens possess legal rights and obligations is a:
   a. bureaucracy
   b. government
   c. nation-state
   d. city-state

8. A hereditary right to rule or a divine right to rule is most likely to be found in a(n)
   a. monarchy.
   b. authoritarian regime.
   c. totalitarian regime.
   d. democracy.

9. The National Socialist (Nazi) party in Germany during World War II is an example of a(n)
    a. monarchy.
    b. authoritarian regime.
    c. totalitarian regime.
    d. democracy.

10. All of the following statements regarding democracy are true, **except**:
    a. Democracy is a political system in which the people hold the ruling power either directly or through elected representatives.
    b. The United Kingdom and Canada have recently attempted direct democracy at the national level.
    c. Representative democracy is not always equally accessible to all people in a nation.
    d. The framers of the Constitution established a system of representative democracy in the United States.

11. According to the pluralist model, power in political systems is
    a. widely dispersed throughout many competing interest groups.
    b. concentrated in the hands of a small group of elites.
    c. comprised of leaders at the top of business, the executive branch of the federal government, and the military.
    d. controlled by members of the ruling class.

12. _____ put political pressure on leaders in order to influence legislation.
    a. Power elites
    b. Elite groups
    c. Special interest groups
    d. PACs

13. Political action committees:
    a. generally have been abolished by reforms in campaign finance laws.
    b. are comprised of people who volunteer their time but not money to political candidates and parties.
    c. are organizations of special interest groups that fund campaigns to help elect candidates based on their stances on specific issues.
    d. encourage widespread political participation by citizens at the grassroots level.

14. The power elite model was developed by:
    a. Emile Durkheim
    b. Thomas R. Dye and Harmon Zeigler.
    c. Joseph Steffan.
    d. C. Wright Mills.

15. The organizations responsible for developing and articulating policy positions, educating voters about issues, and recruiting candidates to run for office are known as:
    a. political parties
    b. political action committees
    c. interest groups
    d. federal election committees

16. _____ is the process by which people learn political attitudes, values, and behavior.
    a. Indoctrination
    b. Military training
    c. Resocialization
    d. Political socialization

17. According to the text, most of the actual functioning of the U.S. government is carried on by
    a. political action committees.
    b. the federal bureaucracy.
    c. the criminal justice system.
    d. political parties.

18. The three-way arrangement in which a private interest group, a congressional committee, and a bureaucratic agency make the final decision on a political issue that is to be decided by that agency is known as:
    a. political subversion
    b. the iron law of oligarchy
    c. the iron triangle of power
    d. the power elite

19. _____ is a societal focus on military ideals and an aggressive preparedness for war.
    a. Militarism
    b. Authoritarianism
    c. Totalitarianism
    d. Warmongerism

20. Which of the following statements is **correct** regarding the U.S. military?
    a. As more women have moved into the ranks of officers in the military, sexual harassment largely has been eliminated.
    b. Women's roles in the military have disturbed the existing world of male militarism.
    c. The percentage of African American women enlisted in the U.S. Army is much smaller than their proportion in the U.S. population.
    d. People of color and poor people from all racial ethnic groups may enlist in the military because of the more limited options available to them in society at large.

21. All of the following statements regarding power are TRUE, except:
    a. power is a social relationship that involves both leaders and followers.
    b. power is a dimension in the structure of social stratification.
    c. persons in positions of power control of valuable resources of society.
    d. the most basic form of power is political influence.

22. Sociologist G. William Domhoff argues that the media
    a. are more powerful than government officials.
    b. are more powerful than corporate leaders.
    c. tends to reflect the biases of those with access to them.
    d. tends to be totally neutral in news coverage.

23. Sociologist Max Weber suggested that most leaders do not want to base their power on force alone; they seek to legitimize their power by turning it into _____, which is the power that people accept as legitimate rather than coercive.
    a. authority
    b. control
    c. dominance
    d. democracy

24. According to sociologist Max Weber, people have a greater tendency to accept authority as legitimate if they are _____ dependent on those who hold power.
    a. economically or politically
    b. culturally or institutionally
    c. institutionally or economically
    d. politically or culturally

25. Sociologist Max Weber outlined three ideal types of authority, which include all of the following, except:
    a. charismatic authority
    b. rational legal authority
    c. traditional authority
    d. coercive authority

26. According to political scientist Zilah Eisenstein, _____, the continual interplay of race and gender reinforces traditional structures of power in contemporary society.
    a. ethnic genderization
    b. class authority
    c. racialized patriarchy
    d. gendered authority

27. Although the U.S. Constitution grants _____ authority to the office of the presidency, a president who fails to uphold the public trust may be removed from office.
    a. coercive
    b. charismatic
    c. tradition
    d. rational-legal

28. A _____ is a unit of political organization that has recognizable national boundaries and whose citizens possess specific legal rights and obligations.
   a. national-state
   b. state
   c. bureaucracy
   d. city-state

29. In dictatorships, power is gained and held by a single individual. Pure dictatorships are rare; all rulers need the support of the military and the backing of business elites to maintain their position. Dictatorships occur in _____ nation-states political systems.
   a. democracies
   b. authoritarianism
   c. monarchies
   d. totalitarianisms

30. In the United States, people have a voice in the government through _____, whereby citizens elect representatives to serve as bridges between themselves and the government.
   a. representative democracy
   b. monarchy
   c. direct democracy
   d. independent involvement democracy

31. Political scientists Thomas Dye and Harmon Zeigler summarized the key elements of pluralism. All of the following were identified, <u>except</u>:
   a. decisions are made on behalf of the people by leaders who engage in a process of bargaining, accommodation, and compromise
   b. veto groups are strictly prohibited
   c. public policy is not always based on majority preference; rather, it reflects a balance among competing interest groups
   d. power is widely dispersed in society

32. Capitalists control the government through special interest groups, lobbying, campaign financing, and other types of "influence peddling" to get legislatures and the courts to make decisions favorable to their class. The state exists only to support the interests of the dominant class. This would exemplify the _____ perspective on class conflict.
   a. instrumental Marxist
   b. structural Marxist
   c. restrictive Marxist
   d. normative Marxist

33. A political party does all of the following, <u>except</u>:
   a. develops and articulates policy positions
   b. educates political candidates about critical social issues
   c. recruits political candidates and helps them win office
   d. holds the political candidates responsible for the party's policy positions

34. People with a _____ perspective tend to focus on equality of opportunity, the need for government regulation and social safety nets.
    a. liberal
    b. socialist
    c. conservative
    d. libertarian

35. People with a _____ perspective are more likely to emphasize economic liberty and freedom from government interference.
    a. liberal
    b. socialist
    c. conservative
    d. radical

36. According to the text's discussion of the federal bureaucracy,
    a. during the nineteenth century the government held a central role in everyday life.
    b. the government bureaucracy has been able to expand more in recent decades.
    c. the federal bureaucracy employs less than 1 million people.
    d. the public expects much less from the government.

37. _____ is a societal focus on military ideals and an aggressive preparedness for war.
    a. Patriarchal complex
    b. Military complex
    c. War
    d. Militarism

38. _____ activities constitute a long-standing political practice – projects designed to bring jobs and public monies to the home state of members of Congress, for which they can take credit.
    a. Cracker barrel
    b. Grease men
    c. Pork barrel
    d. Bread and butter

39. In Libya, Colonel Muammar Qaddafi has provided money and training for terrorist groups such as the Arab National Youth Organization, which was responsible for skyjacking a Lufthansa airplane over Turkey and forcing the German government to free the surviving members of Black September. This is an example of:
    a. repressive terrorism
    b. state-sponsored terrorism
    c. national terrorism
    d. jihad terrorism

40. _____ terrorism is conducted by a government against its own citizens for the purpose of protecting an existing political order.
   a. Repressive
   b. State-sponsored
   c. National
   d. Jihad

TRUE/FALSE QUESTIONS

1. Political sociology is the same as political science.
   T   F

2. Power is a social relationship that involves both leaders and followers.
   T   F

3. Charismatic authority tends to be temporary and relatively unstable.
   T   F

4. Globally, today everyone is born, lives, and dies under the auspices of a nation-state.
   T   F

5. In a representative democracy, elected representatives are expected to keep the "big picture" in mind, and they do not necessarily need to convey the concerns of those they represent in all situations.
   T   F

6. Functionalists suggest that divergent viewpoints lead to a system of political pluralism in which the government functions as an arbiter between competing interests and viewpoints.
   T   F

7. Over the past two decades, special interest groups have become more involved in multiple-issue politics because so many political issues are intertwined with other concerns.
   T   F

8. Sociologist G. William Domhoff believes that the ruling class wields sufficient power to constrain political processes in the United States.
   T   F

9. Structural Marxists contend that the state is a passive instrument of the capitalist class.
   T   F

10. Children do not typically identify with the political party of their parents until they reach college age.
   T   F

11. People in the upper classes tend to vote based on a philosophy of *noblesse oblige*.
   T   F

12. The United States has one of the highest percentages of voter turnout of all Western nations.
   T   F

13. The U.S. governmental bureaucracy has been able to perpetuate itself and expand because many of its employees have highly specialized knowledge and skills and cannot be replaced easily by "outsiders."
   T   F

14. Pork barrel projects are a recent political practice in the United States.
   T   F

15. According to symbolic interactionists, militarism may be related to the social construction of masculinity in societies.
   T   F

16. Historically, the development of manhood and male superiority has been linked to militarism and combat.
   T   F

17. According to Box 14.1, television coverage of political scandals first began in the mid-1990s.
   T   F

18. Some analysts, as depicted in Box 14.2, suggest that the European Union has produced more than fifty years of peace in Europe.
   T   F

19. According to Box 14.3, California Governor Arnold Schwarzenegger has not been able to use media framing to develop favorable responses to his political initiatives.
   T   F

20. According to Box 14.4, recently all twenty-five of top news stories were adequately covered by the U.S. media.
   T   F

FILL-IN-THE-BLANK QUESTIONS

1. Some social analysts refer to the government as the _____, which possesses a legitimate monopoly over the use of force.

2. The main types of political systems are _____, _____, _____, and _____.

3. _____ is the ability of persons or groups to achieve their goals, despite the opposition of others.

4. Power legitimized by law or written rules and regulations is _____.

5. According to C. Wright Mills, the _____ is comprised of political, economic, and military elite.

6. The mutual interdependence of the military establishment and private military contractors is known as the _____.

7. The three types of political terrorism are: _____, _____, and _____.

8. The organized, armed conflict between nations is known as _____.

9. According to sociologist _____, the U.S. actually has a ruling class that wields power to constrain political processes.

10. The two worst terrorist attacks that ever occurred in the U.S. were in the years _____, and _____.

11. According to _____, the three ideal types of authority are traditional, charismatic, and rational-legal.

12. When cities first developed, the _____ became the center of political power.

13. The people who are paid to influence legislation on behalf of special clients are referred to as _____.

14. When military officers seize power from the government, _____ result.

15. According to _____, one purpose of government is to socialize people to be good citizens.

SHORT ANSWER/ESSAY QUESTIONS

1. What is the role of the individual in politics, considering such issues as political socialization, political action, lobbying activities, voter apathy and political elites?
2. What is the importance of political parties? What do they do? Are they good for democracies? Why? Why not?
3. What are the major political systems around the world?
4. What are the three ideal types of authority? In what ways are some more effective than others in achieving certain goals?
5. How do the pluralists and the elite models explain power in the United States? Which do you think best represents the U.S. political system? Why?

STUDENT CLASS PROJECTS AND ACTIVITIES

1. The text notes the low rate of voter participation in our country. You are to attempt to discover some explanations for this condition. You are to conduct an informal survey of a cross section of 20 people of different ages, ethnicity, social class, race, and sex. On your survey forms, you must include the basic background information of your respondents (name, race, sex, address, age ethnicity, occupation, or student

2. status). Obtain responses to the following questions: (1) When was the last time you voted? Why? (2) Did you vote for the President in the last election? (3) Do you usually vote in all elections? Why? Why not? (4) Do you actually vote along political party lines, such as for the Democrats or Republicans? If so, which party? Why? (5) Do other members of your family vote? (6) Do you know at least one elected official? Who? (7) Do you support the present position of the President of the United States on most issues? Explain. Formulate any other question you may want to ask. Provide a conclusion, an analysis, and an evaluation of the data in the writing up of this project. Submit your paper at the appropriate time.

3. Conduct an informal survey of 10 male and 10 female students on campus. Provide the necessary biographical background of your respondents, such as name, address, age, sex, and college classification. The survey is called "What If Women Ran America?" Ask and record all responses to the following questions: If women ran America: (1) Would child care be more available and maternity leave (with pay) guaranteed? Explain. (2) Would government be more attentive to the needy? Explain. (3) Would abortion be legal? Explain. (4) Would there be greater equality for working women? Explain. (5) Would there be greater sexual tolerance? Explain. (6) Would there be stricter gun control? Explain. (7) Would they be tougher on crime? Explain. (8) Would America be a better place? Explain. Summarize the responses and provide an evaluation of this project.

4. Every year (usually in January), Parade Magazine reports on the "Ten Worst Dictators in the World." Locate the latest issue and record each of the ten, summarizing basic information of each, noting the country that each dictates, the length of time of the rule of each, the reason for being on this dubious list and any other pertinent information provided in the article. Next, survey any 20 people and ask these questions: (1) Who do you think are 10 of the worst dictators in the world today? (2) What country do you think they represent? (3) Why do they have the distinction of being among the worst? (4) Do you have any additional information to provide? Provide a summary, if possible, in a pie-chart or graph, the responses to your survey, providing any extra information. Submit your paper, following the specific directions from your professor.

INTERNET ACTIVITIES

1. **The Center for Democracy and Technology**, http://www.cdt.org/, tracks the impact of new information technologies on the constitutional rights and liberties of Americans. Visit this site for more information.

2. Sociologists are frequently interested in surveys and opinion polls to determine how members of a society feel about a particular issue. One organization that is well known for conducting such research is the **Gallup** organization. Search the site for polling data related to this chapter: **http://www.gallup.com/**

3. In addition to the Gallup organization, the **Roper Center for Public Opinion Research**, **http://www.ropercenter.uconn.edu/**. Investigate the Public Opinion Matters section of the web site to find data and articles on several topics that bear relevance to this chapter.

4. Most of the major political parties now have an online presence. After visiting these sites, compare the major similarities or differences among the political parties. The Democratic Party can be found at **http://democrats.org/**. The Republican National Committee can be found at **http://rnc.org/**. The Libertarian Party can be found at **http://www.lp.org/**. The Socialist Party is located at **http://sp-usa.org/**.

5. As discussed in the text, lobbying groups are a powerful force in **politics** in the United States. Each of these organizations provides information about their legislative concerns and activities. For instance, the NRA keeps track of bills and proposed laws that might impact gun ownership. Overall, how do the agendas of these organizations compare? Some of the best-known lobbying groups are: The National Rifle Association, **http://www.nra.org/**, (NRA); The American Association of Retired Persons (AARP) **http://www.aarp.org/**; and The National Education Association (NEA) **http://www.nea.org/**.

6. **The Center for Responsive Politics**, **http://www.opensecrets.org/about/index.asp**, maintains an interesting site with valuable information on both PACs (political action committees) and lobbyist groups. For example, they maintain an online lobbyist report called "Influence, Inc." This provides background data on an issue, analysis of industry spending, and a relevant database. They also profile every political action committee registered in the Federal Election Commission.

INFOTRAC COLLEGE EDITION EXERCISES

Visit the **InfoTrac College Edition** website at: **http://www.wadsworthmedia.com/webtutor/infotrac.htm**. You will arrive at a screen that enables you to search topics.

1. Search for articles and news stories about **voter apathy**. The United Sates has a low voting turnout. Investigate research on this phenomenon.

2. Opponents of the **death penalty** contend that it is cruel and unusual punishment, and that our legal system makes mistakes and wrongfully executes innocent people. Research this issue to find out how often this happens, and what states are doing about it.

3. Hundreds of **interest groups** seek to influence the outcome of elections and specific legislation. Investigate the range of interest groups, or focus on one group in particular.

4. **Campaign reform** is an important current issue, as the amount of money candidates can raise dictates the quality of their campaign, or even if they will be able to run at all. Look at news stories about this issue. See if you can collect information to present a pro and con class discussion.

5. The media attention given to recent school shootings have left many feeling that we should have some type of **gun control**; the NRA opposes this. Investigate this issue.

6. There are an enormous number of articles under the topic **terrorism** on InfoTrac. Be prepared to make an oral presentation, create posters, create a mural, or a write brief report.

SOLUTIONS

MULTIPLE CHOICE QUESTIONS

1. B, p. 454
2. C, p. 454
3. C, p. 456
4. D, p. 457
5. A, p. 457
6. B, p. 458
7. C, p. 459
8. A, p. 459
9. C, p. 461
10. B, p. 463
11. A, p. 463
12. C, p. 464
13. C, p. 464
14. D, p. 465
15. A, p. 468
16. D, p. 470
17. B, p. 472
18. C, p. 475
19. A, p. 476
20. D, p. 478
21. D, p. 456
22. C, p. 456
23. A, p. 456
24. A, p. 457
25. D, p. 457
26. C, p. 457
27. D, p. 458
28. A, p. 459
29. B, p. 461
30. A, p. 463
31. B, p. 464
32. A, p. 467
33. B, p. 468
34. A, p. 468
35. C, p. 468
36. B, p. 472
37. D, p. 476
38. C, p. 476
39. B, p. 478
40. A, p. 479

TRUE/FALSE QUESTIONS

1. F, p. 454
2. T, p. 455
3. T, p. 457
4. T, p. 459
5. F, p. 463
6. T, p. 463
7. F, p. 464
8. T, p. 466
9. F, p. 467
10. F, p. 471
11. T, p. 471
12. F, p. 471
13. T, p. 472
14. F, p. 476
15. T, p. 477
16. T, p. 477
17. F, p. 456
18. T, p. 460
19. F, p. 462
20. F, p. 481

FILL-IN-THE-BLANK QUESTIONS

1. state, p. 454
2. monarchy, p. 459; authoritarian systems, p. 461; totalitarian systems, p. 461; democratic systems, p. 463
3. Power, p. 455
4. rational-legal authority, p. 458
5. power elite, p. 465
6. military-industrial complex, p. 475
7. revolutionary, p. 478; state-sponsored, p. 478; repressive terrorism, p. 479
8. war, p. 479
9. William Domhoff, p. 466
10. 1995, p. 479; 2001, p. 479
11. Max Weber, p. 457
12. city-state, p. 459
13. lobbyists, p. 464
14. military juntas, p. 461
15. Emile Durkheim, p.463

15

FAMILIES AND INTIMATE RELATIONSHIPS

BRIEF CHAPTER OUTLINE

CHAPTER SUMMARY

Families are relationships in which people live together with commitment, form an economic unit, care for any young, and consider their identity to be significantly attached to the group. While the **family of orientation** is the family into which a person is born and in which early socialization usually takes place, the **family of procreation** is the family a person forms by having or adopting children. Sociologists investigate marriage patterns (such as **monogamy** and **polygamy**), descent and inheritance patterns (such as **patrilineal**, **matrilineal**, and **bilateral** descent), familial power and authority (such as **patriarchal**, **matriarchal**, and **egalitarian** families), residential patterns (such as **patrilocal**, **matrilocal**, and **neolocal** residence, and in group or out group marriage patterns (i.e. **endogamy** and **exogamy**). Functionalists emphasize that families fulfill important societal functions, including sexual regulation, socialization of children, economic and psychological support, and the provision of social status. By contrast, conflict and feminist perspectives view the family as a source of social inequality and focus primarily on the problems inherent in relationships of dominance and subordination. Symbolic interactionists focus on family communication patterns and subjective meanings that members assign to everyday events. Postmodern perspectives view the family as *permeable* – one that is subject to modification or change. The nuclear family, therefore, is one of many family forms. Families have changed dramatically in the United States where there have been significant increases in cohabitation, domestic partnerships, dual earner marriages, single parent families, and rates of divorce and remarriage. Housework and child-care responsibilities remain key issues for the family. Violence in the family is more common than violence among groups of people. Problems in the family contribute to the large numbers of children who are in *foster care*, where adults other than a child's own parents or biological relatives serve as caregivers. Divorce has contributed to greater diversity in family relationships, including stepfamilies or blended families and the complex binuclear family. Most people who divorce get remarried. While some never-married singles choose to remain single, others do so out of necessity. Support systems and extended family networks are important in African American, Latino/a, Asian American, and Native American families; however, factors such as age and class may reduce such family ties. Biracial families have greatly increased; today, skin color and place of birth are not reliable indicators of a person's identity and origin. In the future, people's perceptions of family will continue to change, as will the major issues facing the family.

LEARNING OBJECTIVES

After reading Chapter 15, you should be able to:

1. Describe kinship ties and distinguish between families of orientation and families of procreation.

2. Define the social institution of the family.

3. Distinguish between patriarchal, matriarchal, and egalitarian families.

4. Describe the different forms of marriage found across cultures.

5. Explain the differences in residential patterns and note why most people practice endogamy.

6. Describe functionalist, conflict and feminist, symbolic interactionist and postmodernist perspectives on families.

7. Explain the major causes and consequences of divorce and remarriage in the United States.

8. Describe how U.S. families have changed over the past two decades.

9. Discuss the major issues associated with adoption, teenage pregnancies, single-parent households, and two-parent households.

10. Describe the ways that the double shift is experienced by men and women.

11. Compare and contrast extended and nuclear families.

12. Discuss the system of descent and inheritance, and explain why such systems are important in societies.

13. Describe cohabitation and domestic partnerships and note key social and legal issues associated with each.

14. Describe the diversity found in contemporary U.S. families.

15. Describe the major problems faced by people in dual-earner marriages.

16. Explain why it has become increasingly difficult to develop a concise definition of family.

17. Describe the changing definition of the term "interracial marriage" and family.

KEY TERMS

(defined at page number shown and in glossary)

bilateral descent, p. 492
cohabitation, p. 499
domestic partnerships, p. 499
dual earner marriages, p. 601
egalitarian families, p. 493
endogamy, p. 493
exogamy, p. 494
extended family, p. 488
family, p. 486
family of orientation, p. 488
family of procreation, p. 488
homogamy, p. 493
kinship, p.487
marriage, p. 489

matriarchal family, p. 492
matrilineal descent, p. 492
matrilocal residence, p. 493
monogamy, p. 490
neolocal residence, p. 493
nuclear family, p. 489
patriarchal family, p. 492
patrilineal descent, p. 492
patrilocal residence, p. 493
polyandry, p. 491
polygamy, p. 490
polygyny, p. 490
second shift, p. 501
sociology of family , p. 494

KEY PEOPLE

(identified at page number shown)

Jean Baudrillard, p. 497
Peter Berger and Hansfried Kellner, p. 496
Jessie Bernard, p. 496
Francesca Cancian, p. 498
Andrew Cherlin, p. 510
Emile Durkheim , p. 494
David Elkind, p. 496
Arlie Hochschild, p. 501
Alfred C. Kinsey, p. 498
Charlene Miall, p. 503

Margaret Mead, p. 499
Sara McLanahan and Karen Booth, p. 505
Talcott Parsons, pp. 494-495
Brian Robinson, p. 504
Alice Rossi, p. 506
Lillian Rubin, p. 514
Lenore Walker, p. 496
Jane Riblett Wilkie. p. 495
Norma Williams, p. 510

CHAPTER OUTLINE

I. FAMILIES IN GLOBAL PERSPECTIVE
    A. **Families** are relationships in which people live together with commitment, form an economic unit and care for any young, and consider their identity to be significantly attached to the group.
    B. Family Structure and Characteristics
        1. In preindustrial societies, the primary social organization is through **kinship** – a social network of people based on common ancestry, marriage, or adoption.
        2. In industrialized societies, other social institutions fulfill some functions previously taken care of by kinship ties; families are responsible primarily for regulating sexual activity, socializing children, and providing affection and companionship for family members.
        3. Many of us will be members of two types of families: a **family of orientation** – the family into which we are born or adopted and in which early socialization usually takes place, and a **family of procreation** – the family we form by having or adopting children.
        4. Extended and nuclear families
            a. An **extended family** is a family unit composed of relatives (such as grandparents, uncles, and aunts) in addition to parents and children who live in the same household.
            b. A **nuclear family** is a family composed of one or two parents and their dependent children, all of whom live apart from other relatives.
    C. Marriage Patterns
        1. **Marriage** is a legally recognized and/or socially approved arrangement between two or more individuals that carries certain rights and obligations and usually involves sexual activity.
        2. In the United States, **monogamy** – a marriage between two partners, usually a woman and a man – is the only form of marriage sanctioned by law.

3. **Polygamy** is the concurrent marriage of a person of one sex with two or more members of the opposite sex.
   a. The most prevalent form of this marriage pattern is **polygyny** – the concurrent marriage of one man with two or more women.
   b. **Polyandry** – the marriage of one woman with two or more men – is very rare.
D. Patterns of Descent and Inheritance
   1. In preindustrial societies, the most common pattern of unilineal descent is **patrilineal descent** – tracing descent through the father's side of the family – whereby a legitimate son inherits his father's property and sometimes his position upon the father's death.
   2. **Matrilineal descent** traces descent through the mother's side of the family; however, inheritance of property and position usually is traced from the maternal uncle (mother's brother) to his nephew (mother's son).
   3. In industrial societies such as the United States, kinship usually is traced through both parents; **bilateral descent** is a system of tracing descent through both the mother's and father's sides of the family.
E. Power and Authority in Families:
   1. **Patriarchal family**: a family structure in which authority is held by the eldest male (usually the father), who acts as head of household and holds power over the women and children.
   2. **Matriarchal family**: a family structure in which authority is held by the eldest female (usually the mother), who acts as head of household.
   3. **Egalitarian family**: a family structure in which both partners share power and authority equally.
F. Residential Patterns:
   1. **Patrilocal residence**: the custom of a married couple living in the same household (or community) with the husband's family.
   2. **Matrilocal residence**: the custom of a married couple living in the same household (or community) with the wife's parents.
   3. In industrialized nations, most couples hope to live in a **neolocal residence**: the custom of a married couple living in their own residence apart from both the husband's and the wife's parents.
G. Endogamy and Exogamy:
   1. **Endogamy** is the practice of marrying within one's own group. In the U.S. people practice endogamy.
   2. **Exogamy** is the practice of marrying outside one's own social group or category. The family, the church, and the state are the most important sources of positive or negative sanction for practicing exogamy.
   3. Most people engage in **homogamy** – the pattern of individuals marrying those who have similar characteristics, such as race/ethnicity, religious background, age, education, or social class.
II. THEORETICAL PERSPECTIVES ON FAMILIES
A. The **sociology of family** is the subdiscipline of sociology that attempts to describe and explain patterns of family life and variations in family structure.

B. Functionalist Perspectives
  1. The family is important in maintaining the stability of society and the well being of individuals.
  2. According to **Emile Durkheim**, both marriage and society involve a mental and moral fusion of individuals; division of labor contributes to greater efficiency in all areas of life.
  3. **Talcott Parsons** further defined the division of labor in families: the husband/father fulfills the instrumental role (meeting the family's economic needs, making important decisions, and providing leadership) while the wife/mother fulfills the expressive role (doing housework, caring for children, and meeting the emotional needs of family members).
  4. Four key functions of families in advanced industrial societies are:
     a. sexual regulation
     b. socialization
     c. economic and psychological support for members
     d. provision of social status and reputation.
C. Conflict and Feminist Perspectives
  1. Families are a primary source of inequality.
  2. According to some conflict theorists, families in capitalist economies are similar to workers in a factory: women are dominated at home by men the same way workers are dominated by capitalists in factories; reproduction of children and care for family members at home reinforce the subordination of women through unpaid (and devalued) labor.
  3. Some feminist perspectives focus on patriarchy rather than class because men's domination over women existed long before private ownership of property; contemporary subordination is rooted in men's control over women's labor power.
  4. Conflict and feminist perspectives on families focus primarily on the problems inherent in relationships of dominance and subordination.
D. Symbolic Interactionist Perspectives
  1. Interactionists examine the roles of husbands, wives, and children as they act out their own parts and react to the actions of others.
  2. According to **Peter Berger** and **Hansfried Kellner**, interaction between marital partners contributes to a shared reality: newlyweds bring separate identities to a marriage but gradually construct a shared reality as a couple.
  3. According to **Jessie Bernard**, women and men experience marriage differently: there is "his" marriage and "her" marriage.
E. Postmodern Perspectives
  1. In the information age, the postmodern family is *permeable* –capable of being diffused or invaded in such a way that an entity's original purpose is modified or changed.
     a. The nuclear family is only one of many family forms.
     b. Maternal love is transformed into shared parenting.
     c. The individual values autonomy of the individual more than the family unit.

2. Urbanity is another characteristic of the postmodern family.
    a. The boundaries between the workplace and the home become more open and flexible.
    b. New communications technologies integrate and control labor.
3. Some paint a bleak future for families in the age of the "integrated circuit."
    a. There is a growing "digital divide" and a type of "cyber class warfare."
    b. Many families are left out of Internet access.

III. DEVELOPING INTIMATE RELATIONSHIPS AND ESTABLISHING FAMILIES
  A. Love and Intimacy
    1. Although the ideal culture emphasizes romantic love, men and women may not share the same perceptions about love: women tend to express their feelings verbally, while men tend to express their love through nonverbal actions.
    2. Scholars suggest that love and intimacy are closely intertwined.
        a. Perspectives about sexual activities vary according to culture and time period.
        b. **Kinsey's** research provided a model for research on human sexuality for over forty years.
  B. Cohabitation and Domestic Partnerships
    1. **Cohabitation** refers to a couple who live together without being legally married.
    2. Characteristics of persons most likely to cohabit are as follows: under age 45, have been married before, or are older individuals who do not want to lose financial benefits (such as retirement benefits) that are contingent upon not marrying.
    3. Many lesbian and gay couples cohabit because they cannot enter into a legally recognized marriage; some have sought recognition of **domestic partnerships** – household partnerships in which an unmarried couple lives together in a committed sexually intimate relationship and is granted the same benefits as those accorded to married heterosexual couples.
  C. Marriage
    1. Couples marry for reasons such as being "in love," desiring companionship and sex, wanting to have children, feeling social pressure, attempting to escape from their parents' home, or believing they will have greater resources if they get married.
    2. Communication and support are crucial to the success of marriages; problems that cause the most concern are lack of emotional intimacy, poor communication, and lack of companionship.
  D. Housework and Child-care Responsibilities
    1. Over 50% of all U.S. marriages are **dual-earner marriages** – marriages in which both spouses are in the labor force. Over half of all employed women hold full-time, year-round jobs.
    2. Many married women also have a **second shift** – the domestic work that employed women perform at home after they complete their workday on the job; this amounts to an extra month of work for women each year.

3. In families where the wife's earnings are essential to family finances, more husbands have attempted to share some of the household and child-care responsibilities.
    a. Women and men perform different household tasks.
    b. Recurring tasks tend to be more the women's responsibilities while men are more likely to do periodic tasks.

IV. CHILD-RELATED FAMILY ISSUES AND PARENTING
  A. Deciding to have children
    1. Couples deciding not to have children may consider themselves "child-free," while those who do not produce children through no choice of their own may consider themselves "childless."
        a. Advances in birth control techniques now make it possible to determine choice of parenthood, number of children, and the spacing of their births.
        b. Some couples experience *involuntary infertility*; a leading cause is sexually transmitted diseases.
        c. Women who are involuntarily childless engage in "information management" to combat the social stigma associated with childlessness.
  B. Adoption
    1. *Adoption* is a legal process through which rights and duties of parenting are transferred from a child's biological and/or legal parents to new legal parents.
    2. About 6.4 million women become pregnant each year in the United States; of that number, about 44% of pregnancies are intended while 56% are unintended.
  C. Teen pregnancies have decreased over the past three decades; however, they are viewed as a crisis because an increase has occurred in the number of births among unmarried teenagers and the rates have not declined as rapidly as for older married women.
  D. Recently, single- or one-parent households have increased significantly due to divorce and to births outside marriage.
  E. Two-parent households
    1. Parenthood in the United States is idealized, especially for women.
    2. Children in two-parent families are not guaranteed a happy childhood.
V. TRANSITIONS IN FAMILIES AND PROBLEMS IN FAMILIES
  A. Family Violence
    1. Violence between men and women in the home is often referred to as *spouse abuse* or domestic violence.
    2. Physical violence is more common among family members than any other group of people.
    3. A common pattern of *nonintervention* results in slow or little assistance from individuals or law enforcement officials.

B. Divorce
   1. Divorce is the legal process of dissolving a marriage that allows former spouses to remarry if they so choose. Most divorces are based on "*irreconcilable differences*" (there has been a breakdown of the marital relationship for which neither partner is specifically blamed).
   2. Under *no-fault divorce laws*, proof of the blame is generally no longer necessary.
   3. Recent studies have shown that 43 percent of first marriages end in separation or divorce within fifteen years; one in three first marriages ends within ten years, where one in five ends within five years.
   4. Causes of Divorce
      a. At the macrolevel, societal factors contributing to higher rates of divorce include changes in social institutions such as religion and law.
      b. At the microlevel, characteristics that appear to contribute to divorce are:
         i. Marriage at an early age
         ii. A short acquaintanceship before marriage
         iii. Disapproval of the marriage by relatives and friends
         iv. Limited economic resources
         v. Having a high-school education or less
         vi. Parents who are divorced or have unhappy marriages
         vii. The presence of children (depending on gender and age) at the beginning of marriage.
   5. Consequences of Divorce
      a. An estimated 60 percent of divorcing couples have one or more children.
      b. By age 16, about one in every three white and two in every three African American children will experience divorce within their families. Some children experience more than one divorce during their childhood because one or both of their parents may remarry and subsequently divorce again.
      c. Divorce changes relationships for other relatives, especially grandparents.
C. Remarriage
   1. Most people who divorce get remarried: more than 40% of all marriages take place between previously married brides and/or grooms, and about half of all persons who divorce before age 35 will remarry within three years.
   2. Most divorced people remarry others who have been divorced. At all ages, a greater proportion of men than women remarry and often relatively soon after divorce. Among women, those who divorce at younger ages are more likely to remarry than are those who are older; women with a college degree and without children are less likely to remarry.
   3. As a result of divorce and remarriage, some people become part of *blended families*.

VI. DIVERSITY IN FAMILIES

A. Diversity among Singles

1. Some never-married singles choose to remain single because of opportunities for a career (especially for women), the availability of sexual partners without marriage, a belief that the single lifestyle is full of excitement, and a desire to be self-sufficient and have the freedom to change and experiment.

2. Other never-married singles remain single out of necessity; they cannot afford to marry and set up their own household.

3. Among persons age 15 and over, over 43.5% of African Americans have never married, as compared with almost 33.2% of Latino/as, 33.1% of Asian and Pacific Islander Americans, and 24.5% of whites. Among women age 20 and over, the difference is even more pronounced.

B. African American Families

1. A higher proportion of African Americans than whites live in extended family households that may provide emotional and financial support not otherwise available.

2. Among middle and upper-middle class African American families, nuclear families are more prevalent than extended family ties.

C. Latino/a Families

1. Family support systems found in many Latino/a families – "*la familia*" – cover a wide array of relatives including parents, aunts, uncles, cousins, brothers and sisters, and their children.

2. Some sociologists question the extent that *familialism* exists across social classes. **Norma Williams** found that extended family networks are disappearing, especially among advantaged urban Latino/as.

D. Asian American Families

1. While many Asian Americans live in nuclear families, others (especially those residing in Chinatowns) have extended family networks. Some are referred to as semi-extended families because other relatives live in close proximity but not necessarily in the same household.

2. Extended family networks of some Vietnamese Americans are limited because family members died in the war in Vietnam, and others did not migrate to the United States.

E. Native American Families

1. Family ties remain strong among many Native Americans, who have always been known for their strong family ties.

2. Extended family patterns are common among lower income Native Americans living on reservations; most others live in nuclear families.

F. Biracial Families

1. Since the 1970s, there has been a 300 percent increase in marriages between people of different races; when these couples produce offspring, their children are considered to be biracial.

2. Interracial marriage was illegal in sixteen states until 1967.

3. At the beginning of the twenty-first century, skin color and place of birth have ceased to be reliable indicators of a person's identity or origin; the term *interracial marriage* has taken on a much broader interpretation.

VII.   FAMILY ISSUES IN THE FUTURE
    A. Some people believe that the family as we know it is doomed; others believe that a return to traditional family values will save this social institution and create greater stability in society.
    B. Sociologist **Lillian Rubin** suggests that clinging to a traditional image of families is hypocritical in light of our society's failure to support them. Some laws have the effect of hurting children whose families do not meet the traditional model. For example, cutting down on government programs which provide food and medical care for pregnant women and infants will result in seriously ill children rather than model families.
    C. People's perceptions about what constitutes a family will continue to change in the next century: the family may become those persons on whom one can depend for emotional support, who are available in crises and emergencies, or who provide continuing affections, concern, and companionship.

ANALYZING AND UNDERSTANDING THE BOXES

After reading the chapter and studying the outline, re-read the boxes and write down key points and possible questions for class discussion.

Sociology and Everyday Life: How Much Do You Know About Contemporary Trends in U.S. Family Life?

Key Points:

Discussion Questions:

1.

2.

3.

Sociology in Global Perspective: Buffering Financial Hardship: Extended Families in the Global Community

Key Points:

Discussion Questions:

1.

2.

3.

Sociology and Social Policy: Should the U. S. Constitution be Amended to Define "Marriage?"

Key Points:

Discussion Questions:

1.

2.

3.

You Can Make a Difference: Providing Hope and Help for Children

Key Points:

Discussion Questions:

1.

2.

3.

PRACTICE TESTS

MULTIPLE CHOICE QUESTIONS

Select the response that best answers the question or completes the statement.

1. According to the text, traditional definitions of the family
   a. are still highly applicable to today's families.
   b. includes all persons in a relationship who wish to consider themselves a family.
   c. need to be expanded to provide a more encompassing perspective on what constitutes a family.
   d. are indistinguishable from contemporary definitions of the family.

2. A social network of people based on common ancestry, marriage, or adoption is known as:
   a. kinship
   b. a family
   c. a clan
   d. subculture

3. The family of procreation is defined as "the family _____."
   a. into which a person is born
   b. in which people receive their early socialization
   c. that is composed of relatives in addition to parents and children who live in the same household
   d. a person forms by having or adopting children

4. Families that include grandparents, uncles, aunts, or other relatives who live in close proximity to the parents and children are known as a(n):
   a. clan
   b. extended family
   c. nuclear family
   d. family of procreation

5. _____ is the concurrent marriage of one man with two or more women, while _____ is the concurrent marriage of one woman with two or more men.
   a. Monogamy, polygamy
   b. Patriarchy, matriarchy
   c. Polygyny, polyandry
   d. Polyandry, polygyny

6. The most prevalent pattern of power and authority in families is:
   a. matriarchy
   b. monarchy
   c. oligarchy
   d. patriarchy

7. All of the following statements are correct regarding power and authority in families, **except**:
   a. Recently, there has been a trend toward more egalitarian family relationships in a number of countries.
   b. Some degree of economic independency makes it possible for women to delay marriage.
   c. The egalitarian family places a heavier burden on women around the globe than does patriarchal.
   d. Scholars have found no historical evidence to indicate that true matriarchies ever existed.

8. The custom of a married couple living in their own residence apart from both the husband's and the wife's parents is known as:
   a. isolated residence.
   b. neolocal residence.
   c. neutral local residence.
   d. exogamous residence.

9. When a person marries someone who comes from the same social class, racial-ethnic group, and religious affiliation, sociologists refer to this marital pattern as:
   a. endogamy
   b. exogamy
   c. inbreeding
   d. intraclass reproduction

10. In the United States, _____ was a key figure in developing a functionalist model of the family.
    a. C. Wright Mills
    b. Talcott Parsons
    c. Charles H. Cooley
    d. Peter Berger

11. According to functionalists, all of the following are key functions of families, **except**:
    a. provision of social status
    b. economic and psychological support
    c. maintenance of workers so that they can function effectively in the workplace
    d. sexual regulation and socialization of children

12. According to _____ theorists, interaction between marital partners contributes to a shared reality.
    a. feminist
    b. conflict
    c. functionalist
    d. symbolic interactionist

13. According to Lenore Walker, females are socialized to be _____, while males are socialized to be _____.
    a. jealous, possessive.
    b. passive, aggressive.
    c. dominant in the home, dependent in the work force.
    d. passive, communicative.

14. At what point in history did people in the United States come to view home and work as separate spheres?
    a. during colonial times
    b. during the Industrial Revolution
    c. at the end of World War II
    d. in the 1980s when more middle class white women entered the paid workforce

15. Which of the following statements regarding cohabitation is true?
    a. Attitudes about cohabitation have not changed very much in the past two decades.
    b. The Bureau of the Census recently developed a more inclusive definition of cohabitation.
    c. People most likely to cohabit are those who are under age 30 and who have not been married before.
    d. In the United States, many lesbian and gay couples cohabit because they cannot enter into a legally recognized marital relationship.

16. Sociologist _____ coined the term "second shift" to refer to the domestic work that employed women perform at home after they complete their workday on the job.
    a. Arlie Hochschild
    b. Talcott Parsons
    c. Kath Weston
    d. Francesca Cancian

17. Recent studies of teen pregnancy have concluded that:
    a. teenage pregnancies are increasing rapidly in the United States.
    b. teenage mothers often are as skilled at parenting as older mothers.
    c. there has been an increase in teenage pregnancies among unmarried teenagers.
    d. the media no longer focuses on "the problem" of teenage pregnancies.

18. According to sociologist Alice Rossi, men secure their status as adults by _____; women secure their status as adults by _____.
    a. fathering, mothering
    b. their employment, their employment
    c. their sports abilities, their housekeeping skills
    d. their employment, maternity

19. All of the following are cited in the text as primary social characteristics of those most likely to get divorced, **except**:
    a. marriage at a later age and being set in one's ways
    b. a short acquaintanceship before marriage
    c. disapproval of marriage by relatives and friends
    d. parents who are divorced or have unhappy marriages

20. In discussing diversity in families, the text points out that
    a. a higher percentage of whites live in extended family households than do African Americans.
    b. there is no such thing as "the" African American, Latino/a, Asian American, or Native American family.
    c. working-class African American women often are encouraged by their families to choose marriage over education.
    d. extended family patterns are no longer common among lower-income Native Americans living on reservations.

21. Sociologist Lillian Rubin suggests that clinging to a traditional image of families
    a. is needed in today's society
    b. is hypocritical in light of our society's failure to support families.
    c. maintains our faith in family values
    d. would lower the rate of divorce.

22. Many Asian Americans live in nuclear families; however, others (especially those residing in Chinatowns) have extended family networks. In a(n) _____ family, other relatives live in close proximity but not necessarily in the same household.
    a. extended
    b. partially extended
    c. non-extended
    d. semi-extended

23. Today, extended family patterns are common among lower-income Native Americans who
    a. live near the reservation.
    b. live on the reservation.
    c. live in large, urban areas.
    d. live in small, rural areas.

24. Since the 1970s, there has been a _____ increase in marriages between people of different races. When these couples produce offspring, their children are considered to be biracial.
    a. 100 percent
    b. 200 percent
    c. 300 percent
    d. 350 percent

25. Interracial marriage – which is often thought of as marriage between whites and blacks – was illegal in sixteen states until the U.S. Supreme Court overturned _____ in 1967.
    a. miscegenation laws
    b. segregation laws
    c. racial rules
    d. endogamy laws

26. A comparison of Census Bureau data from 1970 to 2000 shows that there has been _____ in the percentage of U.S. households comprising a married couple with their own children under eighteen years of age.
    a. a significant decline
    b. ad significant increase
    c. no change
    d. a doubling

27. The Census Bureau reported that the percentage distribution of nonfamily and family households had changed during the past 30 years. The most noticeable trend is the decline in the number of married-couple households with their own children living with them, which decreased from about 40 percent of all households in 1970 to about _____ percent in 2003.
    a. 17
    b. 23
    c. 32
    d. 35

28. _____ is a legally recognized and/or socially approved arrangement between two or more individuals that carries certain rights and obligations and usually involves sexual activity.
    a. Cohabitation
    b. Love
    c. Marriage
    d. Monogamous habitation

29. In the United States, the only legally sanctioned form of marriage is _____ which refers to a marriage between two partners, usually a woman and a man.
    a. polyandry
    b. homogamy
    c. polygamy
    d. monogamy

30. The _____ refers to family units that are composed of relatives in addition to parents and children who live in the same household.
    a. extended family
    b. nuclear family
    c. conventional family
    d. blended family

31. In horticultural and agricultural societies, _____ are extremely important; having a large number of family members participate in food production may be essential for survival.
    a. families of procreation
    b. extended families
    c. blended families
    d. nuclear families

32. With the advent of industrialization and urbanization, maintaining the _____ family pattern becomes more difficult in societies. Increasingly, young people move from rural to urban areas in search of employment in the industrializing sector of the economy. At that time, the _____ family typically becomes the predominant family pattern in the society.
    a. nuclear, extended
    b. extended, blended
    c. extended, nuclear
    d. nuclear, blended

33. _____ refers to family units composed of one or two parents and their dependent children, all of whom live apart from other relatives.
    a. Married family
    b. Parent-child family
    c. Conventional family
    d. Nuclear family

34. According to the sociologist Norma Williams, extended family networks among Latino/as have been:
    a. disappearing.
    b. growing stronger.
    c. are strongest in urban areas.
    d. are strongest in large cities.

35. In regard to housework, sociologist Arlie Hochschild notes that
    a. domestic work is shared equally by both husbands and wives.
    b. women and men perform different household tasks, but they spend as much time in these activities.
    c. even when husbands share some of the household responsibilities, they typically spend much less time in these activities than do their wives.
    d. husbands now cook as often as their wives.

36. Couples with more _____ ideas about women's and men's roles tend to share more equally in food preparation, housework, and child care.
    a. patriarchal
    b. matriarchal
    c. equal-partner
    d. egalitarian

37. In 2000, Latinas (Hispanic women) had a total fertility rate of 2.6, which was 50 percent above that of white (non-Hispanic) women. Among Latina women, the highest fertility rate is found among _____ women.
   a. Puerto Rican
   b. Cuban American
   c. Mexican American
   d. Latin American

38. According to sociologists Sara McLanahan and Karen Booth, children from mother-only families are more likely than children to two-parent families to have experience all of the following, except:
   a. to have poor academic achievement
   b. to delay marriage and postpone having children
   c. to have higher school absences and dropout rates
   d. to experience more drug and alcohol abuse

39. According to sociologist Brian Robinson, all of the following are myths regarding teenage fathers, except:
   a. they are worldy-wise "superstuds" who engage in sexual activity early and often
   b. they have few emotional feelings for the women they impregnate
   c. they sometimes try to be good fathers
   d. they are "phantom fathers" who rarely are involved in caring for and rearing their children

40. Some single fathers who do not have custody of their children remain actively involved in their children's lives. Others may become "_____" who take their children to recreational activities and buy them presents for special occasions but have a very small part in the children's day-to-day lives.
   a. phantom fathers
   b. Disneyland daddies
   c. big day dads
   d. just for fun fathers

TRUE/FALSE QUESTIONS

1. The standard sociological definition of "family" is still in use today.
   T   F

2. In industrialized societies, other social institutions fulfill some of the functions previously taken care of by the kinship network.
   T   F

3. In the United States, the only legally sanctioned form of marriage is monogamy.
   T   F

4. Today, there are no known societies that practice polyandry.
   T   F

5. In industrial societies, kinship is usually traced through patrilineal descent.
   T   F

6. Functionalists suggest that the erosion of family values may occur when religion becomes less important in everyday life.
   T   F

7. According to Parsons, the husband/father fulfills the expressive role in the family.
   T   F

8. According to some feminist scholars, hostility and violence against women and children may be attributed to patriarchal attitudes, economic hardship, and rigid gender roles in society.
   T   F

9. Postmodernists' perspectives describe the post-modern family as permeable.
   T   F

10. A domestic partnership is a household partnership in which an unmarried couple lives together in a committed, sexually intimate relationship that is granted the same rights and benefits as those accorded to married heterosexual couples.
    T   F

11. Over 50 percent of all marriages in the United States are dual-earner marriages.
    T   F

12. According to research, many U.S. women spend up to one-third of their life attempting to control their reproductivity.
    T   F

13. Women who are involuntarily childless are often thought of as "selfish."
    T   F

14. An estimated 60 percent of divorcing couples have one or more children.
    T   F

15. Interracial marriage was illegal in sixteen states until 1950, when the Supreme Court overturned miscegenation.
    T   F

16. According to Box 15.1, in today's society, people in the U.S. are more inclined to get married than at any other time in history.
    T   F

17. The number of children living with grandparents and with no parent in the home increased by more than 50 percent between 1990 and 2000, according to Box 15.1
    T   F

18. According to Box 15.2, in the global economy, the extended family is primarily a thing of the past.
    T   F

19. According to Box 15.3, in 1996, Congress passed the Defense of Marriage Act, which provides that every U.S. state is required to recognize a same-sex marriage under the laws of another state.
T   F

20. Since its commencement, according to Box 15.4, Hope Meadows has been largely successful in helping children get adopted.
T   F

FILL-IN-THE-BLANK QUESTIONS

1. The family of _____ is the one into which a person is born.

2. The most prevalent form of polygamy is _____.

3. A system of tracing descent through the mother's side is known as _____. descent.

4. The practice of marrying those with similar characteristics is known as _____ and is somewhat similar to the practice of _____.

5. For over forty years, the eminent _____ was considered to be the definite researcher on human sexuality.

6. _____ suggested that the two-stage marriage pattern would best prepare couples for commitment and parenthood.

7. Household partnerships wherein an unmarried couple lives together in a committed sexually intimate relationship and is granted the same benefits as those accorded to married heterosexual couples are defined as _____.

8. According to sociologist Arlie Hoschild, the _____ is the domestic work that employed women perform at home after they complete their work day on the job.

9. Over 50 percent of all marriages in the United States are _____.

10. The "_____" nanny is the best paid and most prestigious home caregiver.

11. The U.S. culture emphasizes _____ love, which refers to a deep emotion, the satisfaction of significant needs, and a caring for an acceptance of the person we love.

12. The sociologist _____ pointed out that women and men experience marriage differently, such as "her" and "his" marriage.

13. Violence between men and women in the home is often called domestic violence or _____.

14. In the U.S. today, most divorces are granted on the grounds of _____.

15. Under _____ laws, proof of spouse blame is generally no longer necessary.

SHORT ANSWER/ESSAY QUESTIONS

1. How do sociologists define family?
2. What are the functionalists, symbolic interactionists, conflict, feminists, and postmodernists perspectives on families?
3. What are some of the problems of the family in the United States today? What would you recommend as some solutions to these problems?
4. How could the family be described as both a "haven in a heartless world" and a "cradle of violence?"
5. What are some of the causes and consequences of divorce?

STUDENT CLASS PROJECTS AND ACTIVITIES

1. Select and then research a topic of unconventional reproduction, such as IN VITRO fertilization, (also known as test-tube fertilization), artificial insemination, or surrogacy. After selecting the topic, (1) trace the history of the development of that specific type of reproduction; (2) provide a description and definition of the reproduction procedure; (3) describe the widespread use of the procedure; and (4) discuss any complications, moral, legal, or medical that have surfaced related to the practice. Additionally, (5) examine and report any court cases that relate to your specific topic that have surfaced. In the writing of your paper, ensure that you provide a conclusion and a personal evaluation of this specific project.
2. Write a paper on the American family in the century ahead. Provide at least four bibliographic sources, (such as the text, other books, periodicals, newspaper articles, etc.). The paper could center on the topic: "The American Family in the Twenty-First Century." Issues to be included are the following: (1) What type of family form will be dominant; has the nuclear family disappeared? (2) As families change, has the type of community changed as well? (3) What is the status of commuter marriage? (4) Describe the dating practice: will the dating game be a "dangerous sport" because of the health risk? (5) What about life-expectancy? Has the life span increased? (6) What about "variations of the theme family"? Are there a variety of new categories, such as gay and lesbian couples with and without children, single women having babies by donor insemination, etc.? (7) What about diet? (8) What type of child care is available?; (9) What is the size of the family?; (10) Who is taking care of aged relatives?; (11) Describe the place of residence. Include any other information that pertains to this topic. In the writing of this project, follow any other specific guidelines and submit your paper at the appropriate time.

3. This is a short assignment that should be typed. Be sure and submit your paper to your instructor on the due date. This paper should be a minimum of five pages. Instructions: construct a "sociological" family tree of your relatives of at least three generations (your grandparents, parents, aunts, uncles, and your own generation). Record names, dates, and information as you gather this information. Examine your genealogy for some of the following norms of marriage: (1) Who married, who did not; (2) Who did your relatives marry?; (3) Age at marriage; (4) Did the norms of homogamy operate?; (5) Did they marry "up" or "down?"; (6) Number of children in the marriage; (7) Number of divorces, if any; and (8) Their religion. You may need to interview several family members, perhaps, to get the correct information for the above. Record all information carefully. After recording the information, examine how closely your family's marital choices follow the patterns of the typical American explained in the text and why variations might have occurred in your family tree. In your paper, include your reactions to your research on your family.

INTERNET ACTIVITIES

1. This **family educational page**, **http://www.familyeducation.com/home/**, contains gateways to a number of sites with information relating to studies of the family, with emphasis on children. Surf these sites and write up a critique or summary about the information each contains.
2. This "**family.com**" site, **http://family.go.com/**, is a resource center for busy families. Can you guess who the primary user of this site might be? Imagine yourself as a working parent. What kinds of information on this site could you use? What information would not be useful?
3. The Economic and Legal Benefits of Marriage, **http://www.religioustolerance.org/mar_bene.htm**, - search the site and present a debate on the pros and cons of cohabitation.
4. Here, **http://www.silcom.com/~paladin/madv/astrid.html**, you will find resources related to domestic violence. Contact one or more references and report back to class on the value of the resource.
5. Use these sites to produce a report on the state of children both here in the United States and globally. Describe the ways that the welfare of children is directly connected to the state of families:
http://www.unicef.org/sowc96/contents.htm?477,233,
http://www.futureofchildren.org/; and **http://www.researchforum.org/**.

INFOTRAC COLLEGE EDITION EXERCISES

Visit the **InfoTrac College Edition** website at:
http://www.wadsworthmedia.com/webtutor/infotrac.htm. You will arrive at a screen that enables you to search topics.

1. Investigate **domestic partnerships** (legal relationships between gay/lesbian couples that grant them some of the same rights as married couples) from a global perspective, and from a state by state pattern in the United States. Which country has the most restrictive policy on these partnerships? Which has the most non-restrictive policy?

2. An increasing number of children are raised in **single parent households**; most of these are female headed. What are the social and social psychological consequences of being without a father? Search the InfoTrac for several responses to this question.

3. Search for the term, **domestic violence**, and learn about measures being taken to combat this social problem. Read the report on African-American resources reported in Ebony magazine.

4. Search the category **polygamy**. Compare research about cross-cultural marriage practices. Apply the functionalist perspective to these kinds of practices.

5. Divide up the subdivisions on **teenage pregnancy** among class members. Construct a collective report on this phenomenon using a contribution from each class member.

6. Search for the keywords **dual career families**. Many students in the class will probably fall into this category. Have students search through these reports and compose a list of good advice that they might put to use in their own future.

7. The number of reported incidents of **child abuse** in this country is very high. Analyze the incidence and types of child abuse, as well as risk factors.

8. Assess the incidence of and reasons for **international adoptions**. From which countries are American parents most likely to adopt and why? The InfoTrac provides a wealth of information on this topic.

9. Assess the social consequences of **cohabitation**, especially from a global perspective. Where is cohabitation most common? Most uncommon?

SOLUTIONS

MULTIPLE CHOICE QUESTIONS

1. C, p. 486
2. A, p. 487
3. D, p. 488
4. B, p. 488
5. C, pp. 490-491
6. D, p. 492
7. C, p. 493
8. B, p. 493
9. A, p. 493
10. B, p. 494
11. C, p. 494
12. D, p. 496
13. B, p. 496
14. B, p. 497

15. D, p. 499
16. A, p. 501
17. C, pp. 504-505
18. D, p. 506
19. A, pp. 507-508
20. B, p. 512
21. B, p. 514
22. D, p. 412
23. B, p. 513
24. C, p. 513
25. A, p. 513
26. A, p. 489
27. B, p. 489
28. C, p. 489

29. D, p. 490
30. A, p. 488
31. B, p. 488
32. C, p. 489
33. D, p. 489
34. A, p. 512
35. C, p. 502
36. D, p. 493
37. C, p. 503
38. B, p. 505
39. C, p. 504
40. B, p. 505

TRUE/FALSE QUESTIONS

1. F, p. 486
2. T, p. 486
3. T, p. 490
4. F, p. 491
5. F, p. 492
6. T, p. 495
7. F, p. 494

8. T, p. 495
9. T, p. 496
10. T, p. 499
11. T, p. 501
12. F, p. 503
13. T, p. 503
14. T, p. 509

15. F, p. 513
16. F, p. 488
17. T, p. 488
18. F, p. 490
19. F, p. 500
20. T, p. 508

FILL-IN-THE-BLANK QUESTIONS

1. orientation, p. 488
2. polygyny, p. 490
3. matrilineal, p. 492
4. homogamy, p. 493; endogamy, p. 493
5. Alfred Kinsey, p. 498
6. Margaret Mead, p. 499
7. domestic partnerships, p. 499
8. sociology of family, p. 494
9. dual earner marriages, p. 501
10. "six-figure", p. 502
11. romantic love, p. 498
12. Jessie Bernard, p. 496
13. spouse abuse, p. 506
14. irreconcilable differences, p. 507
15. no-fault divorce, p. 407

16
EDUCATION

BRIEF CHAPTER OUTLINE

CHAPTER SUMMARY

Education is one of the most significant social institutions in the United States and other high-income nations; however, there is a lack of consensus regarding the effect of social factors upon individuals' access to and differential rewards of academic achievement. **Education** is the social institution responsible for the systematic transmission of knowledge, skills, and cultural values within a formally organized structure. People in preliterate societies acquire knowledge and skills through **informal education**; in preindustrial and industrial societies, people acquire specific knowledge, skills, and thinking processes through **formal education**. **Mass education** refers to providing free public schooling for wide segments of a nation's population. In addition to teaching the basics, U.S. schools today also teach a myriad of topics ranging from computer skills to AIDS prevention. In examining contemporary education in other nations, education in Japan and in Bosnia are contrasted; Japan's system is centralized whereas many schools in Bosnia identify and educate their students by their ethnic background. Functionalists have suggested that education performs a number of essential functions for society; however, conflict theorists emphasize that education perpetuates class, racial-ethnic, and gender inequalities. Symbolic Interactionists point out that education may become a *self-fulfilling prophecy* for students who perform up – or down – to the expectations held for them by teachers. Postmodernists highlight the permeability of both lower and higher education today, with urbanity, autonomy, and consumption, as major characteristics. The U. S. public schools today are a microcosm of many of the problems facing the country: inequality in public vs. private schools, unequal funding of public schools, inequality within public school systems, and racial and ethnic segregation and resegregation. Problems within elementary and secondary schools include school discipline and teaching styles, bullying, teasing, and sexual harassment, and students dropping out. School safety and school violence are major concerns in education. Opportunities and challenges in community colleges and in four-year colleges and universities vary. The soaring cost of a college education, racial and ethnic differences in enrollment, lack of faculty diversity, and the debate over affirmative action remain issues in education today. Much has been written about the issues facing education in the United States and other nations. Some social analysts believe that academic standards are not high enough in U.S. schools. Suggestions for reducing both illiteracy and **functional literacy** have focused on school reforms. As the U.S. becomes more racially, ethnically, and culturally diverse, the issue of *bilingual education* becomes even more critical. In *Lau v. Nichols,* the court affirmed the responsibility of the states and local districts for providing education for minority-language students. Another recent concern in education has been how to provide better educational opportunities for students with disabilities. The debate continues over school vouchers, which give parents the choice of what school their child will attend. As compared to the voucher system, the charter school movement creates public schools that operate under charter contracts negotiated by the schools' organizers and a sponsor. Another alternative educational program is home schooling, which is chosen by some parents who hope to avoid the problems of public schools while providing a quality education for their children. In 2001, Congress passed the "No Child Left Behind Act" intended to challenge the state's public school systems to meet certain education goals. In the United States, education in the future must accommodate the increasing diversity of the U. S. population at all levels of education.

LEARNING OBJECTIVES

After reading Chapter 16, you should be able to:

1. Describe the type of education found in preliterate societies.

2. List and discuss some of the major problems in U. S. elementary and secondary education.

3. Trace the history of education in preindustrial and industrial societies to the present.

4. Describe the functionalist perspective on education.

5. Discuss the postmodernist perspective on education, and discuss the significance of the "students as consumers" model.

6. Trace the history of the debate over affirmative action in higher education, citing the two U. S. Supreme Court decisions and the influence of those decisions.

7. Compare and contrast contemporary education in other nations using Japan and Bosnia as the models.

8. Describe the significance of the self-fulfilling prophecy and labeling on educational achievement.

9. Describe conflict perspectives on education and note how they differ from a functionalist perspective.

10. Differentiate symbolic interactionist perspectives on education from other paradigms.

11. Explain the manifest and latent functions fulfilled by the institution of education.

12. Integrate recent examples of violence in schools with some of the latent functions of education.

13. Use trends in education to formulate ideas about the future of this institution.

14. Describe four recent problems in U. S. higher education and assess possible solutions to these problems.

15. List and critique four alternative approaches for improving education.

KEY TERMS

(defined at page number shown and in glossary)

credentialism, p. 532
cultural capital, p. 529
cultural transmission, p. 520
education, p. 520
formal education, p. 521

functional illiteracy, p. 549
hidden curriculum, p. 531
informal education, p. 520
mass education, p. 522
tracking, p. 530

KEY PEOPLE

(identified at page number shown)

CHAPTER OUTLINE

I. AN OVERVIEW OF EDUCATION:
    A. Education is a powerful and influential institution that imparts values, beliefs, and knowledge considered essential to the social reproduction of individual personalities and entire cultures.
    B. Education is a socializing institution: early socialization primarily takes place in families and friendship networks; later socialization occurs in more formalized organizations created for the purposes of educating people.
II. EDUCATION IN HISTORICAL-GLOBAL PERSPECTIVE
    A. **Education** is the social institution responsible for the systematic transmission of knowledge, skills, and cultural values within a formally organized structure.
    B. **Informal education**: learning that occurs in a spontaneous, unplanned way- is found in preliterate societies.
    C. **Formal education**: learning that takes place within an academic setting such as a school, which has a planned instructional process and teachers who convey specific knowledge, skills, and thinking process to students, is found in preindustrial and industrial societies.
    D. Early Formal Education
        1. Earliest formal education probably occurred in ancient Greece and Rome where philosophers taught elite males to become thinkers and orators.
        2. Between the fall of the Roman Empire and the beginning of the Middle Ages, only the sons of wealthy lords received formal education; other children were trained through apprenticeships in merchant and crafts guilds.
        3. During the Middle Ages, the first colleges and universities were developed under the auspices of the church and the concept of human depravity was introduced into the curriculum.
        4. During the Renaissance, education shifted from a focus on human depravity to the importance of developing well-rounded and liberally educated people.
        5. 5. As societies industrialize, the need for formal education for the masses increases. **Mass education** refers to providing free, public schooling under the leadership of Horace Mann.

6. In addition to teaching the basics, U.S. schools teach a myriad of topics and perform many tasks that previously were performed by other social institutions.
7. Controversy exists over whose values should be taught.

E. Contemporary Education in Other Nations
1. In Japan, when the country began to industrialize during the Meiji Period (1868-1912), public education became mandatory for children.
   a. An educational system and national educational goals were set up.
   b. Education was linked with economic and national development; conformity and nationalism was emphasized.
   c. Today, toddlers are sent to cram schools (*jukus*) to ensure enrollment in good pre-schools.
   d. Middle school academic achievement determines the high school; to enter colleges and universities, students must score well on a variety of college entrance exams prepared by each college or university.
   e. At the college and university level, there is an absence of female students and professors.
2. In Bosnia, in spite of the peace agreement among the Muslims, Eastern Orthodox Serbs, and Roman Catholic Croats, animosities remain.
   a. Each group has taken control over how its children are educated; in certain classes, students are segregated into ethnically distinct classrooms.
   b. Students are taught different versions of history, language, and art depending on their ethnic identity.
   c. Schools with integrated classrooms have increase in conflicts among the different ethnic groups.

III. SOCIOLOGICAL PERSPECTIVES ON EDUCATION
A. Functionalists view education as one of the most important components of society.
1. Education serves five major **manifest functions** – open, stated, and intended goals or consequences of activities within an organization or institution:
   a. Socialization
   b. Transmission of culture
   c. Social control
   d. Social placement
   e. Change and innovation.
2. Education has at least three **latent functions** – hidden, unstated, and sometimes unintended consequences of activities within an organization or institution:
   a. Restricting some activities
   b. Matchmaking and production of social networks
   c. Creation of a generation gap.

B. According to conflict theorists, schools perpetuate class, race-ethnic, and gender inequalities as some groups seek to maintain their privileged position at the expense of others.
  1. Cultural Capital and Class Reproduction: education is a vehicle for reproducing existing class relationships.
     a. According to Pierre Bourdieu, children have less chance of academic success when they lack **cultural capital** – social assets that include values, beliefs, attitudes, and competencies in language and culture.
     b. Children from middle- and upper-income families are endowed with more cultural capital than children from working class and poverty level families.
  2. Tracking and Social Inequality:
     a. Class reproduction also occurs through standardized tests, ability grouping, and tracking – the assignment of students to specific courses and educational programs based on their test scores, previous grades, or both.
  3. Social Class and The Hidden Curriculum:
     a. The **hidden curriculum** is the transmission of cultural values and attitudes, such as conformity and obedience to authority, through implied demands found in rules, routines, and regulations of schools.
     b. Lower class students may be disqualified from higher education and the credentials needed in a society that emphasizes **credentialism** – a process of social selection in which class advantage and social status are linked to the possession of academic qualification.
     c. Credentialism is closely related to *meritocracy* – a social system in which status is assumed to be acquired through individual ability and effort.
  4. Gender Bias in Schools
     a. Studies show that gender bias pervades the overall academic environment and cuts across lines of race and class.
     b. Teachers tend to encourage boys to be the problem solvers; through reading materials, classroom activities, and treatment by teachers and peers, female students learn that they are less important than male students.
     c. Some analysts suggest that girls receive subtle cues from adults that lead them to attribute success to their intelligence and ability. Conversely, girls attribute failure to lack of ability, while boys attribute failure to lack of effort.
  5. Ethnicity, Language and Hidden Curriculum
     a. When teaching English as a second language, instructors frequently find that they not only teach children English, but social skills as well.
     b. The children are exposed to a wide range of beliefs, values, attitudes and behavior expectations that are not directly related to their subject matter.
     c. Conflict theorists argue that inequality is structurally produced and reproduced by formal and informal socialization processes in schools and other educational settings.

C. Symbolic Interactionist Perspectives on Education
  1. Labeling and the Self-Fulfilling Prophecy
    a. For some students, schooling may become a self-fulfilling prophecy – an unsubstantiated belief or prediction that results in behavior which makes the originally false belief come true.
  2. Using Labeling Theory to Examine the IQ Debate
    a. The experiment of Rosenthal and Jacobson demonstrated the effect of teacher expectation upon the resultant IQ scores.
    b. If a teacher (as a result of stereotypes based on the relationship between IQ and race) believes that some students of color are less capable of learning, that teacher (sometimes without even realizing it) may treat them as if they were incapable of learning.
  3. Education and Labeling
    a. IQ testing has resulted in labeling of students (e.g., African American and Mexican American children have been placed in special education classes on the basis of IQ scores when they could not understand the tests).
    b. A *self-fulfilling prophecy* also can result from labeling students as gifted. When some students are labeled as better than others, they may achieve at a higher level because of the label, or they may face discrimination from others (e.g., Asian Americans as super intelligent).
D. Postmodern Perspectives
  1. Postmodernists often highlight differences and inequality in society; therefore education is characterized by its permeability.
  2. Urbanity is reflected in multicultural and anti-bias curricular introduced in early childhood education.
  3. Autonomy is evidenced in policies such as voucher systems.
  4. In higher education, the permeability is exemplified by "educational consumption," the offering of "high-tech" or "wired" campuses, virtual classrooms, and student centers offering extravagant activities and services.
  5. According to Ritzer, "McUniversity" can be thought of as a means of educational consumption that allows students to consume educational services and eventually obtain "goods" such as degrees and credentials.
IV. INEQUALITY AMONG ELEMENTARY AND SECONDARY SCHOOLS
  A. Inequality in Public vs. Private Schools
    1. Almost 90 percent of U. S. elementary and secondary students are educated in public schools; about 9.5 percent of all students are educated in low-tuition private schools, primarily Catholic; 1.5 percent of all students attend high-tuition private school.
    2. Private schools are perceived as better than public schools but, according to some social analysts, there is little to substantiate this claim.
  B. Unequal Funding of Public Schools
    1. Most educational funds are derived from local property taxes and state legislative appropriations.
    2. Children living in affluent suburbs often attend relatively new schools and have access to the latest equipment which students in central city schools and poverty-ridden rural areas lack.

C. Inequality within Public School Systems
   1. Using the New York City public schools as a case study, the research shows that schools are stratified into higher-, middle-, and lower-tiered schools.
   2. Higher-tier schools are highly specialized; admission is limited based upon competitive entrance examination; dropout rates are low. A large number of graduates attend prestigious universities.
   3. Middle-tier schools, located in fairly well-integrated neighborhoods are often referred to as "ed op"(educational option) schools; they offer training for specific careers; some serve as magnet schools, offering a specialized curriculum and enrolling high-achieving students from other neighborhoods.
   4. Lower-tier high schools are very large, overcrowded, and located in highly segregated, deteriorating, and violent neighborhoods. Students have the highest dropout rate, the lowest graduate rates, the worst scores on standardized tests, the poorest attendance patterns, and the worst statistics on assaults and possession of weapons.
D. Racial Segregation and Resegregation of Schools
   1. In many areas of the U.S., schools remain segregated or have become resegregated after earlier attempts at integration failed.
      a. Four decades after the 1954 U.S. Supreme Court ruled in *Brown v. The Board of Education* of Topeka, Kansas, that segregated schools were unconstitutional, racial segregation remains.
      b. Racial segregation is increasing in many school districts and efforts to bring about desegregation or integration have failed in many school districts.
   2. Racially segregated schools often have low retention rates, students with below-grade level reading skills, high teacher-student ratios, less qualified teachers, and low teacher expectations.
      a. Even in supposedly integrated schools, tracking and ability grouping may produce resegregation at the classroom level.
      b. African American and white achievement differences increase with every year of schooling; thus, schools may reinforce rather than eliminate the disadvantages of race and class.
V. PROBLEMS WITHIN ELEMENTARY AND SECONDARY SCHOOLS
   A. School Discipline and Teaching Styles
      1. Leading discipline problems include violence, drug abuse, suicide, robbery, assault, and other forms of aggressive behavior.
      2. Most states now prohibit the use of corporal punishment and some school districts prohibit various forms of disciplinary actions.
      3. Teaching and disciplinary styles may differ across racial and ethnic lines; some argue that color should be noticed.
      4. Jean Anyon suggests that structural inequalities need to be addressed in order to achieve educational reform.

B. Bullying, Teasing, and Sexual harassment
   1. 1. According to research on behalf of the AAUW, 83 percent of girls and 79 percent of boys report experiencing harassment.
   2. Seventy-six percent of students have experienced nonphysical harassment, while 58 percent have experienced physical harassment.
   3. Harassed girls report being negatively affected by harassment more often than do boys.
   4. Substantial numbers of students fear that they will be sexually harassed or hurt in school; for those who are constantly picked on, ridiculed, or harassed, school becomes sheer torture.
C. Dropping Out
   1. Dropout rates vary by race/ethnic, class and regional differences.
   2. Explanations for the group having highest rate of dropouts (Hispanics, or Latino/as) vary:
      a. Labeled as troublemakers, they actually become "dropouts."
      b. They view school as a waste of time.
      c. They may be skeptical about the value of school
   3. From the research of Romo and Falbo, the most common reasons given for dropping out included: lack of interest in school, serious personal problems, serious family problems, poor grades, and alcohol and/or drug problems.
   4. The same research found that a growing number of students are taking the GED (General Equivalent Diploma) exam rather than graduating from high school, which puts many in dead-end situations.
VI. SCHOOL SAFETY AND SCHOOL VIOLENCE
A. Problems such as bullying, harassment, and high school dropout rates are a major concern.
   1. Schools should provide safe and supportive environment so students and teachers can feel secure.
   2. Many officials in schools to colleges and universities are focusing on eliminating problems of safety. Some are creating anger management and peer mediation programs, and other programs such as National Education Association's Safe Schools Programs.
   3. Studies now show that U.S. schools are among the safest places for young people.
   4. Weapons-scanning metal detectors, security guards, magnetic door locks are commonplace.
B. Violence and fear of violence continue to be pressing problems in schools throughout the U.S.
   1. Numerous school, college and university students, faculty, and other victims have been subjected to acts of violence and multiple deaths.
   2. Many students must learn in an academic environment that is similar to a maximum security prison.
   3. Gun control advocates are calling for greater control over the licensing and ownership of firearms, among several other suggestions.

VII. OPPORTUNITIES AND CHALLENGES IN COLLEGES AND UNIVERSITIES
  A. Opportunities and Challenges in Community Colleges
     1. Access to colleges and universities is determined not only by a person's prior academic record but also by the ability to pay.
     2. Community colleges are one of the fastest-growing areas of U.S. higher education today.
     3. They are more affordable and offer more flexible class schedules than the typical four-year-college.
     4. Limited resources are one of the major problems facing the community college today.
  B. Opportunities and Challenges in Four-Year Colleges and Universities
     1. They offer a variety of degrees, ranging from the bachelor's degree to the master's to the doctorate to the professional degree.
     2. Most provide a liberal education by offering a general education curriculum.
     3. Major problems include cost of higher education, problems with students completing a degree program, racial and ethnic differences in enrollment, and lack of faculty diversity.
  C. The Soaring Cost of a College Education
     1. The cost of attending private and public institutions has increased dramatically over the past decade; higher than the overall rate of inflation.
     2. This high cost of a college education reproduces the existing class system: students who lack money may be denied access to higher education, and those who are able to attend college tend to receive different types of education based on their ability to pay.
  D. Racial and Ethnic Differences in Enrollment
     1. Latino/a college enrollment increased from about 5.7 percent to 6.0 percent between 1990 and 2001, accounting for 9.8 percent of all college students, but 14.4 percent of the total U.S. population.
     2. African American enrollment increased somewhat between 1990 and 2001, and remained steady at about 11 percent. Gender differences are evident: women accounted for more than 63 percent of all African American College students in 2001.
     3. Native Americans remained at about 00.9 percent of enrollment from the 1970s to the 2000s; however, two-year community colleges – known as tribal colleges – on reservations have experienced an increase in Native American student enrollment.
     4. At the doctorate degree level, only 18 percent of the doctorates awarded in 1999-2000 were to minority group members.
  E. Lack of Faculty Diversity
     1. African Americans, Latino/as, Asian Americans, and other people of color accounted for only about 12 percent of all faculty members.
     2. Minority faculty members are less likely to be full professors with tenure.
     3. Women are underrepresented at the level of full professor and overrepresented at the assistant professor and instructor levels.

4. Minority faculty members are more likely than their white counterparts to be employed at two-year colleges; for example, 36 percent of Native American faculty members are employed at tribal colleges located on the reservations. Likewise, 39 percent of Latino/a faculty members are at two-year colleges; only 23 percent of white (non-Latino/a) faculty members are employed at two-year institutions.
   5. The lack of Native Americans and Latino/as as college students, especially in Ph.D. programs, perpetuates the problem of diversity among faculty members. Asian American scholars are the exception to this problem; they are more likely to have a terminal degree in their field and to be employed at a research university.
   F. The Debate over Affirmative Action
   1. In 2003, the Supreme Court ruled in *Grutter v. Bollinger* (involving admissions policies of the University of Michigan's law school) and in *Gratz v. Bollinger* (involving the undergraduate admissions policies of the same university) that race can be a factor for universities in shaping their admissions programs, but only within carefully defined limits.
   2. The debate over affirmative action continues; there are more questions on how to achieve greater representation of all racial and ethnic groups, gender and nationality among students, faculty, and administrators.
VIII. FUTURE ISSUES AND TRENDS IN EDUCATION
   A. Academic Standards and Functional Illiteracy
   1. Some social analysts cite the rate of **functional illiteracy**, which is the inability to read and/or write at the skill level necessary for carrying out everyday tasks.
   2. Estimates are that 56 percent of adult Latino/as are functionally illiterate in English, compared with 44 percent of adult African Americans and 16 percent of adult (non-Latino/a) whites.
   3. Most suggestions for reducing both illiteracy and functional illiteracy focus on school reforms such as: more testing of students and teachers, increasing the requirements for high school graduation, and increasing the number of school days per year.
   4. Illiteracy is a global problem as well as a national one; low rates of literacy for women in lower-income nations reflect how different nations view education and gender.
   B. The Debate over Bilingual Education
   1. The issue of bilingual education is not a new one.
      a. During the late nineteenth and early twentieth centuries, children of U. S. immigrants often received classroom instruction from a teacher who spoke the children's native language.
      b. During World War I, members of white ethnic groups, in a demonstration of loyalty to the U. S., advocated that their children should be taught in English.
      c. Little bilingual education was offered until after World War II when a large influx of Mexican and Asian immigrants arrived in the United States, and schools enrolled large numbers of students who did not speak English.

d. Congress mandated in 1968 that public schools must provide non-English speaking students with *bilingual education* – instruction in both a non-English language and in English.

e. The U. S. Supreme Court decision, *Lau v. Nichols* (1974) affirmed the responsibility of the states and local districts for providing appropriate education for minority-language students.

f. Critics of TBE (transitional bilingual education) program believe that TBE programs are less effective than classes that are conducted primarily in English.

g. Supporters of TBE believe that these programs are effective in helping children build competency both in their native language and in English.

C. Equalizing Opportunities for Students with Disabilities

1. *Disability* is regarded as any physical and/or mental condition that limits students' access to, or full involvement, in school life.

2. The Americans with Disabilities Act of 1990 requires schools and government agencies to make their facilities, services, activities, and programs accessible to people with disabilities.

3. The Individuals with Disabilities Education Act (IDEA) mandates that students with disabilities receive free and appropriate education.

4. Many schools have attempted to *mainstream* children with disabilities by *inclusion programs* under which the special education curriculum is integrated with the regular education program, and with *individualized education plans* that provide annual education goals for each child

D. School Vouchers

1. These give parents the choice of a school for the child.

2. Public funds are provided to students to pay for their choice of school, which can be public or private schools.

3. Research findings of a voucher program in Cleveland, Ohio found that parents with children in a voucher school were more satisfied with the voucher system than were the parents having their children in public schools.

E. Charter Schools and "For Profit" Schools

1. These types of schools operate under a charter contract negotiated by the schools' organizers and a sponsor who oversees the provisions of the contract.

2. Charter schools are said to be less bureaucratic and more responsive to the student's needs than public schools in large school districts.

3. Critics argue that these schools take money away from conventional public schools.

4. Proponents argue that charter schools provide a large number of minority students a higher quality education than they would in public schools.

5. The "contracting out" hiring for-profit companies is a more recent approach; companies are contracted to operate public schools. Nationwide, most of the contracting is done by three private companies; critics are concerned about the privatization of education, which they believe undermines the public schools system.

F. Home Schooling
   1. This is a choice by parents who seek to avoid the problems of public school while providing a quality education for their children.
   2. In the 2000s, estimates of have-schooled children rage from 850,000 to 1.7 million in grades K through the 12th grade.
   3. Home-schoolers are more likely than other students to live with two or more siblings in a two-parent family, with only one parent working outside the home.
   4. Typically, these parents are better educated, on average, than other parents, but their income is about the same. Researchers have found that boys and girls are equally likely to be home-schooled.
   5. Proponents believe that their children are receiving a better education at home for various reasons; critics question the knowledge and competence of the typical parent to educate their own children at home.

IX. CONCLUDING THOUGHTS
   1. The education needed for the future is different from that of the past.
   2. Creating a system that encompasses vast cultural differences, but yet respects and preserves those differences, presents a challenge.
   3. President George W. Bush signed into law the "No Child Left Behind Act of 2001," which requires that schools must be accountable for students' learning and sets forth steps for producing an accountable education system.
   4. Will these changes make a difference? How critical are other pressing social problems, such as harassment, and guns and violence in the schools?
   5. Ultimately, the institution of education must be maintained and enhanced in various ways. Spending larger sums of money on education does not guarantee that many of the problems of education will be resolved.

ANALYZING AND UNDERSTANDING THE BOXES
After reading the chapter and studying the outline, re-read the boxes and write down key points and possible questions for class discussion.

Sociology in Everyday Life: How Much Do You Know About U. S. Education?

Key Points:

Discussion Questions:

1.

2.

3.

Framing Education in the Media: "A Bad Report Card" – How the Media Frame Stories About U. S. Schools

Key Points:

Discussion Questions:

1.

2.

3.

You Can Make a Difference: Reaching Out to Youth: College Student Tutors

Key Points:

Discussion Questions:

1.

2.

3.

Sociology and Social Policy: The Ongoing Debate Over School Vouchers

Key Points:

Discussion Questions:

1.

2.

3.

PRACTICE TESTS

MULTIPLE CHOICE QUESTIONS

Select the response that best answers the question or completes the statement:

1. _____ is the social institution responsible for the systematic transmission of knowledge, skills, and cultural values within a formally organized structure.
    a. Religion
    b. Mass media
    c. The government
    d. Education

2. _____ refers to providing free, public schooling for wide segments of a nation's population.
    a. Mass education
    b. Public education
    c. Contemporary education
    d. Federal education

3. During the Middle Ages, the first colleges were developed under the auspices of the
    a. privileged classes.
    b. military.
    c. Catholic church.
    d. national government.

4. The core curriculum consists of all the following courses **except**:
    a. mathematics
    b. English
    c. music education
    d. social sciences

5. According to the text, all of the following statements regarding contemporary U.S. education are true, **except**:
    a. U.S. schools are teaching a more limited number of topics so that they can focus on the "basics."
    b. U.S. schools are attempting to meet the needs of post-industrial society.
    c. Controversy exists over what should be taught.
    d. Many educators believe that their job description now encompasses numerous tasks that previously were performed by other social institutions.

6. According to sociologist _____, education is crucial for promoting solidarity and stability in society.
    a. Emile Durkheim
    b. Pierre Bourdieu
    c. Amitai Etzion
    d. Max Weber

7. Which of the following is a manifest function of education?
    a. creation of a generation gap
    b. restricting some activities
    c. matchmaking and production of social networks
    d. social control

8. _____ functions are hidden, unstated, and sometimes unintended consequences of activities within an organization or institution.
    a. Manifest
    b. Dormant
    c. Latent
    d. Covert

9. Sociologist _____ has suggested that students come to school with differing amounts of cultural capital.
    a. Emile Durkheim
    b. Pierre Bourdieu
    c. Jeannie Oakes
    d. Max Weber

10. The assignment of students to specific courses and educational programs based on their test scores, previous grades, or both is known as:
    a. tracking
    b. the hidden curriculum
    c. equitable assessment
    d. class reproduction

11. According to the _____ perspective, the hidden curriculum affects working class and poverty level students more than it does students from middle- and upper-income families.
    a. functionalist
    b. conflict
    c. symbolic interactionist
    d. feminist

12. The _____ focus on the classroom communication patterns and educational practices, such as labeling, that affect students' self-concept and aspirations.
    a. conflict theorists
    b. functionalists
    c. interactionists
    d. credentialists

13. A teacher who believes (as a result of stereotypes based on the relationship between IQ and race) that some students of color are less capable of learning and treats them accordingly may contribute to:
    a. a self-fulfilling prophecy
    b. the labeling process
    c. tracking
    d. the hidden curriculum

14. About _____ of the U.S. elementary and secondary students are educated in public schools.
    a. 54 percent
    b. 65 percent
    c. 80 percent
    d. 90 percent

15. Head Start programs are funded by
    a. the federal government.
    b. state legislative appropriations.
    c. local property taxes.
    d. special revenue programs such as a state lottery.

16. In public schools in the 1940s, some of the leading discipline problems included
    a. getting out of place in line.
    b. not putting paper in the trash can.
    c. talking without permission
    d. All of these choices were a part of discipline problems.

17. The National Literacy Act of 1991 defines _____ as an individual's ability to read, write, and speak English, and compute and solve problems at levels of proficiency necessary to function on the job and in society.
    a. functional illiteracy
    b. minimum literacy
    c. literacy
    d. functional literacy

18. In _____, the Supreme Court affirmed the responsibility of the states and local districts for providing appropriate education for minority-language students.
    a. *Brown v. Board of Education*
    b. *Lau v. Nichols*
    c. *Hopwood v. State of Texas*
    d. *Bakke v. The University of California at Davis*

19. According to some analysts, in entering college, _____ are more likely enroll in clerical classes, while _____ are more likely to enroll in computer program courses.
    a. Boys, girls
    b. Teachers, principals
    c. Girls, boys
    d. Japanese students, African American students

20. Which of the following statements regarding higher education is true?
    a. The enrollment of low-income students has increased since the 1980s.
    b. There has been an increase in scholarship funds over the past decade.
    c. College students are stratified according to their ability to pay.
    d. Latino/as earned more college doctorates than African Americans in the past two decades.

21. _____ is the process by which children and recent immigrants become acquainted with the dominant cultural beliefs, values, norms, and accumulated knowledge of a society.
    a. Cultural transmission
    b. Education
    c. Socialization
    d. Cultural diffusion

22. _____ societies existed before the invention of reading and writing. These societies have no written language and are characterized by very basic technology and a simple division of labor.
    a. Modern
    b. Preindustrial
    c. Preliterate
    d. Industrial

23. _____ is learning that occurs in a spontaneous, unplanned way.
    a. Spontaneous education
    b. Informal education
    c. Traditional education
    d. Latent education

24. In preliterate societies, parents and other members of a group provide information about how to gather food, find shelter, make weapons and tools, and get along with others. This is an accomplished through direct _____ education.
    a. formal
    b. traditional
    c. macrolevel
    d. informal

25. In many schools in the U.S., faculty, staff, and students are required to
    a. wear a photo ID around their necks.
    b. pass through a metal detector.
    c. have all backpacks inspected.
    d. All of these choices are correct.

26. Many schools have attempted to _____ children with disabilities by providing inclusion programs under which the special education curriculum is integrated with the regular education program.
    a. "mainstream"
    b. "include"
    c. "individualize"
    d. "exclude"

27. _____ means that children with disabilities work with a wide variety of people over the course of a day.
    a. Mainstreaming
    b. Inclusion
    c. Submersion
    d. Individualized education

28. Children with disabilities may interact with their regular education teacher, the special education teacher, a speech therapist, an occupational therapist, a physical therapist, and a resource teacher, depending on the child's individual needs. This is a(n) example of:
    a. mainstreaming.
    b. individualized education.
    c. inclusion.
    d. exclusion.

29. Some elected officials, business leaders, and educators have shifted their focus to other ways of improving education. Some are advocating _____ programs, which give parents the choice of what school their child will attend.
    a. social
    b. magnet
    c. school tracking
    d. school voucher

30. Most suggestions for reducing both illiteracy and functional illiteracy have focused on school reforms, such as all of the following, **except**
    a. more testing of students and teachers.
    b. decreasing the number of school days per year.
    c. increasing the requirements for high school graduation.
    d. increasing the number of school days per year.

31. In 2003, the U.S. Supreme Court ruled in _____ and _____ that race can be a factor for universities in shaping their admission programs within carefully defined limits.
    a. *Brown v. Board of Education, Lau v. Nicholas*
    b. *Hopwood v. State of Texas, Lau v. Nichols*
    c. *Bukke v. U.C. Board, Hopwood v. Texas*
    d. *Grutter v. Bollinger, Gratz v. Bollinger*

32. Along with other provisions, the _____ requires schools and government agencies to make their facilities, services, activities, and programs accessible to people with disabilities.
    a. Individuals with Disabilities Education Act (IDEA)
    b. Americans with Disabilities Act (ADA) of 11990
    c. Handicapped Americans Recovery Program (HARP)
    d. Road to Rehabilitation Reform Law (RRR-L)

33. The law that specifically covers the treatment of children with disabilities is the _____, which mandates that students with disabilities must receive free and appropriate education.
    a. Americans with Disabilities Act (ADA)
    b. Individuals with Disabilities Education Act (IDEA)
    c. Americans with Physical Handicaps Act (APHA)
    d. Individuals with Special Needs Act (ISNA)

34. In the country of _____, national achievement tests are administered by the government, and students must pass them in order to advance to the next level of education.
    a. Japan
    b. United States
    c. Bosnia
    d. England

35. The belief in compulsory education in Germany can be traced back to the theologian and professor _____.
    a. Martin Luther
    b. Max Weber
    c. Karl Menniger
    d. Frederick Engels

36. Among parents whose children attend private secondary schools, an important factor for a majority (51 percent) was the fact that
    a. there is a greater emphasis on religion.
    b. there is a greater emphasis on academics.
    c. there is a greater emphasis on money.
    d. there is a greater emphasis on ethical standards.

37. According to the text's discussion of unequal funding as a source of inequality in education,
    a. of the small proportion of school funding that comes from the federal government, most funds are earmarked for special programs that specifically target disadvantaged students or students with disabilities.
    b. most educational funds are derived from state and federal income taxes.
    c. the property tax base for central city schools has continued to grow in most regions.
    d. recent redistribution of funds has made many schools' resources more equitable that in the past.

38. In the United States, the free public school movement was started in 1848 by _____, who stated that education should be the "great equalizer."
    a. Noah Webster
    b. Horace Mann
    c. Thomas Jefferson
    d. Theodore Roosevelt

39. With the rapid growth of capitalism and factories during the _____, it became necessary for workers to have basic skills in reading, writing, and arithmetic.
    a. Industrial Revolution
    b. Enlightenment
    c. Renaissance
    d. Middle Ages

40. During the _____, the focus of education shifted to the importance of developing well-rounded and liberally educated people.
    a. Industrial Revolution
    b. Enlightenment
    c. Renaissance
    d. Middle Ages

TRUE/FALSE QUESTIONS

1. During the Middle Ages, the first colleges and universities were developed under the auspices of the government.
   T   F

2. *Jukus* are cram schools available only to middle and high school Japanese students.
   T   F

3. In public schools in Bosnia, class content is taught on a highly-structured centralized government approach.
   T   F

4. Manifest functions in education include teaching specific subjects, such as science, history, and reading.
   T   F

5. Meritocracy is the process of social selection in which class advantage and social status are linked to the possession of academic qualifications.
   T   F

6. Labeling students as "gifted" can result in a self-fulfilling prophecy.
   T   F

7. The issue of IQ and race/ethnicity originated with the highly controversial book by Herrnstein and Murray.
   T   F

8. Labeling and self-fulfilling prophecy are unique to U.S. schools.
   T   F

9. According to the postmodernists, higher education is not affected by the "McUniversity" phenomenon.
   T   F

10. Japan has the most decentralized system of public education on the modern industrialized countries.
    T   F

11. Most public educational funds come from the federal government.
    T   F

12. In the New York City school system, any student can attend the higher-tier high schools provided that she/he is a legal resident of the city.
 T   F

13. Most states now prohibit the use of corporal punishment of students in their public schools.
 T   F

14. Boys seldom report harassment from their classmates.
 T   F

15. One major problem of the voucher system is that poor children cannot transfer out of their neighborhood inner-city, lower-tier public school.
 T   F

16. About 11 percent of people between the ages of sixteen and nineteen leave school before earning a high school diploma.
 T   F

17. According to Box 16.2, one of the less frequent media approaches to reporting on education is an inside-the-classroom view of what is actually going on in schools.
 T   F

18. College students tutoring students in local public school appears to be, according to Box 16.3, a lose-lose situation for everyone.
 T   F

19. The original idea of school vouchers, according to Box 16.4, was to prevent a student from entering a private school.
 T   F

20. According to Box 16.4, all states in the U.S. have now accepted school voucher programs.
 T   F

FILL-IN-THE-BLANK QUESTIONS

1. _____ refers to providing free public schooling for wide segments of a nation's population.

2. Intended goals or consequences of education such as socialization, transmission of culture, social control, social placement, and change and innovation are

   _____ .

3. _____ is the assignment of students to specific courses and educational programs based on their test scores, previous grades, or both.

4. About _____ percent of people between the ages of fourteen and twenty-four have left school before earning a high school diploma.

5. According to _____, certain racial-ethnic groups differ in average IQ and are likely to differ in "intelligence genes" as well.

6. _____ offer specialized curriculum that focuses on areas such as science, music, or art, and enroll high-achieving students from other neighborhoods.

7. In the United States, the free public school movement was started in 1848 by _____.

8. Only _____ of white faculty members are employed at two-year institutions of higher learning.

9. Some analysts believe that _____ has fulfilled its purpose and is no longer necessary, other analyst believe that the playing field remains uneven for many white women and people of color.

10. _____ means that children with disabilities work with a wide variety of people; over the course of a day, children may interact with their regular education teacher, the special education teacher, a speech therapist, an occupational therapist, a physical therapist, and a resource teacher, depending on the child's individual needs.

11. The German word _____ is both a German word and a German creation that has been widely adopted around the world.

12. In Germany, the _____ is the highest academic secondary school providing students with a liberal education.

13. About one-half of the nation's undergraduates are educated in _____ colleges.

14. A(n) _____ prophecy may result from the labeling of students by teachers and administrators.

15. Many schools have attempted to _____ children with disabilities by providing inclusion programs with the regular education program.

SHORT ANSWER/ESSAY QUESTIONS

1. Is education a vehicle for decreasing social inequality or is it that education reproduces existing class relationships? How might computer technology influence schools?
2. Compare and contrast the U.S. public schools with those of other countries.
3. What are the major problems in public schools today? In higher education?
4. What do you think are the most significant changes in education over the past twenty years? What changes would you predict for the future?
5. Based on the postmodernist perspectives on education, what educational services and goods will students in both early childhood and higher education consume in the future? What will be the major attractions in the future educational setting?

STUDENT CLASS PROJECTS AND ACTIVITIES

1. A 2000 issue of *Time* magazine provides a special report on the status of public education. *Time* provides an analysis of America's best schools and not-so-good schools. Read the report and respond to the following questions: (1) What are some of the criteria that make for a good school? (2) Do inner-city schools support the voucher system of education? (3) What standards do many of the states advocate? (4) How should "Johnny" learn to read? Is the method of phonics the best program? Or is the "whole-language" method of learning to read the better method? (5) Critique the "success" stories of three schools in three different places. Where are the schools? Why are they considered to be successful?

2. In the same issue, *Time* examines a major issue, "Where does the money go?" This article tracks the flow of cash through a single large urban district. Read the article and answer the following questions: (1) Do some urban schools fail to because they don't have the money to do better? (2) What is meant by the "dance of the lemons?" (3) What are some of the advantages/disadvantages of educational localism? (4) How can "bad" schools be fixed? (5) What are the roles of the federal government and external governmental control? (6) Critique the article.

3. Research the topic of public education in several high-income countries such as France, Japan, New Zealand, the Netherlands, Sweden, England, Norway, Canada, or Germany. In the writing up of your paper, include the following information: (1) a description of their educational system; (2) a description of their specific or unique feature; (3) a description of their curriculum; (4) a description of some of the strengths and weaknesses of their program; (5) a critique of their programs: Which do you think work well? Which do not? (6) any other information that will enhance our understanding of their education; (7) Provide a bibliography of your references; (8) Turn your paper in to your instructor at the appropriate time.

4. The U.S. Congress passed the "No Child Left Behind Act of 2001." As stated in the text, list the major provisions of this act. Next, select 6 states, 2 in the North/Northeastern states, 2 in the South/Southwestern states, and 2 from the West/Midwestern states. From your selected states, research their response to this 2001 act, critique each of the present states, providing vital information from each chosen state. Follow your professor's requirements for this project. Write up your report and submit your findings to your professor.

INTERNET ACTIVITIES

1. The **National Education Association** (NEA), **http://www.nea.org/**, is currently the largest teachers' organization in the United States. Its purpose is to advance the interests of the teaching profession, push for teaching reform, and press for quality education in the United States. When you visit their site, see if you can analyze their core concerns.

2. The **NEA** also addresses the issue of low-performing schools and what can be done to increase their performance: **http://www.nea.org/priorityschools/priority.html**. Analyze this plan using your Sociological Imagination and then brainstorm to create a list of obstacles that might prevent this plan from happening.
3. The **American Federation of Teachers**, **http://www.aft.org/index.html**, (affiliated with the labor union the AFL-CIO) is one of the largest teachers' unions in the United States. Like the NEA, the AFT tries to represent teacher interests and further the quality of education. The AFT, like the NEA, sponsors research and reports on education. Explore and discover the current concerns of this teachers' group.
4. The NCES is the **National Center for Education Statistics**: This organization is the primary federal entity that collects and analyzes educational data for the United States and other countries. This site gives you fast facts topics like the transition from high school to college, various statistics on educational systems, earnings of college graduates, and much more. Search this site: **http://nces.ed.gov/index.asp**.
5. To examine available information on home schooling, you may wish to visit the following sites. These kinds of sites provide information about the home school movement and provide resources for homeschoolers: **http://homeschooling.about.com/mbody.htm**, **http://www.home-school.com**, and **http://www.hslda.org/splash/default.asp**.

INFOTRAC COLLEGE EDITION EXERCISES

Visit the **InfoTrac College Edition** website at: **http://www.wadsworthmedia.com/webtutor/infotrac.htm**. You will arrive at a screen that enables you to search topics.

1. More and more states are turning to **standardized tests** to assess proficiency as a measure of school accountability and to determine whether students should pass to the next grade. What are some the costs and benefits of this approach?
2. **Home schooling** has become increasingly popular. What kinds of research questions are sociologists and educators asking? Compose a list of five research questions from articles listed on InfoTrac.
3. In the aftermath of Columbine and other school shootings, parents, students, and teachers are increasingly fearful of **school violence**. Investigate this issue with respect to its incidence and social policy implications.
4. Many states allow parents vouchers that permit them to send their child to a school outside their district. Investigate the **school voucher** system and see if you can learn what are the costs and benefits.
5. **Charter schools** operate in accordance with a "bottom line" that is specified by their charter. In exchange for meeting their educational goals, they are exempt from meeting traditional state rules and restrictions. Research the trend towards charter schools. What makes one school succeed and another fail?
6. There is currently an explosion of **on-line education** at the college level. On-line courses offer convenience to the student, but at the same time, does not offer the same type of classroom interaction, even if the professor and students interact in a "chat room." Assess this trend and its significance for American education.

SOLUTIONS

MULTIPLE CHOICE QUESTIONS

1. D, p. 520
2. A, p. 522
3. C, p. 521
4. C, p. 522
5. A, p. 522
6. A, p. 526
7. D, p. 527
8. C, p. 527
9. B, p. 529
10. A, p. 520
11. B, p. 531
12. C, p. 533
13. A, p. 533
14. D, p. 536

15. A, p. 537
16. D, p. 540
17. C, p. 549
18. B, p. 550
19. C, p. 532
20. C, p. 546
21. A, p. 520
22. D, p. 520
23. B, p. 520
24. D, p. 520
25. D, p. 543
26. A, p. 551
27. B, p. 551
28. C, p. 551

29. D, p. 551
30. B, p. 550
31. D, p. 549
32. B, p. 551
33. B, p. 551
34. D, p. 536
35. A, p. 524
36. B, p. 536
37. A, p. 537
38. B, p. 521
39. A, p. 521
40. C, p. 521

TRUE/FALSE QUESTIONS

1. F, p. 521
2. F, p. 523
3. F, p. 523
4. T, p. 526
5. F, p. 532
6. T, p. 533
7. F, p. 534

8. F, p. 534
9. F, p. 535
10. F, p. 536
11. F, p. 537
12. F, p. 537
13. T, p. 540
14. F, p. 542

15. F, p. 552
16. T, p. 542
17. T, p. 541
18. F, p. 552
19. F, p. 554
20. F, p. 554

FILL-IN-THE-BLANK QUESTIONS

1. Mass education, p. 522
2. manifest functions, pp. 526-527
3. Tracking, p. 530
4. eleven percent, p. 542
5. Hernstein and Murray, p. 534
6. Magnet school, p. 537
7. Horace Mann, p. 521
8. twenty-three percent, p., 547

9. affirmative action, p. 548
10. Inclusion, p. 551
11. Kindergarten, p. 524
12. gymnasium, p. 525
13. community, p. 544
14. self-fulfilling, p. 533
15. mainstream, p. 551

17
RELIGION

BRIEF CHAPTER OUTLINE

CHAPTER SUMMARY

Religion is one of the most significant social institutions in society. It can be a highly controversial topic; one group's deeply-held or cherished religious practices may be a source of irritation to another. It is a source of both stability and conflict not only in the United States but throughout the world. **Religion** is a system of beliefs, symbols, and rituals, based on some sacred or supernatural realm that guides human behavior, gives meaning to life, and unites believers into a moral community. According to Durkheim, sacred refers to the extraordinary or supernatural aspects of life, while those things that people do not set apart as sacred are referred to as **profane**. Although religion seeks to answer important questions such as why we exist, why people suffer and die, and what happens when we die, increases in scientific knowledge have contributed to

secularization – the process by which religious beliefs, practices, and institutions lose their significance in sectors of society and culture. According to functionalists, religion provides meaning and purpose to life, promotes social cohesion and a sense of belonging, and provides social control and support for the government. The conflict perspective suggests that religion can have negative consequences: the capitalist class uses religion as a tool of domination to mislead workers about their true interests and retard social change. However, Max Weber believed that religion could be a catalyst for social change. Symbolic interactionists examine the meanings that people give to religion and the meanings they attach to religious symbols in their everyday life. Religions vary around the world. *Hinduism*, one of the world's oldest religions, is complex and diverse, comprising numerous varieties of beliefs and practices. *Buddhism* first emerged in India about 2,500 years ago. It is based on the teachings of Buddha, who perceived that life is suffering and pain; the answer to world's problems is achieved through mediation which leads to personal transformation and a spiritual existence. *Confucianism*, based upon the teachings of Confucius, was the official religion of China until the Communist takeover and establishment of the People's Republic of China. Order in human relationships and a strict code of moral conduct are some of the basic concepts. Despite centuries of religious hatred and discrimination, *Judaism* persists as one of the world's most influential religions. The three key components of Judaism are: God (the deity), Torah (God's teachings), and Israel (the community or holy nation). *Islam* is based on the teachings of Muhammad, a prophet, not a deity; followers must adhere to the five Pillars of Islam. Almost one-third of the world's population is *Christian*, the dominant world religion today. The Christian deity is a sacred Trinity comprised of God the Father, Jesus the Son, and the Holy Spirit. Contemporary religious organizations may be categorized as **ecclesiae**, **churches**, **denominations**, **sects**, and **cults**. Religion in the United States is diverse; pluralism and religious freedom are among the cultural values most widely espoused. Social science research continues to identify significant patterns between social class and religious denominations. With the increase of **secularization**, there has been a resurgence of religious **fundamentalism,** as well as debates over religious beliefs and values. New religious forms, such as **liberation theology,** the **Goddess movement**, and **religious nationalism** have recently emerged. Maintaining an appropriate balance between religion and other aspects of social life will be an important challenge for the United States and other nations of the world in the future.

LEARNING OBJECTIVES

After reading Chapter 17, you should be able to:

1. Define and discuss the four main categories of religion, and link them to the types of societies in which they tend to occur.

2. List the six major world religions and state their basic beliefs.

3. Define the concept of religion and indicate how the sociological investigation of religion is different than a theological investigation.

4. Summarize the controversy about teaching intelligent design in the classroom.

5. Describe the ways that a group may move from one type of religious organization to another over time.

6. Discuss the relationship between religion and questions about the meaning of life.

7. Describe the functionalist perspective on religion including its major functions in societies.

8. Describe symbolic interactionist perspectives on religion and explain how women and men may view religion differently.

9. Discuss the effects of race and class on central-city and suburban churches in the U.S.

10. Explain Durkheim's ideas about religion focusing on the sacred and the profane.

11. Describe conflict perspectives on religion and distinguish between the approaches of Karl Marx and Max Weber.

12. Describe the relationship between religion and social inequality.

13. Distinguish between the different types of religious organization.

14. Compare and contrast civil religion with other forms of religion.

15. Explain the relationship between the growth of secularization and the rise of religious fundamentalism.

16. Describe the future of religion in the U.S. and in other nations of the world.

17. Appraise the major arguments for and against the theory of secularization.

KEY TERMS

(defined at page number shown and in glossary)

animism, p. 563
church, p. 581
civil religion, p. 568
cult, p. 583
denomination, p. 582
ecclesia, p. 581
faith, p. 562
fundamentalism, p. 582
liberation theology, p. 586
monotheism, p. 564

nontheism, p. 564
polytheism, p. 564
profane, p. 562
religion, p. 560
rituals, p. 563
sacred , p. 562
sect, p. 562
secularization, p. 564
simple supernaturalism, p. 563
theism, p. 564

KEY PEOPLE

(identified at page number shown)

Nancy Ammerman, p. 584
Robert Bellah, p. 568
Peter Berger, p. 562
Randall Collins, p. 563
Emile Durkheim , p. 565
Clifford Geertz, p. 563
Lester Kurtz, p. 560
Karl Marx, pp. 568-569
Meredith B. McGuire, p. 568
Keith A. Roberts, p. 561
Wade Clark Roof, p. 565
Max Weber, p. 569

CHAPTER OUTLINE

I. THE SOCIOLOGICAL STUDY OF RELIGION
   A. Religion and the Meaning of Life
      1. **Religion** is a system of beliefs, symbols, and rituals, based on some sacred or supernatural realm, that guides human behavior, gives meaning to life, and unites believers into a community.
      2. Religion seeks to answer important questions such as why we exist, why people suffer and die, and what happens when we die.
         a. According to Peter Berger, religion is a *sacred canopy*, a sheltering fabric hanging over people providing security and answers to questions of life.
         b. This sacred canopy requires **faith**, unquestioning belief that does not require proof or scientific evidence.
      3. Sacred and Profane
         a. According to Emile Durkheim, **sacred** refers to those aspects of life that are extraordinary or supernatural; those things that are set apart as "holy."
         b. Those things people do not set apart as sacred are referred to as **profane** – the everyday, secular or "worldly," aspects of life.
      4. In addition to beliefs, religion also is comprised of symbols and **rituals** – symbolic actions that represent religious meanings – that range from songs and prayers to offerings and sacrifices.
      5. Religions have been classified into four main categories based on their dominant belief:
         a. **Simple supernaturalism** is the belief that supernatural forces affect people's lives either positively or negatively.
         b. **Animism** is the belief that plants, animals, or other elements of the natural world are endowed with spirits or life forces having an impact on events in society.
         c. Theism is a belief in one or more god or gods.

i. **Monotheism** is a belief in a single, supreme being or god who is responsible for significant events such as the creation of the world. Examples include Christianity and Judaism.

ii. **Polytheism** is a belief in more than one god. Examples include Hinduism, Buddhism, and Shinto.

d. **Nontheistic religion** is based on a belief in divine spiritual forces such as sacred principles of thought and conduct, rather than a God or gods.

B. Religion and Scientific Explanation

1. During the Industrial Revolution, rapid growth in scientific and technological knowledge gave rise to the idea that science ultimately would answer questions that previously had been in the realm of religion.

2. Many scholars believed that scientific knowledge would result in **secularization** – the process by which religious beliefs, practices, and institutions lose their significance in sectors of society and culture – but others point out a resurgence of religious beliefs and an unprecedented development of alternative religions in recent years.

II. SOCIOLOGICAL PERSPECTIVES ON RELIGION

A. Functionalist Perspectives on Religion

1. According to Emile Durkheim, all religions share in common three elements: (a) beliefs held by adherents; (b) practices (rituals) engaged in collectively by believers; and (c) a moral community which results from the groups shared beliefs and practices pertaining to the sacred.

2. Religion has three important functions in any society:

a. providing meaning and purpose to life

b. promoting social cohesion and a sense of belonging

c. providing social control and support for the government.

3. The informal relationship between religion and the state is referred to as **civil religion**, the set of beliefs, rituals, and symbols that make sacred the values of the society and place the nation in the context of the ultimate system of meaning.

B. Conflict Perspectives on Religion

1. According to Karl Marx, the capitalist class uses religious ideology as a tool of domination to mislead the workers about their true interests; thus, religion is the "opiate of the people."

2. By contrast, Max Weber argued that religion could be a catalyst to produce social change.

a. In *The Protestant Ethic and the Spirit of Capitalism*, Weber linked the teachings of John Calvin with the rise of capitalism.

b. John Calvin emphasized the doctrine of *predestination* – the belief that all people are divided into two groups – the saved and the damned – and only God knows who will go to heaven (the elect) and who will go to hell, even before they are born.

c. Because people cannot know whether they will be saved, they look for signs that they are among the elect. As a result, people work hard, save their money, and do not spend it on worldly frivolity; instead, they reinvest it in their land, equipment, and labor.

      d. As people worked ever harder to prove their religious piety, structural conditions in Europe led to the industrial revolution, free markets, and the commercialization of the economy, which worked hand in hand with Calvinist religious teachings.

  C. Symbolic Interactionist Perspectives on Religion

    1. For many people, religion serves as a reference group to help them define themselves. Religious symbols, for example, have a meaning to large bodies of people (e.g., the Star of David for Jews; the crescent moon and star for Muslims; and the cross for Christians).

    2. *Her* Religion and *His* Religion: all people do not interpret religion in the same way. Women and men may belong to the same religions, but their individual religion will not necessarily be a carbon copy of the group's entire system of beliefs.

    3. Religious symbolism and language typically create a social definition of the roles of men and women.

III. WORLD RELIGIONS

  A. Hinduism

    1. Hinduism is one of the world's oldest current religions originating in Pakistan between 3,500 and 4,500 years ago.

    2. Hinduism does not have a specific founder or worship of a single god; it is considered by many to be polytheistic.

    3. Because this religion does not have a sacred book inspired by a god or gods, it is not based on the teachings of any one person. Some religion scholars refer to it as an *ethical religion* – a system of beliefs that calls upon adherents to follow an ideal way of life, which is achieved by adhering to the duties of one's caste.

    4. Individuals pass through cycles of life, death, and rebirth until the soul earns liberation; the soul's acquisition of each new body is tied to the law of *karma*; the ultimate goal is to enter the state of *nirvana* – becoming liberated from the world by uniting the individual soul with the universal soul.

    5. Of the more than 700 million Hindus, 95 percent reside in India. Over 80 percent of India's population is Hindu.

  B. Buddhism

    1. Buddhism first emerged in India about 2,500 years ago, but expanded into other nations in various forms.

    2. The founder, Siddhartha Gautama, spent his life teaching others how to obtain Enlightenment and how to reach nirvana.

    3. The Four Noble Truths and the Eightfold Path to Nirvana are incorporated in some form in the three major branches of Buddhism today.

    4. Today, Buddhism is one of the fastest-growing Eastern religions in the United States.

  C. Confucianism

    1. Confucianism, which means the "family of scholars," existed before its eventual leader, Confucius, was born. He emerged as a teacher about the same time that the Buddha became a significant figure in India.

2. Important teachings are: order in human relationships, a strict code of moral conduct, respect for others, benevolence and reciprocity. Humans are by nature good and learn from good role models.
3. Confucius established the social hierarchy of Five Constant Relationships: ruler-subject, husband-wife, elder brother-younger brother, elder friend-junior friend, and father-son.
4. Confucianism was the official religion of China until the Communist takeover and the establishment of the People's Republic of China.

D. Judaism
1. About 18 million Jews reside in about 134 countries; the majority reside in the U.S. or in Israel.
2. Jewish belief is monotheism; a single God, called Yahweh, made a covenant with his chosen people following a critical element in Judaism, the Exodus from Egypt in the eighteenth century B.C.E.
3. Believers of Judaism find the source of their beliefs in the Ten Commandments, and in the moral, ceremonial, and cultural laws contained in four books of the Torah: Exodus, Leviticus, Numbers and Deuteronomy. The Torah, also called the Pentateuch, is the sacred book of contemporary Judaism.
4. Jews worship in synagogues on the Sabbath, which is observed from sunset Friday to sunset Saturday. They celebrate a set of holidays that are distinct from U. S. dominant cultural religious celebrations, and believe that one day the Messiah will come to earth ushering in an age of peace and justice for all.
5. Throughout their history, Jews have been the object of prejudice and discrimination. The Holocaust that took place in Nazi Germany between 1933 and 1945 took the lives of six million Jews.
6. The three branches of Judaism – Orthodox, Reformed, and Conservative – provide for a variety of cultural practices and teachings.

E. Islam
1. Islam, whose followers are known as Muslims, is based on the life and teachings of Muhammad, who was born about 570 C.E. in the city of Mecca, located today in Saudi Arabia.
2. Followers must adhere to the five Pillars of Islam.
3. Allah is the one God of Islam; the Qur' an is their holy book and was revealed to the Prophet Muhammad through the Angel Gabriel at the command of God.
4. A core belief is the idea of *jihad*, meaning struggle; the Greater Jihad is the internal struggle against sin; the Lesser Jihad is the external struggle that takes place in the world, including violence and war. The term *jihad* is typically associated with religious fundamentalism.
5. Some social analysts believe that Islamic fundamentalism is uniquely linked to the armed struggles of some groups.
6. Because of a high recent wave of migration and a relatively high rate of conversion, Islam is one of the fastest-growing religions in the United States.

F. Christianity
   1. As with Judaism and Islam, Christianity follows the Abrahamic tradition; although Jews and Christians share some common scriptures in what Christians call the "Old Testament," they interpret them differently.
      a. God offers to Christians a new covenant, to the followers of Jesus Christ.
      b. The central themes in the teachings of Jesus are the Kingdom of God and standards of personal conduct.
      c. Central teachings of Christianity are: the death and then the resurrection of Jesus – establishing that He is the son of God – and the ascension of Jesus into heaven forty days after His resurrection with an eventual "second coming" marking the end of the world as it is now known.
   2. The Christian deity is a sacred Trinity comprised of God the Father, Jesus the Son, and the Holy Spirit.
   3. Almost one-third of the world's population refers to themselves as Christian; it is the world's dominant religion.
IV. TYPES OF RELIGIOUS ORGANIZATION
   A. Some countries have an official or state religion known as an **ecclesia** – a religious organization that is so integrated into the dominant culture that it claims as its membership all members of a society. Examples include: the Anglican Church (the official Church of England), the Lutheran Church in Sweden and Denmark, the Catholic Church in Italy and Spain, and Islam in Iran and Pakistan.
   B. The Church Sect Typology.
      1. A church is a large, bureaucratically organized religious organization that tends to seek accommodation with the larger society in order to maintain some degree of control over it.
      2. Midway between the church and the sect is a **denomination** – a large organized religion characterized by accommodation to society but frequently lacking in ability or intention to dominate society.
      3. A **sect** is a relatively small religious group that has broken away from another religious organization to renew what it views as the original version of the faith.
   C. A **cult** is a religious group with practices and teachings outside the dominant cultural and religious traditions of a society
      1. Some major religions (including Judaism, Islam, and Christianity) and some denominations (such as the Mormons) started as cults.
      2. Cult leadership is based on charismatic characteristics of the individual, including an unusual ability to form attachments with others.
      3. Over time, some cults undergo transformation into sects or denominations.
V. TRENDS IN RELIGION IN THE UNITED STATES
   A. Religion and Social Inequality
      1. Religion in the United Sates is diverse; pluralism and religious freedom are among the cultural values.
      2. Social scientists identify patterns between social class and religious denominations.

      a. Denominations with affluent members and high social status are the mainline liberal churches; the earliest members were immigrants from Britain who helped establish the U. S. government and capitalist economy.

      b. Denominations that adhere to conservative religious beliefs are often in the middle to lower socioeconomic status and tend to discourage women from working outside the home.

  B. Race and Class in Central-City and Suburban Churches

    1. As more middle and upper-income individuals and families moved to the suburbs, some churches relocated to the suburbs.

    2. The former building site is often replaced by a minority congregation, working class, and older members living on fixed income, or recent immigrant groups.

  C. Secularization and the Rise of Religious Fundamentalism

    1. *Secularization* – the decline in the significance of the sacred in daily life – has occurred, causing an increase of religious fundamentalism among some groups.

    2. **Fundamentalism** is a traditional religious doctrine that is conservative, typically opposed to modernity, and rejects worldly pleasures in favor of other worldly spirituality.

      a. In the United States, it originally appealed to people from low-income, rural, southern backgrounds.

      b. The "new" fundamentalist movement has a wider appeal to people from all socioeconomic levels, geographical areas, and occupations.

      c. Some political leaders vow to bring the Christian religion "back" into schools and public life.

VI. RELIGION IN THE FUTURE

  A. New religious forms are being created, while traditional forms of religious life are experiencing dramatic revitalization; liberation theology is one example.

    1. **Liberation theology** is the Christian movement that advocates freedom from political subjugation within a traditional perspective and advocates the need for social transformation to benefit the poor and powerless.

    2. Some feminist movements have turned to pagan religions and witchcraft to counter the patriarchal structure and content of the world's religions.

      a. The *Goddess movement* acknowledges the legitimacy of female power and encompasses a variety of counterculture beliefs based on paganism and feminism.

    3. In other nations, the rise of religious nationalism has led to the blending of strongly held religious and political beliefs; it is strong in the Middle East, where Islamic nationalism has spread, strongly affecting women and children.

    4. In the U. S., the influence of religion will be evident in ongoing battles over school prayer, abortion, gay rights, and women's issues.

  B. On some fronts, religion may unify people; on others it may contribute to confrontations among individuals and groups.

  C. Maintaining an appropriate balance between religion and other aspects of social life will be an important challenge in the future – the debate continues over what religion it is and what it should do.

ANALYZING AND UNDERSTANDING THE BOXES

After reading the chapter and studying the outline, re-read the boxes and write down key points and possible questions for class discussion.

Sociology in Everyday Life: How Much Do You Know About the Effect of Religion on U. S. Education?

Key Points:

Discussion Questions:

1.

2.

3.

Framing Religion in the Media: Shaping the Intersections of Science and Religion

Key Points:

Discussion Questions:

1.

2.

3.

Sociology and Social Policy: Should Prayer Be Allowed in Public Schools? Issues of Separation of Church and State

Key Points:

Discussion Questions:

1.

2.

3.

You Can Make a Difference: Understanding and Tolerating Religious and Cultural Differences

Key Points:

Discussion Questions:

1.

2.

3.

PRACTICE TESTS

MULTIPLE CHOICE QUESTIONS

Select the response that best answers the question or completes the statement:

1. The famous court case that dealt with the teaching of evolution in the public school classroom was
    a. the Scopes trial.
    b. *Brown v. Board of Education*.
    c. *Plessey v. Ferguson*.
    d. the Hopwood trial.

2. _____ study specific religious doctrines or belief systems, including answers to questions such as the nature of God or the gods, and the relationships between the sacred and human beings.
    a. Priests
    b. Prophets
    c. Theologians
    d. Millennialists

3. Sociologist _____ referred to religion as a sacred canopy.
   a. Emile Durkheim
   b. Clifford Geertz
   c. Peter Berger
   d. Robert Bellah

4. According to Emile Durkheim, _____ refers to those aspects of life that are extraordinary or supernatural.
   a. religion
   b. sacred
   c. profane
   d. superhuman

5. The belief that land, animals, or other elements of the natural world are endowed with spirits or life forces that have an impact on events in society is known as:
   a. polytheism
   b. animism
   c. theism
   d. monotheistic religion

6. Three of the major world religions – Christianity, Judaism, and Islam – are characterized as
   a. simple supernaturalism.
   b. animism.
   c. polytheism.
   d. monotheism.

7. Which of the following is not one of the elements Durkheim believed to be common to all religions?
   a. a liberating theology
   b. a system of beliefs
   c. a moral community
   d. practices or rituals

8. According to the functionalist perspective, religion offers meaning for the human experience by
   a. providing an explanation for events that create a profound sense of loss on both an individual and a group basis.
   b. offering people a reference group to help them define themselves.
   c. reinforcing existing social arrangements, especially the stratification system.
   d. encouraging the process of secularization.

9. Celebrations on Memorial Day and the Fourth of July are examples of:
   a. religious tolerance
   b. civil religion
   c. patriotic ethnocentrism
   d. separation of church and state

10. According to _____, the capitalist class uses religious ideology as a tool of domination.
    a. Emile Durkheim
    b. C. Wright Mills
    c. Karl Marx
    d. Max Weber

11. In regard to religion, Max Weber asserted that
    a. church and state should be separated.
    b. religion could be a catalyst to produce social change.
    c. religion retards social change.
    d. the religious teachings of the Catholic church were directly related to the rise of capitalism.

12. "Her religion" and "his religion" have been examined from a(n) _____ perspective.
    a. functionalist
    b. neo Marxist
    c. conflict
    d. interactionist

13. The ideas of dharma, karma, and nirvana are associated with
    a. animism.
    b. Buddhism.
    c. Confucianism.
    d. Hinduism.

14. The "Four Noble Truths" and the practice of the "Eightfold path" are today recognized as teachings of
    a. Hinduism.
    b. Buddhism.
    c. Confucianism.
    d. Islam.

15. All of the following are one of the "Five Constant Relationships of Confucianism" **except:**
    a. mother-daughter
    b. father-son
    c. husband-wife
    d. elder brother-younger brother

16. Which one of the following is not associated with Judaism?
    a. People are responsible for making ethical choices.
    b. The Messiah will come to earth one day.
    c. God rested on the seventh day after He created the world.
    d. The Messiah has already come to earth.

17. Which of the following is not associated with Islam?
    a. Muslim
    b. the Qur' an
    c. The Greater Jihad
    d. Four Noble Truths

18. The Anglican church in England and the Lutheran church in Sweden are examples of a(n):
    a. ecclesia.
    b. sect.
    c. denomination.
    d. secular institution.

19. In a _____, membership is largely based on birth, and children of members typically are baptized as infants.
    a. church
    b. sect
    c. denomination
    d. cult

20. According to the text, religious nationalism – the blending of strongly held religious and political beliefs – is especially strong today in
    a. the United States.
    b. Middle Eastern nations.
    c. Japan.
    d. Brazil.

TRUE/FALSE QUESTIONS

1. According to Peter Berger, the sacred canopy is the institution of the family.
   T   F

2. Across cultures and in different eras, a wide variety of things have been considered sacred.
   T   F

3. Secularization is the process by which religious beliefs, practices, and institutions lose their significance in sectors of society and culture.
   T   F

4. Marx used the phrase, the "opiate of the masses" to refer to the organized activity of religious people seeking a change in society.
   T   F

5. Karl Marx wrote *The Protestant Ethic and the Spirit of Capitalism* to explain how religion may be used by the powerful to oppress the powerless.
   T   F

6. Ninety-five percent of all Hindus in the world today reside in Pakistan.
   T   F

7. The Dalai Lama is associated with Hinduism.
   T   F

8. Today, Confucianism is the official religion of the People's Republic of China.
   T   F

9. Because it has fewer adherents worldwide than some other major religions, the influence of Judaism is weak in western culture today.
   T   F

10. In the United States today, Muslims outnumber Presbyterians and Episcopalians.
    T   F

11. More people in the world today refer to themselves as Christians than any other religion.
    T   F

12. Even though Christianity spread outward from Europe to other cultures, it changed very little, utilizing one highly integrated body of integrated religious beliefs.
    T   F

13. Denominations tend to be more tolerant and less likely than churches to expel or excommunicate members.
    T   F

14. "New" fundamentalists have encouraged secular humanism in U.S. schools.
    T   F

15. Liberation theology initially emerged in Latin America.
    T   F

16. In the United States, traditional fundamentalism primarily appealed to people from lower income, rural, and southern backgrounds.
    T   F

17. According to Box 17.1, the U.S. Constitution originally specified that religion should be taught in the public schools.
    T   F

18. According to Box 17.2, world-renown physicist Albert Einstein suggested that "true religion" and scientific knowledge are incompatible.
    T   F

19. According to Box 17.3, the First Amendment to the Constitution mandates the separation of church and state.
    T   F

20. According to Box 17.4, Islam, Christianity, and Judaism believe in one God, as rooted in the same part of the world, and share some holy sites.
    T   F

FILL-IN-THE-BLANK QUESTIONS

1. Those things that are considered secular or "worldly" are referred to as _____.

2. _____ is the belief that supernatural forces affect people's lives either positively or negatively.

3. The belief in one or more gods is known as _____.

4. The process by which religious beliefs lose their significance in sectors of society is known as _____.

5. According to anthropologist _____, religion is a set of cultural symbols that establishes powerful motivation.

6. According to _____, religion is the opiate of the masses.

7. Max Weber wrote in his classical book, _____, that religion could be a catalyst to produce social change.

8. In the United States today, one of the fastest-growing eastern religions is _____.

9. According to table 17.1, the largest Christian denomination in the United States is the _____.

10. The informal relationship between religion and the state is referred to as _____, the set of beliefs, rituals, and symbols that make sacred the values of the society and place the nation in the context of the ultimate system of meaning.

11. _____ is the unquestioning belief that does not require proof or scientific evidence.

12. A religious organization that claims as its membership all members of a society is known as a(n) _____.

13. The _____ encompasses a variety of countercultural beliefs based on paganism and feminism and acknowledgement of the legitimacy of female power.

14. The rise of _____ has led to the blending of strongly held religious and political beliefs, especially in the Middle East.

15. The Christian movement _____ advocates freedom from political subjugation within a traditional perspective and to benefit the poor and downtrodden.

SHORT ANSWER/ESSAY QUESTIONS

1. What are some of the central beliefs of the six largest world religions?
2. What is religion and how do three major sociological perspectives – functionalist, conflict, and symbolic interactionist – differ in their perspectives on religion?

3. What are some of the relationships among race, class gender, and religion?
4. What is meant by civil religion? In what ways are people affected in their everyday lives by civil religion?
5. What are some of the current trends and changes in worldwide religion? Do you think there will be a continuing tendency toward secularization, or toward fundamentalism? Will religion continue as a major social institution?

STUDENT CLASS PROJECTS AND ACTIVITIES

1. Compare and contrast the religious books of law of at least two different religions. The Torah, the Qur' an, or the New Testament are some suggestions. Answer the following questions: (1) What are some similarities of the two books? (2) What are some unique differences? (3) How do they treat the issue of gender, and gender roles? (4) What are the expectations of a believer dealing with a non-believer? (5) What are some of the ethical laws, such as: (a) How do the believers relate to the poor, or the rich? (b) How do the believers relate to those who have disobeyed the rules? (c) What are the expected rites and rituals of the believer? (d) What are the "sacreds" (as defined by Durkheim) of the religion? (6) What are the major beliefs? and (7) What happens to believers after their deaths? Provide bibliographical information in preparing your paper. Summarize your findings and provide a personal evaluation of this project. Submit your paper to your instructor at the appropriate time, following any instructions given to you.
2. Research a particular cult found in the United States or elsewhere. You should, in this research project, provide in your paper: (1) a definition of the cult; (2) the purpose of the cult; (3) a general description of the cult which should include: (a) its origin; (b) its membership and method of acquiring members; (4) the nature of its group life, including advantages and disadvantages of membership in the cult. After providing this basic information, include a two-page paper on a day in the life of the cult. Pretend to be a member and take the membership point of view in the writing up of a "day in the life of this cult." This day could be a very special day, such as your wedding day, or a typical school, work, or ceremonial day. Submit your papers to your instructor on the appropriate date, following any additional instructions you have been given.
3. Conduct an informal survey of any twenty people in order to obtain their opinion of the ordination of women in our society. You must provide this basic information about your respondents: (1) their name, age, and address; (2) if possible, their religious affiliation (if any); (3) their gender. Provide responses to the following questions: (1) Do you support the entry of women into the seminary schools? Why? Why not? (2) Do you support the ordination of women? Why? Why not? (3) Should women be allowed to serve as full-time professional clergy or rabbi or cantor? Why? Why not? (4) What do you think should be the primary role of women in the church? (5) What do you think should be the role of women who are clergy (pastor, rabbi, cantor, and priest)? (6) Do you support the decision of the Anglican Church of England to ordain women? (7) Do you support the position of the Roman Catholic Church to not allow women to become priests? Why? Why not? (8) Do you have other statements you would like to make regarding the issue of women and the church? Summarize your findings and provide a conclusion statement per each respondent. Include any raw data that you have collected. Submit your paper to your instructor at the appropriate time.

INTERNET ACTIVITIES

1. Peggy Wehmeyer, a reporter for abcnews.com, contends that the Internet is having much the same effect on transforming religion as printing the Bible did. Explore a comparative religious community that provides information on the major world religions, information on morality and culture, family and prayer circles. **http://www.beliefnet.com/**
2. To illustrate the extent to which the Internet is shaping religious experience, go to **http://www.hollywoodjesus.com/** and explore the uniquely American interplay between religion and entertainment.
3. The **American Atheists** frequently asked questions (FAQ), **http://www.atheists.org/schoolhouse/faqs.prayer.html**, about prayer in public school web page. Use information from this site for in-class discussions about the interconnections between religion and education.
4. Explore information on the Islamic faith, news, and social organizations: **http://www.islamworld.net/**; and, **http://www.islamicity.com/.**
5. Compare and contrast Internet sites of insiders and outsiders on the subject of Islam and Terrorism: **http://www.submission.org/terrorism.html**, **http://www.islam101.com/terror/, http://www.islam-guide.com/ch3-11.htm**, **http://groups.colgate.edu/aarislam/response.htm**, and **http://answering-islam.org/Terrorism/.**

INFOTRAC COLLEGE EDITION EXERCISES

Visit the **InfoTrac College Edition** website at: **http://www.wadsworthmedia.com/webtutor/infotrac.htm**. You will arrive at a screen that enables you to search topics.

1. The Southern Baptist Convention, the largest Protestant denomination in the U.S., is deeply divided over many issues. One issue is that of the status of women serving in the ministry. Research the issue of the role of **women in religious life**.
2. To what extent can you find evidence of how religious Americans are? Use the keyword **religiosity** (the sociological concept for religious belief and behavior).
3. Many churches and religious groups explicitly condemn **homosexuality**. Frame the debate among the various major world religions. Where do the various religious groups stand on this issue?
4. Fundamentalist religious organizations continue to grow, while other religious organizations have experienced declining memberships. Research **religious fundamentalism** and why it appeals to new adherents.
5. Many Islamic groups advocate a "holy war" against westernization. What does this term really mean to them? Does a holy war necessarily mean resorting to terrorist tactics? Look up the words: **jihad**, **holy war**, **religious conflict** and **terrorism**. Don't limit your search to stories from the Middle East and Islam. Investigate the religious war that has been going on in Northern Ireland for decades. Look up articles on Bosnia-Herzegovina as well.
6. The issue of prayer in schools continues to be important as some groups contend that **school prayer** would improve the moral fiber of children, while others contend that it violates the premise of separation of church and state.

SOLUTIONS

MULTIPLE CHOICE QUESTIONS

1. A, p. 559
2. C, p. 561
3. C, p. 562
4. B, p. 562
5. B, p. 563
6. D, p. 564
7. A, p. 565
8. A, p. 565
9. B, p. 568
10. C, p. 569
11. B, p. 569
12. D, p. 572
13. D, p. 573
14. B, p. 575

15. A, p. 576
16. D, p. 578
17. D, p. 575
18. A, p. 581
19. A, p. 581
20. B, p. 586
21. A, p. 580
22. B, p. 580
23. A, p. 585
24. D, p. 583
25. A, p. 562
26. B, p. 562
27. D, p. 563
28. C, p. 562

29. B, p. 563
30. B, p. 564
31. B, p. 564
32. D, p. 564
33. D, p. 569
34. D, p. 569
35. C, p. 573
36. B, p. 573
37. C, p. 575
38. A, p. 575
39. C, p. 576
40. A, p. 576

TRUE/FALSE QUESTIONS

1. F, p. 562
2. T, p. 562
3. T, p. 564
4. F, p. 569
5. F, p. 569
6. F, p. 573
7. F, p. 575

8. F, p. 573
9. F, p. 576
10. T, p. 579
11. T, p. 580
12. F, p. 581
13. T, p. 582
14. F, p. 586

15. T, p. 586
16. T, p. 585
17. F, p. 562
18. F, p. 566
19. T, p. 571
20. T, p. 587

FILL-IN-THE-BLANK QUESTIONS

1. profane, p. 562
2. Simple supernaturalism, p. 563
3. theism, p. 564
4. secularization, p. 564
5. Clifford Geertz, p. 563
6. Karl Marx, p. 569
7. The Protestant Ethic and the Spirit of Capitalism, p. 569
8. Buddhism, p. 575
9. Roman Catholic Church, p. 583
10. civil religion, p. 568
11. Faith, p. 562
12. ecclesia, p. 581
13. Goddess movement, p. 586
14. religious nationalism, p. 586
15. liberation theology, p. 586

18

HEALTH, HEALTH CARE, AND DISABILITY

BRIEF CHAPTER OUTLINE

CHAPTER SUMMARY

Although many people may think of health as simply being the absence of disease, the World Health Organization defines **health** as a state of complete physical, mental and social well-being. **Health care** is any activity intended to improve health. The **infant mortality rate**, the number of deaths of infants under one year of age per 1,000 live births in a given year, and **life expectancy**, the estimate of the average lifetime of people born in a specific year, are two indicators of the well-being of people in a society. A vital part of health care is **medicine**, an institutionalized system for the scientific diagnosis, treatment, and prevention of illness. In analyzing health from a global perspective, we find that health and health care expenditures vary widely among nations. **Social epidemiology** is the study of the causes and distribution of health, disease, and impairment throughout a population. Social factors such as age, sex, race/ethnicity, social class, lifestyle choices, and drug use and abuse affect the health and longevity of people in a given area. Sexual activity, usually associated with good physical and mental health, may involve health-related hazards, such as transmission of sexually transmitted diseases. The fee-for-service method continues in the United States, which is expensive; few restrictions are placed on medical fees. The United States and the Union of South Africa are the only high-income nations without some form of universal health coverage for all citizens. Private health insurance, public health insurance (consisting of Medicare and Medicaid), health maintenance organizations, and managed care compromise the major approaches for paying for U.S. health care. Advances in high-tech medicine is becoming a major part of overall health care; however, many people are turning to holistic medicine – is an approach to health care that focuses on prevention of illness and disease and is aimed at treating the whole person-body and mind-rather than just the part or parts in which symptoms occur. Many practitioners of **alternative medicine** – healing practices inconsistent with dominant medical practice – take a holistic approach. According to a **functionalist** approach, if a society is to function as a stable system, people must be healthy in order to contribute to their society. People who assume the **sick role** are unable to fulfill their necessary social roles, contributing to an inability to fulfill their functions. Unequal access to good health and health care is a major issue of concern to **conflict** theorists. The **medical-industrial complex**-that which encompasses both local physicians and hospitals as well as global health-related industries- produces and sells medicine as a commodity; people below the poverty level and those just above it have great difficulty gaining access to medical care. **Symbolic Interactionists**, in explaining the social construction of illness, focus on the meaning that social actors give their illness or disease and how that meaning affects their self-concept and their relationships with others. According to a postmodernist approach, doctors have gained power through the clinical gaze, and gained prestige through the classification of disease. **Mental illness**, a condition in which a person has a severe **mental disorder** requiring extensive treatment, differs from a mental disorder, a condition that makes it difficult or impossible for a person to cope with everyday life. **Disability**, a physical or health condition that stigmatizes or causes discrimination, is often defined in terms of work from a business perspective, or in terms of organically-based impairments from a medical professional perspective. Health care in the future is critically affected by health care technologies; however, health in the future will somewhat be up to each of us – how we safeguard ourselves against illness, and how we help others who are victims of diseases and disabilities.

LEARNING OBJECTIVES

After reading chapter 18, you should be able to:

1. Define social epidemiology and its role in society.

2. Describe the rise of scientific medicine in the U.S.

3. Define health, health care and medicine and explain the importance of these issues for individuals and the whole of society.

4. Describe the key events and ideas associated with the rise of scientific medicine and professionalism.

5. Summarize three major social implications of advanced medical technology.

6. Illustrate functionalist, conflict, symbolic interactionist, and postmodernist perspectives on health and medicine.

7. Classify four methods of paying for health care and controlling health care costs in the U.S.

8. Describe the relationships of race, class, gender and mental disorders.

9. Compare health in a global perspective to health in the U.S.

10. Compare and contrast how age, sex, race/ethnicity and social class affect health and mortality.

11. Explain how lifestyle choices affect health, disease, and impairment.

12. Compare and contrast holistic medicine and alternative medicine with traditional, or orthodox medical treatment

13. Distinguish between health care in the U.S. and that of other countries.

14. Define and explain mental illness and mental disorder and explain the treatment of each in the U.S.

15. Critique inequalities related to disability.

16. Evaluate the impact of stereotypes, prejudice, and discrimination on persons with disabilities.

KEY TERMS

(defined at page number shown and in glossary)

acute diseases, p. 596
chronic diseases, p. 596
deinstitutionalization, p. 615
demedicalization, p. 613
disability, p. 618
drug, p. 597
health, p. 592
health care, p. 592
health maintenance organization
(HMO), p. 604
holistic medicine, p. 610

infant mortality rate, p. 592
life expectancy, p. 592
managed care, p. 604
medical-industrial complex, p. 612
medicalization, p. 613
medicine, p. 592
sick role, p. 611
social epidemiology, p. 595
socialized medicine, p. 607
universal health care, p. 606

KEY PEOPLE

(identified at page number shown)

Ronald Angel, p. 621
Joe R. Feagin and Melvin Sikes, p. 617
Deborah Findlay and Leslie Miller,
p. 613
Abraham Flexner , p. 601
Michel Foucault , p. 614

Eliot Freidson , p. 620
Erving Goffman, p. 616
Talcott Parsons, p. 611
Thomas Pettigrew, p. 617
Paul Starr, p. 603
Ingrid Waldron, p. 597

CHAPTER OUTLINE

I. HEALTH IN GLOBAL PERSPECTIVE
   A. The World Health Organization defines **health** as a state of complete physical, mental and social well-being.
   B. Medical sociology, as a subdiscipline, is the study of health, healing, and illness.
   C. **Health care** is any activity intended to improve health.
   D. Some measures of health and well-being are:
      1. The **infant mortality rate**, the number of deaths of infants under one year of age per 1,000 live births in a given year.
      2. **Life expectancy**, the estimate of the average lifetime of a people born in a specific year.
      3. **Medicine**, an institutionalized system for the scientific diagnosis, treatment, and prevention of illness; it is a vital part of health care.
   E. Health care and health care expenditures vary widely among nations. The WHO reports that babies born in low-income countries will not live as long compared with children in high-income countries. Almost 14 percent of all children born in low-income countries die before they reach their first birthday.
II. HEALTH IN THE UNITED STATES
   A. **Social epidemiology**, the study of the causes and distribution of health, disease, and impairment throughout a population, provides some explanations of health and disease.
   B. Social epidemiologists investigate disease agents, the environment, and the human host as sources of illness.
   C. Younger people are more likely to have **acute diseases**, illnesses that strike suddenly and cause dramatic incapacitation and sometimes death, while older people are more likely to have **chronic diseases**, illnesses that are long-term and develop gradually or are present from birth.
   D. Today, women on average now live longer than men; at birth, females have lower mortality rates both in the prenatal stage and during the first month of life.
   E. Some research points out that social class is a greater determinant of health than race/ethnicity; however, illness is also related to a person's occupation.

F. Three (of many) **lifestyle factors** relate to health:
1. **Drugs**, any substances other than food and water that, when taken into the body, alter its functioning in some way, affect health in many ways.
    a. Long-term heavy use of alcohol can damage the brain and other parts of the body, cause nutritional deficiencies, cardiovascular problems, and alcoholic cirrhosis.
    b. **Nicotine** (tobacco) is a toxic, dependency-producing psychoactive drug that is more addictive than heroin. Smoking tobacco is linked to cancer and other serious diseases. Environmental tobacco smoke – the smoke in the air that nonsmokers inhale from other people's tobacco smoking, is hazardous for nonsmokers.
    c. The use of **illegal drugs**, like cocaine and marijuana, affects the lifestyle and the health of individuals.
2. Sexual activity may involve health-related hazards such as transmission of certain **sexually transmitted diseases**.
    a. The most prevalent STDs are HIV/AIDS, gonorrhea, syphilis, and genital herpes. All of these affect the health of individuals; however, the most serious of these is AIDS.
    b. On a global basis, the number of people infected with HIV or AIDS is increasing; ten percent of all new cases worldwide are children; nearly two-thirds of all people with HIV/AIDS live in sub-Saharan Africa; 14 percent live in south and southeast Asia.
    c. HIV is transmitted through unprotected sexual intercourse with an infected partner; sharing a hypodermic needle with someone who is infected; by exposure to contaminated blood or blood products; or by an infected woman passing the virus on to a child during pregnancy, childbirth, or breast feeding.
3. Staying healthy: diet and exercise
    a. Lifestyle choices can include positive actions such as maintaining a healthy diet and good exercise program.
    b. Both of these have contributed to a significant decrease in heart disease and some cancers.

III. HEALTH CARE IN THE UNITED STATES
A. The rise of scientific medicine resulted from several significant discoveries during the nineteenth century in areas such as bacteriology and anesthesiology; at the same time, advocates of reform in medical education were promoting scientific medicine.
1. The **Flexner Report**, complied by Abraham Flexner, included a model of medical education; those medical schools that did not fit the "model" were closed. Only two of the African American medical schools survived, and only one of the medical schools for women survived.
2. With professionalization resulting from the Flexner Report, licensed medical doctors gained control over the entire medical establishment; this continues today and may continue into the future.

B. Medicine Today
   1. In the U.S., patients are billed individually for each service they receive.
      a. This includes treatment by doctors, laboratory work, hospital visits, other health-related expenses, and prescriptions.
      b. This payment is expensive; few restrictions are placed on the fees that doctors, hospitals, and medical providers can charge.
      c. This method, representing the "true spirit" of capitalism, has led to advances in medicine.
      d. However, this fee-for-service medicine results in its inequality of distribution.
C. Paying for medical care in the United States
   1. Private health insurance expansion in the United States led to escalated costs.
      a. Third party providers pay large portions of doctor and hospital bills for insured patients.
      b. Private health insurance plans are organized like other large-scale for-profit corporations.
   2. Public health insurance is subsidized by federal and state taxes; the U.S. has two nationwide public health insurance programs: Medicare, for those 65 or older who are eligible; and Medicaid, a jointly funded federated-state program to make health care more available to the poor. Both programs are in financial difficulty today.
   3. **Health Maintenance Organizations (HMOs)** were created to provide workers with health coverage by keeping costs down.
      a. **HMOs** provide, for a set monthly fee, total care with an emphasis on prevention to avoid costly treatment later.
      b. Supporters applaud the emphasis on preventative care.
      c. Critics charge that some HMOs require their physicians to withhold vital information and medical treatment from their patients in order to reduce potential large sums of money for medical procedures or hospitalization.
   4. **Managed care**, any system of cost containment that closely monitors and controls health care providers' decisions about medical procedures, diagnostic tests, and other services that should be provided to patients.
      a. Patients choose a primary-care physician from a list of participating doctors.
      b. Doctors must get approval before they perform any procedures or admit a patient to a hospital; if they fail to obtain advance approval, the insurance company has the right to refuse to pay for the treatment or the hospital stay.
   5. The **uninsured and the underinsured** comprise one-third of all U. S. citizens.
      a. Children comprise a large amount not covered by insurance.
      b. The working poor constitutes a large portion; they earn too little to afford health insurance, too much to qualify for Medicaid, and their employers do not provide health insurance.

D. Paying for Medical Care in Other Nations
1. Canada has a **universal health care system**, a health care system in which all citizens receive medical services paid for by tax revenues. In Canada these revenues are supplemented by insurance premiums paid by all taxpaying citizens. It does not constitute what is referred to as **socialized medicine**, a health care system in which the government owns the medical care facilities and employs the physicians; the physicians are not government employees.
2. Great Britain, in 1946, passed the National Health Service Act, which provided for all health care services to be available at no charge to the entire population. The government sets health care policies, raises funds and controls the medical care budget, owns health care facilities, and directly employs physicians and other health personnel.
3. China adopted innovative strategies after the Civil War in order to improve the health of its populace. Physician extenders, known as *street doctors* in urban areas and *barefoot doctors* in rural areas, had little formal training and worked under the supervision of trained physicians. Today, medical training is more rigorous; all doctors receive training in both Western and traditional Chinese medicine. Doctors who work in hospitals receive a salary; all other doctors are on a fee-for-service basis. The cost of health care generally remains low, but the cost of hospital care has risen. Chinese who can afford it purchase health care insurance to cover the rising cost of hospitalization. The health of its citizens is only slightly below that of most industrialized nations.
E. Social implications of advanced medical technology include:
1. The new technologies create options for people and society, but options that alter human relationships.
2. The new technologies increase the cost of medical care (such as the computerized axial tomography, CT or CAT scanner).
3. The new technologies raise provocative questions about the very nature of life, such as duplicating mammals from adult DNA.
F. **Holistic** medicine and alternative medicine provide variations from traditional (or orthodox) medical treatment.
1. **Holistic medicine** focuses on prevention of illness and disease and aims at treating the whole person-body and mind, rather than just the part or parts in which symptoms occur.
2. Alternative medicine involve healing practices inconsistent with dominant medical practice which take a holistic approach; many people today are utilizing this in addition to or in lieu of traditional medicine.
IV. SOCIOLOGICAL PERSPECTIVES ON HEALTH AND MEDICINE
A. A functionalists perspective on health: the **sick role**
1. According to Talcott Parsons, the sick role is the set of patterned expectations that define the norms and values appropriate for individuals who are sick and for those who interact with them.
2. The sick role has four characteristics:
a. A person's sickness is not deliberate
b. A sick person is exempted from responsibilities

   c. A sick person must want to get well

   d. A sick person must seek competent help from a medical profession

  3. Illness is dysfunctional for individuals and the larger society; sick people are unable to fulfill their necessary social roles; it is important for the society to maintain social control over people who enter the sick role.

B. A conflict perspective: inequalities in health and health care

  1. Conflict theorists emphasize the political, economic, and social forces that affect health and the health care delivery system.

  2. Medicine is a commodity that is produced and sold by the **medical-industrial complex** which encompasses both local physicians and hospitals as well as global health-related industries such as insurance companies and pharmaceutical and medical supply companies that deliver health care today.

  3. Medical care is linked to people's ability to pay and their position within the class structure; the affluent or those who have good medical insurance most likely receive care; the medically indigent, those who cannot qualify for Medicaid, but do not earn enough, cannot afford private medical care.

  4. Physicians hold a legal monopoly over medicine. They can charge inflated fees; clinics, pharmacies, laboratories, hospitals, insurance companies, and many other corporations derive excessive profits from the existing system of payment in medicine.

C. A symbolic interactionist perspective: the social construction of illness

  1. Symbolic interactionists focus on the meaning that actors give their illness or disease and how this will affect their self-concept and their relationship with others.

  2. In addition to the objective criteria for determining medical conditions, the subjective component is very important; the term **medicalization** refers to the process whereby nonmedical problems become defined and treated as illness or disorders.

  3. Medicalization may occur on three levels: (a) the conceptual level; (b) the institutional level; and (c) the interactional level. Medicalization is typically the result of a lengthy promotional campaign, often culminating in legislation or other social policy changes that institutionalize a medical treatment of a new "disease."

  4. **Demedicalization** refers to a problem that no longer retains its medical definition; an example is the work of women's health advocates seeking to redefine menopause as a natural process rather than as an illness or psychological disorder.

  5. Symbolic interactionists provide new insights on how illness may be socially constructed, and not strictly determined by medical criteria.

D. A postmodernist perspective: the clinical gaze

  1. According to Michel Foucault, the formation of clinical medicine led to the power that doctors gained over other medical personnel and everyday people.

2. Doctors gain power through the clinical gaze, which they use to gather information.
   a. As doctors "diagnose", they become experts.
   b. As a definition network of disease classification, new rules and new tests developed, the dominance of doctor's wisdom became enhanced.
3. Foucault's analysis is not limited to doctors who treat bodily illness, but also psychiatrists and the treatment of insanities.

V. MENTAL ILLNESS
A. **Mental illness**, a condition in which a person has a severe mental disorder requiring extensive treatment, differs from a **mental disorder**, a condition that makes it difficult or impossible for a person to cope with everyday life.
B. According to the National Comorbidity Survey, nearly 50 percent of respondents between the ages of 15 and 54 years had been diagnosed with a mental disorder sometime in their lives.
C. Treatment of mental illness
   1. Substance-related disorders are the leading cause of hospitalization for men between the ages of 15 and 44 and the second leading cause for women in that age group.
   2. The introduction of new psychoactive drugs to treat mental disorders and the **deinstitutionalization** movement have created dramatic changes in how people with mental disorders are treated; deinstitutionalization refers to the practice of rapidly discharging patients from mental hospitals into the community.
   3. **Deinstitutionalization** is now viewed as a problem by many social scientists; in the 1960s, it was believed that many state mental patients had been deprived of their civil rights, housed in what Goffman called a *total institution*; they were released, many having no place to go, except to reside on the streets and become part of the homeless population.
   4. Involuntary commitment, used as a social control mechanism to keep mentally ill people off the streets, does little to treat underlying medical and social conditions that contribute to mental disorders.
D. Race, Class, Gender, and Mental Disorders
   1. Most studies examining race/ethnicity and mental disorders have compared African Americans and white Americans and have uncovered no significant differences in diagnosable mental illness.
   2. In a qualitative study about the effects of racism, Feagin and Sikes concluded that repeated personal encounters with racial hostility deeply affect the psychological well-being of most African Americans, regardless of their level of education or social class.
   3. According to Pettigrew, racism in all it forms constitutes a mentally unhealthy situation, wherein people do not achieve their full potential.
   4. Researchers agree that social class has a significant impact on mental disorders; as social class increases, rates of mental disorders decrease.
   5. The rate of diagnosable depression is about twice as high for women as for men; this gender difference typically emerges in puberty and increases in adulthood as women and men enter and live out their unequal adult statuses.

VI. DISABILITY
    A. **Disability**, a physical or health condition that stigmatizes or causes discrimination, is often defined in terms of work from a business perspective, or in terms of organically-based impairments from a medical professional perspective.
    B. An estimated 49.7 million persons in the U.S. have one or more physical or mental disabilities, and the number is increasing as medical advances make it possible for those who would have died from an accident or illness to survive, but with an impairment, and as life expectancies increases. Environment, lifestyle, and working conditions may contribute to disability.
    C. Many disability rights advocates argue that persons with a disability are kept out of the mainstream of society, being denied equal opportunities in education by being consigned to special education classes or schools.
    D. Sociological Perspectives on Disability
        1. Functionalist Talcott Parsons focused on how people who are disabled fill the sick role. This is a *medical model* of disability; people with disabilities become chronic patients.
        2. Symbolic interactionists examine how people are labeled as a result of disability; Eliot Freidson determined that the particular label results from (1) the person's degree of responsibility for the impairment, (2) the apparent seriousness of the condition, and (3) the perceived legitimacy of the condition.
        3. From a conflict perspective, persons with disabilities are a subordinate group in conflict with persons in positions of power in the government, the health care industry, and the rehabilitation business. When people with disabilities are defined as a social problem and public funds are spent to purchase goods and services for them, rehabilitation becomes big business.
    E. Social inequalities based on disability include prejudice and discrimination.
        1. Stereotypes of persons with a disability fall into two categories: (1) deformed individuals who also may be horrible deviants; or (2) persons who are to be pitied. Even positive stereotypes become harmful to people with a disability, those that excel despite the impairment. Disability rights advocates note that such stereotypes do not reflect the daily struggle of most people with disabilities.
        2. Employment, poverty, and disability are related; people may become economically disadvantaged as a result of chronic illness or disability; on the other hand, poor people are less likely to be educated and more likely to be malnourished and have inadequate access to health care.
        3. Disability has a stronger negative effect on women's labor participation than it does on men's.
        4. The cost of "mainstreaming" persons with disabilities closely relates to current expenditures that provide greater access to education and jobs.
VII.   HEALTH CARE IN THE FUTURE
    A. In the future, advanced health care technologies for high-income countries will provide more accurate and quicker diagnosis, effective treatment techniques, and increased life expectancy; however, new ethical concerns, the inability of the poor of the world, (including the U.S.) to have access to health care, the growth of managed-care companies in the U.S. making decisions formerly reserved for

physicians and hospital personnel and ethical concerns of technological advances all indicate a change in the future of health and medicine.

B. Ultimately, healthcare in the future will to some degree be up to each of us – what measures we take to safeguard ourselves against illness and disorders, and how we help others who are victims of diseases and disabilities.

ANALYZING AND UNDERSTANDING THE BOXES

After reading the chapter and studying the outline, re-read the boxes and write down key points and possible questions for class discussion.

Sociology in Everyday Life: How Much Do You Know About Health, Illness, and Health Care?

Key Points:

Discussion Questions:

1.

2.

3.

Sociology and Social Policy: Medicare and Medicaid: the Pros and Cons of Government-Funded Medical Care

Key Points:

Discussion Questions:

1.

2.

3.

Framing Health Issues in the Media: Death and Dying as a Television Spectacle

Key Points:

Discussion Questions:

1.

2.

3.

You Can Make a Difference: Joining the Fight Against Illness and Disease!

Key Points:

Discussion Questions:

1.

2.

3.

PRACTICE TESTS

MULTIPLE CHOICE QUESTIONS

Select the response that best answers the question or completes the statement:

1. The World Health Organization (WHO) defines health as the
   a. absence of disease.
   b. complete physical, mental, and social well-being.
   c. body in a state of equilibrium, with all parts in balance.
   d. absences of sickness, viruses, and pains.

2. _____ refers to the positive sense of complete well being; while _____ refers to an interference with health.
   a. Healing, disease
   b. Health, healing
   c. Health, illness
   d. Health care, disease

3. Health care in the United States
   a. is positively linked to money spent on health care and people's physical, mental, and social well-being.
   b. equates positively with longevity.
   c. is more expensive, but more productive for individuals.
   d. is any activity intended to improve health.

4. _____ is an institutionalized system for the scientific diagnosis, treatment, and prevention of illness.
   a. The medical-industrial complex
   b. Universal health care
   c. Medicine
   d. Disease

5. The World Health Organization reports that almost fourteen percent of all babies born in low-income countries die before they reach the
   a. first year of life.
   b. first month of life.
   c. first six weeks of life.
   d. first nine months of life.

6. Social epidemiologists are people who study the causes and distribution of:
   a. health, disease, and impairment in a population.
   b. epidemics that occur throughout a population.
   c. mental illness and behavior disorders in a population.
   d. the role of degenerative diseases in a population.

7. _____ are illnesses that strike suddenly, cause dramatic incapacitation, and sometimes death.
    a. Chronic diseases
    b. Disabilities
    c. Epidemics
    d. Acute diseases

8. _____ live longer than _____, but have higher rates of _____ illness.
    a. Men, women, chronic
    b. Women, men, acute
    c. Women, men, chronic
    d. Men, women, acute

9. _____ is a toxic, dependency-producing drug that is more addictive than _____.
    a. Heroin, nicotine
    b. Nicotine, heroin
    c. Nicotine, marijuana
    d. Marijuana, alcohol

10. Until the 1960s, _____ and _____ were the principal STDs in this country.
    a. genital herpes, syphilis
    b. gonorrhea, genital herpes
    c. gonorrhea, syphilis
    d. genital herpes, HIV/AIDS

11. Nearly 70 percent of all people with HIV/AIDS live in
    a. sub-Saharan Africa.
    b. South Asia.
    c. Southeast Asia.
    d. the United States.

12. The Flexner Report stated in 1910 that
    a. new medical schools should be developed for women.
    b. more medical schools should be developed for African Americans.
    c. new medical schools should be developed in rural areas.
    d. most existing medical schools were inadequate for teaching.

13. Throughout most of the twentieth century, medical care in the United States was paid for
    a. by HMOs.
    b. on a fee-for-service basis.
    c. by third party providers.
    d. managed care.

14. _____ is a jointly funded federal-state local health care program for the poor.
    a. Medicare
    b. Medical
    c. Medicaid
    d. Managed care

15. The only high-income nations without some form of universal health coverage for all its citizens are:
    a. the United States and the Union of South Africa
    b. Canada and the United States
    c. Japan and the United States
    d. Great Britain and the United States

16. Individuals who had little medical formal training in China who worked in rural communities, emphasizing health and health care and treating illness and disease are known as:
    a. barefoot doctors
    b. Red Guard doctors
    c. street doctors
    d. communist herb doctors

17. An approach to health care that treats the whole person, rather than just symptoms that occur, is known as _____.
    a. mainstream medicine
    b. managed care
    c. holistic medicine
    d. medicalization

18. People in the sick role are expected to
    a. be responsible for their condition.
    b. desire to get well.
    c. continue their normal roles and obligations.
    d. ignore competent medical assistance.

19. The practice of rapidly discharging patients from mental hospitals into the community is referred to as _____.
    a. reinstitutionalization
    b. demedicalization
    c. deconstruction
    d. deinstitutionalization

20. In contemporary industrial societies, disability often can be attributed to:
    a. epidemics related to poor sanitation and overcrowding
    b. urban density and poverty
    c. environment, lifestyle, and working conditions
    d. employment in high stress jobs in the primary tier of the labor market

21. Although it is estimated that only _____ percent of U.S. citizens will die prior to age 60, health experts estimate that in low-income nations such as Zambia, _____ percent of the people are not expected to see their 60th birthday.
    a. 5, 85
    b. 10, 90
    c. 13, 80
    d. 19, 75

22. The _____ is the number of deaths of infants under 1 year of age per 1,000 live births in a given year.
    a. child mortality rate
    b. infant mortality rate
    c. first year of life rate
    d. baby mortality

23. Sweden spends _____ per person on health care, but has a _____ infant mortality rate compared to the U. S.
    a. more, lower
    b. less, lower
    c. more, higher
    d. less, higher

24. _____ is the study of the causes and distribution of health, disease, and impairment throughout a population.
    a. Sociology of medicine
    b. Social epidemiology
    c. Social health
    d. Social medicine

25. In relation to social epidemiology, _____ include(s) biological agents such as insects, bacteria, and viruses that carry or cause disease.
    a. the human host
    b. the environment
    c. disease agents
    d. health coefficients

26. Sociologist Ingrid Waldron notes that gender roles and gender socialization contribute to the differences in life expectancy. All of the following are TRUE, **except**:
    a. males are more likely than females to engage in risky behavior
    b. men are more likely to be employed in dangerous occupations
    c. men are more reluctant to consult doctors than women
    d. women are less likely than men to use the health care system

27. Two of the most common sources of chronic disease and premature death are:
    a. tobacco and marijuana use
    b. AIDS and heart disease
    c. heart disease and alcohol abuse
    d. tobacco use and alcohol abuse

28. The Census Bureau projects that, by the year 2005, about _____ percent of the U.S. population will be at least 65 years old.
    a. 12
    b. 18
    c. 20
    d. 30

29. Recent research suggests that _____ may be a more significant factor(s) in health and mortality than is _____.
    a. social class, race or ethnicity
    b. gender, social class
    c. race or ethnicity, gender
    d. race or ethnicity, social class

30. _____ use occurs when a person takes a drug for no purpose other than for achieving a pleasurable feeling or an altered psychological state.
    a. Recreational
    b. Chronic
    c. Therapeutic
    d. Acute

31. High doses of marijuana smoked during pregnancy can
    a. disrupt the development of the fetus.
    b. result in congenital abnormalities of the fetus.
    c. result in neurological disturbance of the fetus.
    d. cause all of these health problems.

32. People who use cocaine over extended periods of time have _____ than do non-users.
    a. higher rates of heart problems
    b. higher rates of infection
    c. higher rates of cardiovascular disorders
    d. all of the above

33. When attempting to treat gonorrhea and syphilis, penicillin
    a. can cure most cases of both STDs, if the disease has not spread.
    b. cannot cure most cases of either STD.
    c. is ineffective in treating gonorrhea, but effective in treating syphilis.
    d. is ineffective in treating syphilis, but effective in treating gonorrhea.

34. All of the following regarding AIDS are true, **except**:
    a. AIDS reduces the body's ability to fight diseases
    b. no one actually dies of AIDS
    c. worldwide, the number of AIDS cases has dropped
    d. AIDS almost inevitably ends in death

35. With professionalization, _____ gained control over the entire medical establishment.
    a. medical schools
    b. medical insurance companies
    c. the American Medical Association
    d. licensed medical doctors

36. Despite public and private insurance programs, about _____ of all U.S. citizens are without health insurance or had difficulty getting or paying for medical care at some time in the last year.
    a. one-fourth
    b. one-third
    c. one-fifth
    d. one-half

37. An estimated _____ million people in the U. S. had no health insurance in 2005.
    a. 35.5
    b. 40.2
    c. 46.6
    d. 50.2

38. A _____ is a document stating a person's wishes regarding the medical circumstances under which his or her life should be terminated.
    a. living will
    b. living license
    c. living terms
    d. living file

39. Examples of _____ include the removal of certain behaviors from the list of mental disorders compiled by the American Psychiatric Association and the deinstitutionalization of mental health patients.
    a. demedicalization
    b. remedicalization
    c. medicalization
    d. non-medicalization

40. _____ is a condition in which a person has a severe mental ailment requiring extensive treatment with medication, psychotherapy, and sometimes hospitalization.
    a. Mental disorder
    b. Comorbidity
    c. Mental illness
    d. Mental dysfunctioning

TRUE/FALSE QUESTIONS

1. Sociologist Ingrid Waldron believes that life expectancy is linked to gender roles and gender socialization.
   T   F

2. Class is generally more of a factor for health problems than race or ethnicity.
   T   F

3. Therapeutic use of drugs occurs when a person takes a drug for a specific purpose, such as reducing fever or a cough.
   T   F

4. Cocaine is the most extensively used illegal drug in the United States.
   T   F

5. HIV can be transmitted by an infected woman breastfeeding her child.
   T   F

6. As a result of the Flexner report, medical schools for African Americans and women were expanded.
   T   F

7. According to Paul Starr, third-party health insurance providers are the main reason for medical inflation.
   T   F

8. Managed care allows doctors more individual freedom in choosing the medical treatment for their patients.
   T   F

9. The Canadian health care system is an example of socialized medicine.
   T   F

10. According to conflict theorists, physicians hold a legal monopoly over medicine.
    T   F

11. The subjective component of medicalization and demedicalization reflects the major concern of the interactionist perspective of health.
    T   F

12. According to Foucault, doctors gain power through the "clinical gaze."
    T   F

13. According to researchers, the prevalence of mental disorders in the U.S. is much less than most analysts had previously believed.
    T   F

14. Less than 15 percent of persons with a disability today were born with it.
    T   F

15. On average, male workers with a severe disability make 50% of what their co-workers without disabilities earn.
T   F

16. According to Box 18.1, the medical-industrial complex has operated in the U.S. with virtually no regulation.
T   F

17. The field of epidemiology, according to Box 18.1, focuses primarily on how individuals acquire disease and bodily injury.
T   F

18. According to Box 18.2, Medicare is considered to be an entitlement program, not a welfare program.
T   F

19. The "biggest thing going in American," according to Box 18.3, is mortality – the more graphic, the better.
T   F

20. According to Box 18.4, Beverly Barnes did not initially intend to start the program now known as Patient Pride.
T   F

FILL-IN-THE-BLANK QUESTIONS

1. The state of complete physical, mental, and social well-being is defined as _____.

2. The average lifetime of people born in a specific year is known as the _____.

3. The _____ report changed the practice of medical education in the U.S.

4. A(n) _____ care system is one wherein all citizens receive medical services paid for by tax revenues.

5. The _____ is the set of patterned expectations that defines the norms and values appropriate for individuals who are sick and for those who interact with them.

6. _____ is an approach to health care that focuses on prevention of illness and disease and is aimed at treating the whole person.

7. The _____ perspective focuses on the meaning that social actors give their disease or illness.

8. The _____ perspective on health and illness states that the myth of the wise doctor was supported by the development of disease classification systems and new tests.

9. The research of _____ suggested that about 15 percent of whites have such high levels of racial prejudice that they tend to exhibit symptoms of serious mental illness.

10. A(n) _____ is a physical or mental health condition that stigmatizes or causes discrimination.

11. A(n) _____ is any substance, other than food or water, that, when taken into the body, alters its functioning in some way.

12. _____ provide, for a set monthly fee, total health care with an emphasis on prevention to avoid costly treatment later.

13. _____ is a health care program for persons age 65 or older who are covered by Social Security or who are eligible and "buy into" the program by paying a monthly premium.

14. In a(n) _____ health care program, doctors must get approval before they perform certain procedures or admit a patient to a hospital.

15. A health care system in which all citizens receive medical services paid for by tax revenues is referred to as a(n) _____.

SHORT ANSWER/ESSAY QUESTIONS

1. What is it meant by health, health care, and disability and why are these important concerns for both individual and entire societies?
2. How is health care paid for in the United States? In some other countries? Which system do you think is better for people in a society? Explain your response.
3. What is holistic medicine and how does it differ from traditional health care? What is meant by alternative health care?
4. What is meant by disability, and what are three theoretical perspectives on disabilities?
5. What are some of the major issues and concerns of health care in the future?

STUDENT CLASS PROJECTS AND ACTIVITIES

1. Collect at least 10 articles dealing with current issues of health and health care found in the text, such as alcoholism, concerns about diet and exercise, AIDS, current causes of death, the American health care system, the profession of medicine (including doctors, nurses, hospitals), health insurance, prepaid health care, managed health care, and health care in other countries. The articles can come from newspapers, professional journals, magazines, etc. You are to: (1) collect the articles; (2) write a summary explaining the message of each article; (3) write a personal reaction to each article; (4) provide a conclusion and personal evaluation of the project, and (5) provide a bibliographic reference for each article selected. Submit the articles at the appropriate time.

2. Investigate a health care system found in another modern industrial country that has a reputation of having a fairly good, reliable health care delivery system. Some suggestions are: Great Britain, Canada, Germany, Sweden, Norway, and Japan. You are to describe in your paper: (1) the specific type(s) of health care system; (2) who receives the health care; (3) how the program is funded; (4) if personal choice is allowed–even for the higher income; (5) who the professionals are in the system; (6) their training; (7) how serious illness or disease is treated; (8) any other information pertinent to this research. In the writing of your paper, include a bibliography, a summary, a conclusion, and a personal evaluation of this project. Submit your paper to your instructor at the appropriate time.

3. The October 12, 2007 issue of *AARP* listed the 50 top hospitals in the United States, as cited by *Consumer's Checkbooks*, a nonprofit consumer education organization. This article summarizes the major findings of the publication *Consumer's Checkbook's Guide to Hospitals*, which rates more than 4,500 hospitals nationwide. Read this article and summarize the major criteria and the major findings of the report. Are there any surprises? What constitutes the accreditation score, the physician's rating, and special offerings/programs/research of the top-ranked hospitals? List the top 25 hospitals in the nation. Next, discuss the top-ranked hospital, noting the explanations for its ranking. Write up your research following any specific directions provided by your instructor.

4. According to an article in *Time* magazine (November 8, 1999), what were the ten causes of death and disabilities in the 1990s? What are the projections for the top ten causes of death and disability in the year 2020? Select one of these causes of the year 2020 and research that specific cause. Why is it a projected cause of death or disability? Why is it listed as a top ten projected cause of death or disability? What should be done to diminish the devastating effects of that cause? Submit your paper following any directions provided by your instructor.

INTERNET ACTIVITIES

1. The **National Institute of Health, http://www.nih.gov/**, is one of eight agencies that are part of the Department of Health and Human Services. In turn, it consists of over twenty-five agencies such as the National Institute of Allergy and Infectious Diseases, the National Institute on Alcohol Abuse and Alcoholism, and the National Institute of Drug Abuse. Investigate the wealth of information on this site.

2. To explore one of the major medical and **health** research institutions in the U.S., visit the web site for the **Centers for Disease Control, http://www.cdc.gov/**, and read about the CDC, its facilities, mission, people, budget, data and statistics. The CDC collects data on a wide range of topics, including cancer rates, AIDS rates, violence, and rare illnesses (e.g., Ebola and Hanta viruses).

3. **Health** problems in other countries are often linked to poor sanitation and hygiene. To learn more about these kinds of diseases, including the guinea worm, which causes internal injuries and permanent, disfiguring scarring, visit the web site for the **World Health Organization, http://www.who.int/home-page/**. Be sure to note some of the differences between the kinds of diseases the CDC focuses on compared to the World Health Organization.

4. The **Department of Health and Human Services, http://www.os.dhhs.gov/**, is a federal agency that provides information on numerous government programs, including Medicare and Medicaid, and the State Children's Health Insurance Program, as well as medical fraud.

5. The **National Institute of Mental Health, http://www.nimh.nih.gov/** (also belongs to the NIH), provides information on mental illness, child and adolescent violence, and rural mental health.

INFOTRAC COLLEGE EDITION EXERCISES

Visit the **InfoTrac College Edition** website at: **http://www.wadsworthmedia.com/webtutor/infotrac.htm**. You will arrive at a screen that enables you to search topics.

1. Search for articles (both professional and news) related to **Health Maintenance Organizations**. How effective are **HMO**s? Are they more interested in cutting costs than in providing quality health care?

2. How pervasive is **Medicare** fraud or **medical fraud** in general? How does this impact our society? Search for news articles that expose these kinds of problems. What are the social issues involved?

3. A general search using the topic **disability** will allow you to get a sense of how this condition touches on a broad range of social life.

4. After searching for periodical references for **holistic medicine**, examine the range of journals in which this term appears. Talk about the different "backdoors" that these nontraditional methods often have to take.

SOLUTIONS

MULTIPLE CHOICE QUESTIONS

1. B, p. 592
2. C, p. 592
3. D, p. 592
4. C, p. 592
5. A, p. 592
6. A, p. 595
7. D, p. 596
8. C, p. 597
9. B, pp. 598-599
10. C, p. 599
11. A, p. 600
12. D, p. 601
13. B, p. 602
14. C, p. 603

15. A, p. 603
16. A, p. 607
17. C, p. 610
18. B, p. 611
19. D, pp. 615-616
20. C, pp. 618-619
21. C, p. 512
22. B, p. 592
23. B, p. 595
24. B, p. 595
25. C, p. 596
26. D, p. 597
27. D, p. 596
28. C, p. 596

29. A, p. 597
30. A, p. 598
31. D, p. 599
32. D, p. 599
33. A, p. 599
34. C, p. 600
35. D, p. 602
36. B, p. 604
37. A, p. 605
38. A, p. 608
39. A, p. 613
40. C, pp. 614-615

TRUE/FALSE QUESTIONS

1. T, p. 597
2. T, p. 597
3. T, p. 597
4. F, p. 599
5. T, p. 600
6. F, p. 601
7. T, p. 603

8. F, p. 604
9. F, p. 607
10. T, p. 612
11. T, p. 613
12. F, p. 614
13. F, p. 615
14. T, p. 618

15. T, p. 621
16. F, p. 594
17. F, p. 594
18. T, p. 605
19. T, p. 608
20. T, p. 622

FILL-IN-THE-BLANK QUESTIONS

1. health, p. 592
2. life expectancy, p. 592
3. Flexner, p. 601
4. universal health, p. 606
5. sick role, p. 611
6. Holistic medicine, p. 610
7. symbolic interaction, p. 612
8. postmodern, p. 614
9. Thomas Pettigrew, p. 617

10. disability, p. 618
11. drug, p. 597
12. Health maintenance organizations, p. 604
13. Medicare, p. 603
14. managed care, p. 604
15. universal health care system, p. 606

19

POPULATION AND URBANIZATION

BRIEF CHAPTER OUTLINE

DEMOGRAPHY: THE STUDY OF POPULATION
    Fertility
    Mortality
    Migration
    Population Composition
POPULATION GROWTH IN GLOBAL CONTEXT
    The Malthusian Perspective
    The Marxist Perspective
    The Neo-Malthusian Perspective
    Demographic Transition Theory
    Other Perspectives on Population Change
A BRIEF GLIMPSE AT INTERNATIONAL MIGRATION THEORIES
URBANIZATION IN GLOBAL PERSPECTIVE
    Emergence and Evolution of the City
    Preindustrial Cities
    Industrial Cities
    Postindustrial Cities
PERSPECTIVES ON URBANIZATION AND THE GROWTH OF CITIES
    Functionalist Perspectives: Ecological Models
    Conflict Perspectives: Political Economy Models
    Symbolic Interactionist Perspectives: The Experience of City Life
PROBLEMS IN GLOBAL CITIES
URBAN PROBLEMS IN THE UNITED STATES
    Divided Interests: Cities, Suburbs, and Beyond
    The Continual Fiscal Crisis of the Cities
RURAL COMMUNITY ISSUES IN THE UNITED STATES
POPULATION AND URBANIZATION IN THE FUTURE

CHAPTER SUMMARY

Demography is the study of the size, composition, and distribution of the population. Population growth is the result of **fertility** (births), **mortality** (deaths), and **migration**. The **population composition** – the biological and social characteristics of a population is affected by changes in fertility, mortality, and migration. One measure of population composition is **sex ratio** – the number of males for every hundred females in a given population. For demographics, sex and age are significant characteristics; they are key indicators of **fertility** and **mortality** rates. The current distribution of a population by sex and age can be depicted in a population pyramid. Over two hundred years ago, Thomas Malthus warned that overpopulation would result in major global problems such as poverty and starvation. According the Marxist perspective, overpopulation occurs because of capitalists' demands for a surplus of workers to suppress wages and heighten workers' productivity. More recently, neo-Malthusians have re-emphasized the dangers of our population, depicting earth as a dying planet with too many people, too little food, compounded by environmental degradation. **Demographic transition** is the process by which some societies have moved from high birth and death rates to relatively low birth and death rates as a result of technological development. Other perspectives on population change include *rational choice theory*, the epidemiological transition, economic development, and the process of "westernization." In explaining international migration theories, the *neoclassical economics approach*, the new household's economics of migration approach, conflict and world systems theory, all add to our knowledge of the way people migrate. *Urban sociology* is the study of social relationships and political and economic structures in the city. Cities are a relatively recent innovation when compared to the length of human existence. Because of their limited size, preindustrial cities tend to provide a sense of community and a feeling of belonging. The Industrial Revolution changed the size and nature of the city; people began to live close to the factories and to one another, which led to overcrowding and poor sanitation. In postindustrial cities, some people live and work in suburbs or outlying edge cities. Functionalist perspectives (ecological models) of urban growth include the *concentric zone model*, the *sector model*, and the *multiple nuclei model*. According to the political economy models of conflict theorists, urban growth is influenced by capital investment decisions, power and resource inequality, class and class conflict, and government subsidy programs. Feminist theorists suggest that cities have *gender regimes*; women's lives are affected by both public and private patriarchy. Symbolic interactionists focus on the positive and negative aspects of peoples' experiences in the urban settings. Rapid population growth in many global cities is producing a wide variety of urban problems including overcrowding, environmental pollution and disappearance of farmland as well as creating a limit in the availability of basic public services. Urbanization, suburbanization, **gentrification**, and the growth of *edge cities* have had a dramatic impact on the U.S. population. Many central cities continue to experience fiscal crises that have resulted in cuts in services, lack of maintenance of the infrastructure, as well as loss of resources because of terrorist attacks. Many cities and large urban areas have created a "disabling" environment for many people; access is critical in order for persons with a disability to become productive members of the community. Some of the traditional issues of the rural community include financial and emotional, proliferation of superstores, and increase in tourism. Rapid global population growth is inevitable in the future. The urban population will triple as increasing numbers of people in lesser developed and developing nations migrate from rural areas to mega cities that contain a high percentage of a region's population.

LEARNING OBJECTIVES

After reading Chapter 19, you should be able to:

1. Describe the study of demography and define the basic demographic concepts.

2. Define the concept of zero population growth.

3. Trace the historical development of cities.

4. Identify the major characteristics of preindustrial, industrial, and postindustrial cities.

5. Explain the Malthusian perspective on population growth.

6. Discuss functionalist perspectives on urbanization and outline the major ecological models of urban growth.

7. Describe global patterns of urbanization in core, peripheral, and semiperipheral nations.

8. Summarize the key assumptions of the major urban theorists.

9. Describe the neo-Malthusian perspective on population growth

10. Present the major ideas behind demographic transition theory.

11. Explain the symbolic interactionist perspective on urban life.

12. Discuss the Marxist perspective on population growth and compare it with the Malthusian perspective.

13. Compare and contrast conflict and functionalist perspectives on urban growth.

14. Construct a comprehensive outlook on global migration using the new households economics of migration approach, the neoclassical economic approach, network theory and institutional theory.

15. Determine why demographic transition theory may not apply to population growth in all societies.

16. Discuss the major problems facing urban areas in the United States today and in the future.

17. Assess the major problems and potential solutions for undocumented workers in the United States.

KEY TERMS

KEY PEOPLE

CHAPTER OUTLINE

I. DEMOGRAPHY: THE STUDY OF POPULATION
   A. **Demography** is a subfield of sociology that examines population size, composition, and distribution.
   B. **Fertility** is the actual level of childbearing for an individual or a population.
      1. The **crude birth rate** is the number of live births per 1,000 people in a population in a given year.
      2. In most areas of the world, women are having fewer children; women who have six or more children tend to live in agricultural regions where children's labor is essential to the family's economic survival and child mortality rates are very high.

C. A decline in **mortality** – the incidence of death in a population – has been the primary cause of world population growth in recent years.
  1. The **crude death rate** is the number of deaths per 1,000 people in a population in a given year.
  2. The **infant mortality rate** is the number of deaths of infants under 1 year of age per 1,000 live births in a given year.
D. **Migration** is the movement of people from one geographic area to another for the purpose of changing residency.
  1. While **immigration** is the movement of people into a geographic area to take up residency, **emigration** is the movement of people out of a geographic area to take up residency elsewhere.
  2. The **crude net migration rate** is the net number of migrants (total in migrants minus total out migrants) per 1,000 people in a population in a given year.
E. **Population composition** is the biological and social characteristics of a population, including age, sex, race, marital status, education, occupation, and income.
  1. The **sex ratio** is the number of males for every hundred females in a given population; a sex ratio of 100 indicates an equal number of males and females.
  2. A **population pyramid** is a graphic representation of the distribution of a population by sex and age.
II. POPULATION GROWTH IN GLOBAL CONTEXT
  A. The Malthusian Perspective
  1. According to **Thomas Robert Malthus**, the population (if left unchecked) would exceed the available food supply; population would increase in a geometric progression (2, 4, 8, 16 . . .) while the food supply would increase only by an arithmetic progression (1, 2, 3, 4 . . .).
  2. This situation could end population growth and perhaps the entire population unless positive checks (such as famines, disease and wars) or preventive checks (such as sexual abstinence and postponement of marriage) intervened.
  B. The Marxist Perspective
  1. According to Karl Marx and Frederick Engels, food supply does not have to be threatened by overpopulation; through technology, food for a growing population can be produced.
  2. Overpopulation occurs because capitalists want a surplus of workers (an industrial reserve army) to suppress wages and force employees to be more productive.
  3. Overpopulation will lead to the eventual destruction of capitalism; when workers become dissatisfied, they will develop class consciousness because of shared oppression.
  C. The Neo-Malthusian Perspective
  1. Neo-Malthusians (or "New Malthusians") reemphasized the dangers of overpopulation and suggested that an exponential growth pattern is occurring.

2. Overpopulation and rapid population growth result in global environmental problems, and people should be encouraging **zero population growth** – the point at which no population increase occurs from year to year because the number of births plus immigrants is equal to the number of deaths plus emigrants.

D. Demographic Transition Theory
  1. **Demographic transition** is the process by which some societies have moved from high birth and death rates to relatively low birth and death rates as a result of technological development.
  2. Demographic transition is linked to four stages of economic development:
     a. Stage 1: Preindustrial Societies – little population growth occurs, high birth rates are offset by high death rates.
     b. Stage 2: Early Industrialization – significant population growth occurs, birth rates are relatively high while death rates decline.
     c. Stage 3: Advanced Industrialization and Urbanization – very little population growth occurs, both birth rates and death rates are low.
     d. Stage 4: Postindustrialization – birth rates continue to decline as more women are employed full-time and raising children becomes more costly; population growth occurs slowly, if at all, due to a decrease in the birth rate and a stable death rate.
     e. Other perspectives on population change.
        i. Other perspectives on population change include rational choice theory, the epidemiological transition, economic development and the process of "westernization."
        ii. Critics suggest that demographic transition theory may not accurately explain population growth in all societies; this theory best explains growth in Western societies.

III. A BRIEF GLIMPSE AT INTERNATIONAL MIGRATION THEORIES
  A. In explaining international migration theories, the neoclassical economics approach, the new households economics of migration approach, conflict and world systems theory, all add to our knowledge of the way people migrate.
  B. Network theory, push or pull factors, and institutional theory provide further explanation of global migrations.

IV. URBANIZATION IN GLOBAL PERSPECTIVE
  A. Urban sociology is a subfield of sociology that examines social relationships and political and economic structures in the city.
  B. Emergence and Evolution of the City
    1. Cities are a relatively recent innovation as compared with the length of human existence. According to Gideon Sjoberg, three preconditions must be present in order for a city to develop:
       a. A favorable physical environment
       b. An advanced technology that could produce a social surplus
       c. A well-developed political system to provide social stability to the economic system.

2. Sjoberg places the first cities in the Mesopotamian region or areas immediately adjacent to it at about 3500 B.C.E.; however, not all scholars agree on this point.
C. Preindustrial Cities
 1. The largest preindustrial city was Rome.
 2. Preindustrial cities were limited in size because of crowded housing conditions, lack of adequate sewage facilities, limited food supplies, and lack of transportation to reach the city.
 3. Many preindustrial cities had a sense of community – a set of social relationships operating within given spatial boundaries that provide people with a sense of identity and a feeling of belonging.
D. Industrial Cities
 1. The nature of the city changed as factories arose and new forms of transportation and agricultural production made it easier to leave the countryside and move to the city.
 2. Between 1700 and 1900, the population of many European cities mushroomed; London increased to about 6.5 million.
 3. New York City became the first U.S. *metropolis* – one or more central cities and their surrounding suburbs that dominate the economic and cultural life of a region.
 4. The countries of Japan and Russia industrialized later and the pattern of urbanization began to move quickly.
E. Postindustrial Cities
 1. Since the 1950s, postindustrial cities have emerged as the U.S. economy has gradually shifted from secondary (manufacturing) to tertiary (service and information processing) production.
 2. Postindustrial cities are dominated by "light" industry, such as computer software manufacturing, information processing services, educational complexes, medical centers, retail trade centers, and shopping malls.
V. PERSPECTIVES ON URBANIZATION AND THE GROWTH OF CITIES
A. Functionalist Perspectives: Ecological Models
 1. **Robert Park** based his analysis of the city on human ecology – the study of the relationship between people and their physical environment – and found that economic competition produces certain regularities in land use patterns and population distributions.
 2. Concentric zone model
  a. Based on Park's ideas, **Ernest W. Burgess** developed a model that views the city as a series of circular zones, each characterized by a different type of land use, that developed from a central core: (1) the central business district and cultural center; (2) the zone of transition – houses where wealthy families previously lived that have now been subdivided and rented to persons with low incomes; (3) working class residences and shops, and ethnic enclaves; (4) homes for affluent families, single family residences of white-collar workers, and shopping centers; and (5) a ring of small cities and towns comprised of estates owned by the wealthy and houses of commuters who work in the city.

b. Two important ecological processes occur: **invasion** is the process by which a new category of people or type of land use arrives in an area previously occupied by another group or land use, and **succession** is the process by which a new category of people or type of land use gradually predominates in an area formerly dominated by another group or activity.

c. **Gentrification** is the process by which members of the middle and upper-middle classes, especially whites, move into the central city area and renovate existing properties.

3. Sector Model

a. **Homer Hoyt**'s sector model emphasizes the significance of terrain and the importance of transportation routes in the layout of cities.

b. Residences of a particular type and value tend to grow outward from the center of the city in wedge-shaped sectors with the more expensive residential neighborhoods located along the higher ground near lakes and rivers, or along certain streets that stretch from the downtown area.

c. Industrial areas are located along river valleys and railroad lines; middle class residences exist on either side of wealthier neighborhoods; lower class residential areas border the central business area and the industrial areas.

4. Multiple Nuclei Model

a. According to **Chauncey Harris** and **Edward Ullman**, cities have numerous centers of development; as cities grow, they annex outlying townships.

b. In addition to the central business district, other nuclei develop around activities such as an educational institution or a medical complex; residential neighborhoods may exist close to or far away from these nuclei.

5. Contemporary Urban Ecology.

a. **Amos Hawley** viewed urban areas as complex social systems in which growth patterns are based on advances in transportation and communication.

b. Social area analysis examines urban populations in terms of economic status, family status, and ethnic classification.

B. Conflict Perspectives: Political Economy Models

1. According to Marx, cities are arenas in which the intertwined processes of class conflict and capital accumulation take place; class consciousness is more likely to occur in cities where workers are concentrated.

2. Three major themes are found in political economy models:

a. Patterns of urban growth and decline are affected by: (1) economic factors such as capitalist investments; and (2) political factors, including governmental protection of private property and promotion of the interests of business elites and large corporations.

b. Urban space has both an exchange value and a use value: (1) exchange value refers to the profits industrialists, developers, and bankers make from buying, selling, and developing land and buildings; (2) use value is the utility of space, land, and buildings for family and neighborhood life.

c. Structure and agency are both important in understanding how urban development takes place: (1) structure refers to institutions such as state bureaucracies and capital investment circuits that are involved in the urban development process; and (2) agency refers to human actors who participate in land use decisions, including developers, business elites, and activists protesting development.
3. According to political economy models, urban growth is influenced by capital investment decisions, power and resource inequality, class and class conflict, and government subsidy programs.
4. Gender Regimes in Cities
   a. According to feminist perspectives, urbanization reflects the workings of the political economy and patriarchy.
   b. Different cities have different *gender regimes* – prevailing ideologies of how women and men should think, feel, and act; how access to positions and control of resources should be managed; and how women and men should relate to each other.
   c. Gender intersects with class and race as a form of oppression, especially for lower-income women of color who live in central cities.
C. Symbolic Interactionist Perspectives: The Experience of City Life
1. Simmel's View of City Life
   a. According to **Georg Simmel**, urban life is highly stimulating; it shapes people's thoughts and actions.
   b. However, many urban residents avoid emotional involvement with each other and try to ignore events taking place around them.
   c. City life is not completely negative; urban living can be liberating – people have opportunities for individualism and autonomy.
2. Urbanism as a Way of Life
   a. **Louis Wirth** suggested that urbanism is a "way of life." Urbanism refers to the distinctive social and psychological patterns of city life.
   b. Size, density, and heterogeneity result in an elaborate division of labor and in spatial segregation of people by race/ethnicity, class, religion, and/or lifestyle; a sense of community is replaced by the "mass society" – a large scale, highly institutionalized society in which individuality is supplanted by mass media, faceless bureaucrats, and corporate interests.
3. Gans's Urban Villagers
   a. According to **Herbert Gans**, not everyone experiences the city in the same way; some people develop strong loyalties and a sense of community within central city areas that outsiders may view negatively.
   b. Five major categories of urban dwellers are: (1) cosmopolites – students, artists, writers, musicians, entertainers, and professionals who choose to live in the city because they want to be close to its cultural facilities;   (2) unmarried people and childless couples who live in the city because they want to be close to work and entertainment; (3) ethnic villagers who live in ethnically segregated neighborhoods; (4) the deprived – individuals who are very poor and see few future prospects; and (5) the trapped – those who cannot escape the city, including downwardly mobile persons, older persons, and persons with addictions.

    4. Gender and City Life
       a. According to **Elizabeth Wilson**, some men view the city as *sexual space* in which women, based on their sexual desirability and accessibility, are categorized as prostitutes, lesbians, temptresses, or virtuous women in need of protection.
       b. More affluent, dominant group women are more likely to be viewed as virtuous women in need of protection while others are placed in less desirable categories.
    5. Cities and Persons with a Disability
       a. Many cities and urban areas create a "disabling" environment for many people.
       b. **Harlan Hahn** suggests that historical patterns in the dynamics of capitalism contributed to discrimination against persons with disabilities, and this legacy remains today.

VI. PROBLEMS IN GLOBAL CITIES
  A. Natural increases in population account for two-thirds of new urban growth, and rural-to-urban migration accounts for the remainder.
  B. Rapid global population growth in Latin American and other regions is producing a wide variety of urban problems, including overcrowding, environmental pollution and the disappearance of farmland.
  C. Most cities in Africa, South America, and the Caribbean are in peripheral nations.
  D. Cities in semiperipheral nations, such as India, Iran, and Mexico, are confronted with unprecedented population growth.

VII. URBAN PROBLEMS IN THE UNITED STATES
  A. Divided Interests: Cities, Suburbs, and Beyond
    1. Since World War II, the U.S. population has shifted dramatically as many people have moved to the suburbs.
    2. Suburbanites rely on urban centers for employment and some services but pay property taxes to suburban governments and school districts; some affluent suburbs have state of the art school districts and infrastructure while central city services and school districts lack funds.
    3. Race, Class, and Suburbs
       a. The intertwining impact of race and class is visible in the division between central cities and suburbs.
       b. Most suburbs are predominantly white; many upper middle and upper class suburbs remain virtually all white; people of color who live in suburbs often are resegregated.
    4. Beyond the Suburbs
       a. *Edge cities* initially develop as residential areas beyond central cities and suburbs; then retail establishments and office parks move into the area and create an unincorporated edge city.
       b. Corporations move to edge cities because of cheaper land and lower utility rates and property taxes.

5. Likewise, Sunbelt cities grew in the 1970s, as millions moved from the north and northeastern states to southern and western states where there were more jobs and higher wages, lower taxes, pork barrel programs funded by federal money that created jobs and encouraged industry, and the presence of high technology industries.

B. The Continual Fiscal Crisis of the Cities
1. The largest cities in the United States have faced periodic fiscal crisis for many years, intensified by higher employee health care and pension costs, declining revenue, and increased expenditures for public safety and homeland safety.
2. Many cities have cut back on spending in areas other than public safety.
3. Even if the U.S. economy improves significantly in the near future, analysts believe that the positive effects of such a rebound will not improve the budgetary problems of our cities and towns for a number of years.

VIII. RURAL COMMUNITY ISSUES IN THE UNITED STATES
A. About 25 percent of the U.S. population resides in rural areas, identified as communities of 2,500 or less by the U.S. Census Bureau.
1. Rural communities today are more diverse; recently, more people from large urban areas and suburbs have moved into rural areas.
2. These recent immigrants to rural areas do not face some traditional problems experienced by long-term rural residents.

B. Individuals in rural areas whose livelihood is farming or other agricultural endeavors have experienced several difficulties.
1. crop failures
2. loss of small businesses
3. problems of divorce, alcoholism, and abuse
4. loss of farm
5. limited economic opportunities
6. limited health care

C. Other influencing factors
1. Proliferation of superstores—putting local small businesses out of business.
2. Increase in tourism in rural America.

IX. POPULATION AND URBANIZATION IN THE FUTURE
A. Rapid global population growth is inevitable: although death rates have declined in many low income nations, there has not been a corresponding decrease in birth rates.
B. In the future, low income countries will have an increasing number of poor people. The world's population will double, the urban population will triple as people migrate from rural to urban areas.
C. At the macrolevel, we may be able to do little about population and urbanization; at the microlevel, we may be able to exercise some degree of control over our communities and our own lives.

ANALYZING AND UNDERSTANDING THE BOXES

After reading the chapter and studying the outline, re-read the boxes and write down key points and possible questions for class discussion.

Sociology and Everyday Life: How Much Do You Know About U. S. Immigration?

Key Points:

Discussion Questions:

1.

2.

3.

You Can Make a Difference: Creating a Vital Link Between College Students and Immigrant Children

Key Points:

Discussion Questions:

1.

2.

3.

Framing Immigration in the Media: Media Framing and Public Opinion

Key Points:

Discussion Questions:

1.

2.

3.

Sociology in Global Perspective: Urban Migration and the "Garbage Problem"

Key Points:

Discussion Questions:

1.

2.

3.

PRACTICE TESTS

MULTIPLE CHOICE QUESTIONS

Select the response that best answers the question or completes the statement.

1. Demography is a subfield of sociology that examines
   a. population size.
   b. population composition.
   c. population distribution.
   d. all of these choices

2. _____ is the actual level of childbearing for an individual or a population, while _____ is the potential number of children that could be born if every woman reproduced at her maximum biological capacity.
   a. Birth rate, fertility rate
   b. Fertility rate, birth rate
   c. Fertility, fecundity
   d. Fecundity, fertility

3. The primary cause of world population growth in recent years is a(n)
   a. increase in the birth rate.
   b. decline in the death rate.
   c. decline in all infectious diseases.
   d. increase in post baby boom birth rates.

4. The average lifetime in years of people born in a specific year is known as:
   a. life expectancy.
   b. the crude mortality rate.
   c. the longevity table.
   d. age specific death rates.

5. _____ is the movement of people out of a geographic area to take up residency elsewhere.
    a. Immigration
    b. Emigration
    c. Transmigration
    d. Ex-migration

6. According to Thomas Malthus's perspective on population,
    a. the population would increase in a geometric progression while the food supply would increase in an arithmetic progression.
    b. the population would increase in an arithmetic progression while the population would increase in a geometric progression.
    c. the food supply is not threatened by overpopulation because technology makes it possible to produce the food and other goods needed to meet the demands of a growing population.
    d. societies move through a process of demographic transition.

7. According to Karl Marx and Frederick Engels's perspective on population,
    a. the population would increase in a geometric progression while the food supply would increase in an arithmetic progression.
    b. the population would increase in an arithmetic progression while the food supply would increase in a geometric progression.
    c. the food supply is not threatened by overpopulation because technology makes it possible to produce the food and other goods needed to meet the demands of a growing population.
    d. societies move through a process of demographic transition.

8. According to the demographic transition theory, significant population growth occurs because birth rates are relatively high while death rates decline in the _____ stage of economic development.
    a. preindustrial
    b. early industrial
    c. advanced industrial
    d. postindustrial

9. All of the following statements regarding preindustrial cities are true, **except**:
    a. the largest preindustrial city was Rome
    b. preindustrial cities were limited in size because of crowded housing conditions and a lack of adequate sewage facilities
    c. food supplies were limited in preindustrial cities
    d. preindustrial cities lacked a sense of community

10. The *Gemeinschaft* and *Gesellschaft* typology originated with:
    a. Emile Durkheim
    b. Gideon Sjoberg
    c. Max Weber
    d. Ferdinand Tönnies

11. _____ refers to one or more central cities and their surrounding suburbs that dominate the economic and cultural life of a region.
   a. Urban sprawl
   b. Megalopolis
   c. Metropolis
   d. Urbanization

12. Postindustrial cities are characterized by:
   a. "light" industry, information processing services, educational complexes, retail trade centers, and shopping malls
   b. the growth of the factory system
   c. agricultural production
   d. "heavy" industry, such as automobile manufacturing

13. Ecological models of urban growth are based on a _____ perspective.
   a. functionalist
   b. conflict
   c. neo Marxist
   d. interactionist

14. All of the following are ecological models of urban growth, except the
   a. concentric zone model.
   b. sector model.
   c. urban sprawl model.
   d. multiple nuclei model.

15. An upper-middle class doctor who moves her family from the suburbs into the central city to renovate an older home is an example of:
   a. succession
   b. gentrification
   c. ex-suburbanization
   d. downward immigration

16. According to political economy models, urban growth is
   a. influenced by terrain and transportation.
   b. based on the clustering of people who share similar characteristics.
   c. linked with peaks and valleys in the economic cycle.
   d. influenced by capital investment decisions, power and resource inequality, and government subsidy programs.

17. _____ refers to the tendency of some neighborhoods, cities, or regions to grow and prosper while others stagnate and decline.
   a. Invasion
   b. Succession
   c. Gentrification
   d. Uneven development

18. The United States, Japan, and Germany are _____ nations.
    a. core
    b. peripheral
    c. semiperipheral
    d. globalized

19. Sociologist _____ has argued that urbanism is a "way of life."
    a. Herbert Gans
    b. Georg Simmel
    c. Louis Wirth
    d. Elizabeth Wilson

20. According to sociologists, edge cities
    a. initially develop as industrial parks.
    b. drain taxes from central cities and older suburbs.
    c. have existed since World War II.
    d. always correspond to municipal boundaries.

21. As used by demographers, a _____ is a group of people who live in a specified geographic area.
    a. society
    b. culture
    c. sect
    d. population

22. Changes in population occur as a result of three processes. All of the following processes are applicable, **except**:
    a. fertility.
    b. mortality.
    c. social disorder.
    d. migration.

23. The potential number of children that could be born if every woman produced at her maximum biological capacity is defined as _____.
    a. reproduction
    b. fecundity
    c. fertility
    d. birthing

24. The most basic measure of fertility is the _____ rate.
    a. fecundity
    b. age-specific birth
    c. crude birth rate
    d. reproduction

25. The primary cause of world population growth in recent years has been a decline in _____ – the incidence of death in a population.
    a. the crude death rate
    b. the total death rate
    c. the death quota
    d. mortality

26. The crude birth rate in the United States in 2004 was _____ as compared with an all time high rate of 27 per 1000 in 1947.
    a. 10.0 per 1,000
    b. 6.0 per 1,000
    c. 14.0 per 1,000
    d. 17.0 per 1,000

27. The _____ is an important reflection of a society's level of preventive (prenatal) medical care, maternal nutrition, childbirth procedures, and neonatal care for infants.
    a. infant death syndrome
    b. child death index
    c. infant mortality rate
    d. underage mortality rate.

28. _____ is the number of people living in a specific geographic area.
    a. Distribution
    b. Migration
    c. Emigration
    d. Density

29. _____ is the movement of people out of a geographic area to take up residency elsewhere.
    a. Distribution
    b. Migration
    c. Immigration
    d. Emigration

30. _____ factors at the international level include violence, war, famine, and political unrest.
    a. Pull
    b. Push
    c. Discharge
    d. Hinge

31. _____ was one of the first scholars to systematically study the effects of population. He argued that "the power of population is infinitely greater than the power of the earth to produce subsistence (food) for man."
    a. Thomas Malthus
    b. Max Weber
    c. Karl Marx
    d. Emile Durkheim

32. One measure of population composition is the _____, which is the number of males for every hundred females in a given population.
    a. sex ratio
    b. rate of gender composition
    c. population composition
    d. refined gender rate

33. Thomas Malthus argued that the population would increase in a geometric (exponential) progression while the food supply would increase only by an arithmetic progression; thus, a _____ occurs.
    a. doubling effect
    b. positive check
    c. preventive check
    d. demographic transition

34. The largest preindustrial city was:
    a. Jericho
    b. Athens
    c. Rome
    d. Babylon

35. Based on his preconditions for the development of cities, sociologist Gideon Sjoberg places the first cities in
    a. Egypt.
    b. Italy.
    c. Sicily.
    d. the Middle Eastern region of Mesopotamia.

36. A _____ refers to one or more central cities and their surrounding suburbs that dominate the economic and cultural life of a region.
    a. metropolis
    b. megalopolis
    c. conglomeration
    d. census district

37. Sociologist Robert Park based his analysis of the city on _____, which is the study of the relationship between people and their physical environment.
    a. demography
    b. urban sociology
    c. human ecology
    d. physical sociology

38. Urban ecologist _____ revitalized the ecological tradition by linking it more closely with functionalism.
    a. Herbert Gans
    b. Louis Wirth
    c. Amos Hawley
    d. Joe Fagin

39. According to _____, many residents develop strong loyalties and a sense of community in central-city area that outsiders may view negatively.
    a. Georg Simmel
    b. Herbert Gans
    c. Homer Hoyt
    d. Ernest W. Burgess

40. A lending practice of banks includes the _____ of certain properties so that acquiring a loan is virtually impossible.
   a. redlining
   b. stonewalling
   c. barricading
   d. edging

TRUE/FALSE QUESTIONS

1. Demographers analyze fertility, mortality, and migration rates.
   T   F

2. The world's population is increasing by more than 76 million people per year.
   T   F

3. In most areas of the world, women are having more children.
   T   F

4. The primary cause of world population growth in recent years has been a decline in mortality.
   T   F

5. Pull factors of migration include political unrest and war.
   T   F

6. Sex ratio is the number of females per 100 males in a given population.
   T   F

7. Population pyramids are graphic representations of the distribution of a population by sex and age.
   T   F

8. According to the Marxist perspective, overpopulation occurs because capitalists desire to have a surplus of workers so as to suppress wages and increase workers' productivity.
   T   F

9. About 50 percent of the world's population lives in cities.
   T   F

10. The sector model views the city as a series of circular areas or zones, each characterized by a different type of land use that developed from a central core.
    T   F

11. According to conflict theorists, cities grow and decline by chance.
    T   F

12. According to feminist theorists, public patriarchy may be perpetuated by cities through policies that limit women's access to paid work and public transportation.
    T   F

13. Symbolic interactionists examine the experience of urban life rather than the political economy of the city.
T   F

14. Herbert Gans has suggested that almost all city dwellers live in urban areas by choice.
T   F

15. Nationally, most suburbs are predominantly white.
T   F

16. According to W. Parker Frisbie and J. D. Kasarda, the influence of human ecology in the field of urban sociology is still very strong today.
T   F

17. According to Box 19.1, unauthorized immigrants from Mexico and Latin America represent slightly less than 50 percent of the unauthorized population in the United States.
T   F

18. According to Box 19.3, Mexican immigrants working in the United States in mostly low-paying jobs send more of their monies back home, which is more than Mexico earns from tourism or foreign investment.
T   F

19. In some global cities, according to Box 19.4, less than 10 percent of the population has regular collection of household waste.
T   F

20. As depicted in the Photo Essay, when we look at the faces of the people around our country today, we see a wide, wide diversity of human beings living together positively and peacefully.
T   F

FILL-IN-THE-BLANK QUESTIONS

1. The biological and social characteristics of a population is known as the
   _____.

2. A graphic representation of a population by sex and age is defined as a
   _____.

3. According to _____, limits to fertility are defined as _____.

4. The point at which no population increases occurs from year to year is defined as
   _____.

5. According to Lynn Appleton, different kinds of cities have different _____ which are prevailing ideologies of how men and women should think.

6. According to the sociologist Herbert Gans, _____ are students, artists, entertainers, and professionals who choose to live in the city.

7. The _____ model of urban development cities has numerous centers of development.

8. According to _____, urban living could have a liberating effect on people because they have opportunities for individualism and autonomy.

9. _____ is the movement of people into a geographic area to take up residency.

10. Things that attract people to move into another country are known as _____ factors.

11. The incidence of death in a population is defined as _____.

12. The first metropolis in the U. S. was _____.

13. For Thomas Malthus, the only acceptable preventative check was _____.

14. The study of the relationship between people and their physical environment is defined as _____.

15. _____ is the process by which members of the middle and upper-middle classes (especially whites), move into the central-city area and renovate existing properties.

SHORT ANSWER/ESSAY QUESTIONS

1. What is demography and how are people affected by demographic changes?
2. What is the Malthusian perspective? Do the views of Karl Marx support this? Explain.
3. Explain the three functionalist models of urban growth.
4. What are the stages in the demographic transition theory?
5. How have recent acts of global terrorism affect U.S. cities and other cities about the world?

STUDENT CLASS PROJECTS AND ACTIVITIES

1. Utilizing the most current U.S. Census data, construct a population pyramid of the United States, your home state, and the city in which you live or reside, or of any city in the United States. Provide the pyramids, interpretations of the statistics, a summary of the data, a conclusion from each pyramid constructed, and bibliographical references. Submit your project to your instructor at the appropriate time.

2. Construct a demographic analysis of the state in which your college or university is located and compare this with an analysis of the entire United States. Present data on the size of the population, the numbers and percentage of increase or decrease from the past twenty years, the birth rate, the death rate, sex ratio, infant mortality, cause of death, and life expectancy. Provide a summary of the data, a conclusion, and an evaluation of this project. Submit your project at the appropriate time to your instructor.

3. Select and submit 10 articles from various sources on any topic covered in this chapter Population and Urbanization. Because this is a topic of interest and concern, there are many sources of information on this subject. You are to provide a copy of each article with your paper and you must: (1) provide the title of the article; (2) provide the data and source of the article; (3) provide a summary of the article; (4) provide two or three issues that each article raises; (5) provide at least two to three questions per each article that are generated from reading the article; (6) provide a personal conclusion and evaluation of this project. Submit your paper to your instructor at the appropriate time, following any additional instructions you may have received.

4. Construct a plan of a model city. In the plan, provide for all necessary public facilities, such as: an efficient transportation system, an efficient governmental system, the most efficient use of land, provision for industrial and commercial bases, provision for education, medical, recreational, and religious facilities, provision for optimum housing, promotions for attracting an equitable balance of ethnic, age, and income groups, and provision for long-range planning for optimum population size and city size. Include any other amenities in the planning of this optimum model city. Submit this project to your instructor at the appropriate time.

INTERNET ACTIVITIES

1. Explore the **HUD** site, **http://www.hud.gov/**, and learn about the history of the government's involvement in the life of cities. The mission of the Department of Housing and Urban Development is to provide "a decent, safe, and sanitary home and suitable living environment for every American."

2. If you are interested in more detailed information on the population crisis in Africa you should access the following sites. Demographic information collected by the World Health Organization, the United Nations and the U.S. Census Bureau are available. **http://www.un.org/esa/africa/, http://www.afro.who.int/**

3. Surf **http://www.un.org/popin/** to learn about world population trends. It is possible from this site to obtain statistical reports, United Nations documents, wall charts, and records of population policies of governments and world organizations.

4. **Population.com, http://www.population.com/**, is a huge resource for studies on population, **migration** and urbanization. This site has an extensive report as well as charts on population trends around the world. You can apply some of the concepts you have learned in this chapter to a specific region of the globe with a distinct history and culture.

5. This is the homepage for the **National Urban League, http://www.nul.org/**. The Urban League movement was founded in 1910. The National Urban League, headquartered in New York City, spearheads this nonprofit, nonpartisan, community-based movement. The heart of the Urban League movement is the professionally staffed Urban League affiliates in over 100 cities in 34 states and the District of Columbia. Search through the site and discover what kinds of issues urban leaders are facing.

INFOTRAC COLLEGE EDITION EXERCISES

Visit the **InfoTrac College Edition** website at:
http://www.wadsworthmedia.com/webtutor/infotrac.htm. You will arrive at a screen that enables you to search topics.

1. Investigate **global population** from a variety of perspectives. Determine the key issues based on these articles.

2. A key component in understanding **demographic transitions** is the **sex ratio** of a society. You will need to weed out articles about fig wasps and turtles, but there are a number of very informative reports and research articles on this phenomenon. Interesting comparisons can be made between biological studies of animals and those about humans' social behavior.

3. Conduct a keyword search for **zero population growth**. Find the article entitled: *Allowing fertility decline: 200 years after Malthus's essay on population*. The economic opportunity model in contrast to the **demographic transition** model is presented as a way to reduce population growth. Analyze this new alternative in light of what you have learned from reading chapter 15.

4. Search InfoTrac for articles about **urbanization**. There are a number of interesting articles that you can read, such as:
   a. How progress affects the United States
   b. The Urbanization of opera: Music theater in Paris in the nineteenth century
   c. You can't keep a good myth down
   d. Urbanization and the decline of witchcraft: an examination of London
   e. Relationships between health services, socioeconomic variables and inadequate weight gain among Brazilian children
   f. Asphalt Jungle: A global land rush is gobbling up the space and resources needed for agriculture and wildlife. Only the cockroaches are thrilled.

SOLUTIONS

MULTIPLE CHOICE QUESTIONS

1. D, p. 629
2. C, p. 629
3. B, p. 631
4. A, p. 631
5. B, p. 632
6. A, p. 637
7. C, p. 637
8. B, p. 639
9. D, p. 643
10. D, p. 643
11. C, p. 644
12. A, p. 644
13. A, p. 645
14. C, p. 645

15. B, p. 646
16. D, p. 647
17. D, p. 647
18. A, p. 652
19. C, p. 649
20. B, p. 655
21. D, p. 629
22. C, p. 629
23. B, p. 629
24. C, p. 630
25. D, p. 631
26. C, p. 630
27. C, p. 631
28. D, p. 632

29. D, p. 632
30. B, p. 632
31. A, p. 636
32. A, p. 636
33. A, p. 637
34. B, p. 643
35. D, p. 643
36. A, p. 643
37. C, p. 645
38. C, p. 646
39. B, p. 649
40. A, p. 654

TRUE/FALSE QUESTIONS

1. T, p. 629
2. T, p. 628
3. F, p. 630
4. T, p. 631
5. F, p. 632
6. F, p. 636
7. T, p. 636

8. T, p. 637
9. T, p. 642
10. F, p. 646
11. F, p. 647
12. T, p. 648
13. T, p. 648
14. F, p. 649

15. T, p. 654
16. T, p. 646
17. T, p. 630
18. T, p. 641
19. T, p. 653
20. T, p. 634

FILL-IN-THE-BLANK QUESTIONS

1. population composition, p. 633
2. population pyramid, p. 636
3. Thomas Malthus, p. 637 ; preventative checks, p. 637
4. zero population growth, p. 639
5. gender regimes, p.648
6. cosmopolites, p. 649
7. multiple nuclei, p. 646

8. Georg Simmel, p. 649
9. Immigration, p. 632
10. pull, p. 637
11. mortality, p. 631
12. New York City, p. 644
13. moral restraint, p. 637
14. human ecology, p. 645
15. Gentrification, p. 646

20

COLLECTIVE BEHAVIOR, SOCIAL MOVEMENTS, AND SOCIAL CHANGE

BRIEF CHAPTER OUTLINE

CHAPTER SUMMARY

Social change is the alteration, modification, or transformation of public policy, culture, or social institutions over time. Such change usually is brought about by **collective behavior** – voluntary, often spontaneous activity that is engaged in by a large number of people and typically violates dominant group norms and values. A **crowd** is a relatively large number of people who are in one another's immediate vicinity. Five categories of crowds have been identified: (1) casual crowds are relatively large gatherings of people who happen to be in the same place at the same time; (2) conventional crowds are comprised of people who specifically come together for a scheduled event and thus share a common focus; (3) expressive crowds provide opportunities for the expression of some strong emotion; (4) acting crowds are collectivities so intensely focused on a specific purpose or object that they may erupt into violent or destructive behavior; and (5) protest crowds are gatherings of people who engage in activities intended to achieve specific political goals. Protest crowds sometime participate in **civil disobedience** – nonviolent action that seeks to change a policy or law by refusing to comply with it. Explanations of crowd behavior include contagion theory, social unrest and circular reaction, convergence theory, and emergent norm theory. Examples of **mass behavior** – collective behavior that takes place when people respond to the same event in much the same way – include rumors, gossip, mass hysteria, fads, fashions, and public opinion. The major types of **social movements** – organized groups that act consciously to promote or resist change through collective action – are reform movements, revolutionary movements, religious movements, alternative movements, and resistance movements. Sociological theories explaining social movements include relative deprivation theory, value-added theory, resource mobilization theory, social constructionist theory; frame analysis, political opportunity theory and new social movement theory. Social change produces many challenges that remain to be resolved: environmental problems, changes in the demographics of the population, and new technology that benefits some – but not all – people. As we head into the future, we must use our sociological imaginations to help resolve the issues of the twenty-first century.

LEARNING OBJECTIVES

After reading Chapter 20, you should be able to:

1. Define collective behavior and describe the conditions necessary for such behavior to occur.

2. Distinguish between crowds and masses. Identify casual, conventional, expressive, acting, and protest crowds.

3. Identify the stages in social movements.

4. Define mass behavior and describe the most frequent types of this behavior.

5. Describe the difference between rumors and gossip and between fads and fashions.

6. 6. State the key assumptions of resource mobilization theory.

7. State the key assumptions of the frame analysis approach to understanding social movements.

8. Describe social movements and note when and where they are most likely to develop.

9. Describe new social movement theory and the diverse array of social movements on which it focuses.

10. Show how the current concern with global climate change can develop into a social movement.

11. Distinguish among mobs, riots, and panics.

12. Distinguish the key elements of these four explanations of collective behavior: contagion theory, social unrest and circular reaction, convergence theory, and emergent norm theory.

13. Compare relative deprivation theory and value-added theory as explanations of why people join social movements.

14. Determine why social movements may be an important source of social change.

15. Differentiate among the five major types of social movements based on their goals and the amount of change they seek to produce.

16. Integrate Simmel's, Veblen's and Bourdieu's perspectives on fashion.

17. Evaluate the most effective types of social movements when the goal is long term social change.

18. Assess the predictions about social change presented in the conclusion of this chapter.

KEY TERMS

(defined at page number shown and in glossary)

civil disobedience, p. 667
collective behavior, p. 662
crowd, p. 665
environmental racism, p. 681
gossip, p. 671
mass, p. 666
mass behavior, p. 669
social movement, p. 673

mob, p. 666
panic, p. 667
propaganda, p. 673
public opinion, p. 673
riot, p. 667
rumors, p. 670
social change, p. 661

KEY PEOPLE

(identified at page number shown)

Herbert Blumer, p. 666
Pierre Bourdieu, p. 671
Lory Britt, p. 675
Steven M. Buechler, p. 662
Stella M. Capek, p. 681
Rachel Carson, p. 662

Steven E. Clayman, p. 669
Riley E. Dunlap, p. 673
Kai Erikson, p. 674
William A. Gamson, p. 678
Lois Gibbs, p. 674
Gustave Le Bon, p. 668

CHAPTER OUTLINE

I. COLLECTIVE BEHAVIOR
  A. **Social change** is the alteration, modification, or transformation of public policy, culture, or social institutions over time; such change usually is brought about by **collective behavior** – relatively spontaneous, unstructured activity that typically violates established social norms.
  B. Conditions for Collective Behavior
    1. Collective behavior occurs as a result of some common influence or a stimulus that produces a response from a collectivity – a relatively large number of people who mutually transcend, bypass, or subvert established institutional patterns and structures.
    2. Major factors that contribute to the likelihood that collective behavior will occur are:
      a. Structural factors that increase the chances of people responding in a particular way
      b. Timing
      c. A breakdown in social control mechanisms and a corresponding feeling of normlessness
      d. A common stimulus, for example the publication of the stimulating book, *Silent Spring*
  C. Dynamics of Collective Behavior
    1. People may engage in collective behavior when they find that their problems are not being solved through official channels; as the problem appears to grow worse, organizational responses become more defensive and obscure.
    2. People's attitudes are not always reflected in their political and social behavior.
    3. People act collectively in ways they would not act singly due to:
      a. The noise and activity around them.
      b. A belief that it is the only way to fight those with greater power and resources.
  D. Distinctions Regarding Collective Behavior
    1. People engaging in collective behavior may be a:
      a. **Crowd** – a relatively large number of people who are in one another's immediate face-to-face presence; or
      b. **Mass** – a number of people who share an interest in a specific idea or issue but who are not in one another's immediate physical vicinity.
    2. Collective behavior also may be distinguished by the dominant emotion expressed (e.g., fear, hostility, joy, grief, disgust, surprise, or shame).

E. Types of Crowd Behavior
1. **Herbert Blumer** divided crowds into four categories:
    a. Casual crowds – relatively large gatherings of people who happen to be in the same place at the same time; if they interact at all, it is only briefly.
    b. Conventional crowds – people who specifically come together for a scheduled event and thus share a common focus.
    c. Expressive crowds – people releasing their pent up emotions in conjunction with others who experience similar emotions.
    d. Acting crowds – collectivities so intensely focused on a specific purpose or object that they may erupt into violent or destructive behavior. Examples:
        i.   A **mob** – a highly emotional crowd whose members engage in, or are ready to engage in, violence against a specific target which may be a person, a category of people, or physical property.
        ii. A **riot** – violent crowd behavior fueled by deep-seated emotions, but not directed at a specific target.
        iii. A **panic** – a form of crowd behavior that occurs when a large number of people react with strong emotions and self destructive behavior to a real or perceived threat.
2. To these four types of crowds, Clark McPhail and Ronald T. Wohlstein added *protest* crowds – crowds that engage in activities intended to achieve specific political goals.
    a. Protest crowds sometimes take the form of **civil disobedience** – nonviolent action that seeks to change a policy or law by refusing to comply with it.
    b. At the grassroots level, protests often are seen as the only way to call attention to problems or demand social change.
F. Explanations of Crowd Behavior
1. According to contagion theory, people are more likely to engage in antisocial behavior in a crowd because they are anonymous and feel invulnerable; Gustave Le Bon argued that feelings of fear and hate are contagious in crowds because people experience a decline in personal responsibility.
2. According to **Robert Park**, social unrest is transmitted by a process of *circular reaction* – the interactive communication between persons in such a way that the discontent of one person is communicated to another whom, in turn, reflects the discontent back to the first person.
3. *Convergence* theory focuses on the shared emotions, goals, and beliefs many people bring to crowd behavior.
    a. From this perspective, people with similar attributes find a collectivity of like-minded persons with whom they can release their underlying personal tendencies.
    b. Although people may reveal their "true selves" in crowds, their behavior is not irrational; it is highly predictable to those who share similar emotions or beliefs.

4. According to Ralph Turner's and Lewis Killian's *emergent norm theory*, crowds develop their own definition of the *situation and establish* norms for behavior that fits the occasion.
   a. Emergent norms occur when people define a new situation as highly unusual or see a long-standing situation in a new light.
   b. Emergent norm theory points out that crowds are not irrational; new norms are developed in a rational way to fit the needs of the immediate situation.

G. Mass Behavior
   1. **Mass behavior** is collective behavior that takes place when people (who often are geographically separated from one anther) respond to the same event in much the same way. The most frequent types of mass behavior are:
   2. **Rumors** – unsubstantiated reports on an issue or subject – and **gossip** – rumors about the personal lives of individuals.
   3. *Mass hysteria* is a form of dispersed collective behavior that occurs when a large number of people react with strong emotions and self destructive behavior to a real or perceived threat; many sociologists believe this behavior is best described as a panic with a dispersed audience.
   4. Fads and Fashions
      a. A *fad* is a temporary but widely copied activity enthusiastically followed by large numbers of people.
      b. *Fashion* is a currently valued style of behavior, thinking or appearance. Fashion also applies to art, music, drama, literature, architecture, interior design, and automobiles, among other things.
   5. **Public opinion** consists of the attitudes and beliefs communicated by ordinary citizens to decision makers (as measured through polls and surveys based on interviews and questionnaires).
      a. Even on a single topic, public opinion will vary widely based on characteristics such as race, ethnicity, religion, region of the country, urban or rural residence, social class, education level, gender, and age.
      b. As the masses attempt to influence elites and visa versa, a two-way process occurs with the dissemination of **propaganda** – information provided by individuals or groups that have a vested interest in furthering their own cause or damaging an opposing one.

II. SOCIAL MOVEMENTS
A. A **social movement** is an organized group that acts consciously to promote or resist change through collective action.
B. Types of Social Movements
   1. *Reform* movements seek to improve society by changing some specific aspect of the social structure.
   2. *Revolutionary* movements seek to bring about a total change in society. **Terrorism** is the calculated unlawful use of physical force or threats of violence against persons or property in order to intimidate or coerce a government, organization, or individual for the purpose of gaining some political, religious, economic or social objective.
   3. *Religious* movements seek to produce radical change in individuals and typically are based on spiritual or supernatural belief systems.

4. *Alternative* movements seek limited change in some aspect of people's behavior (e.g., a movement that attempts to get people to abstain from drinking alcoholic beverages).
5. *Resistance* movements seek to prevent or to undo change that already has occurred.

C. Stages in Social Movements
1. In the *preliminary* stage, widespread unrest is present as people begin to become aware of a threatening problem. Leaders emerge to agitate others into taking action.
2. In the *coalescence* stage, people begin to organize and start making the threat known to the public. Some movements become formally organized at local and regional levels.
3. In the *institutionalization* stage, an organizational structure develops, and a paid staff (rather than volunteers) begins to lead the group.

III. SOCIAL MOVEMENT THEORY
A. *Relative deprivation* theory asserts that people who suffer relative deprivation are likely to feel that a change is necessary and to join a social movement in order to bring about that change.
B. B.   According to **Neal Smelser**'s *value-added theory*, six conditions are necessary and sufficient to produce social movements when they combine or interact in a particular situation:
1. structural conduciveness
2. structural strain
3. spread of a generalized belief
4. precipitating factors
5. mobilization for action
6. social control factors
C. *Resource mobilization theory* focuses on the ability of a social movement to acquire resources (money, time and skills, access to the media, etc.) and mobilize people to advance the cause.
D. Social Constructionist theory is based on the assumption that social movements are interactive, symbolically defined, and a negotiated process.
1. This process involves participants, opponents, and bystanders.
2. Research based on this perspective often investigates how problems are framed and what names they are given.
3. Distinct frame alignment processes occur in social movements:
   a. frame bridging
   b. frame amplification
   c. frame extension
   d. frame transformation
E. Political opportunity theory is based upon the assumption that opportunities for potential protestors and movement organizers exist within the political system at any given point.
1. Opportunity refers to options for collective action, with chances and risks attached to them that depend on factors outside the mobilizing group.

    2. People will choose options for collective action that are most readily available to them and that will produce the most favorable outcome for their cause.

    3. This approach highlights the interplay of opportunity, mobilization, and influence in determining when certain types of behavior may occur.

F. New social movement theory looks at a diverse array of collective actions and the manner in which those actions are based in politics, ideology, and culture.

    1. It incorporates sources of identity, including race, class, gender, and sexuality as sources of collective action and social movement.

    2. Examples of already existing "new social movements" include ecofeminism and environmental justice movements.

       a. According to ecofeminists, patriarchy is a root cause of environmental problems because it contributes to a belief that nature is to be possessed and dominated, rather than treated as a partner.

       b. Environmental justice movements focus on the issue of **environmental racism** – the belief that a disproportionate number of hazardous facilities (including industries such as waste disposal/treatment and chemical plants) are placed in low income areas populated primarily by people of color.

IV. SOCIAL CHANGE IN THE FUTURE

A. The Physical Environment and Change: Changes in the physical environment often produce changes in the lives of people; in turn, people can make dramatic changes in the physical environment, over which we have only limited control.

B. Population and Change: Changes in population size, distribution, and composition affect the culture and social structure of a society and change the relationships among nations.

C. Technology and Change: Advances in communication, transportation, science, and medicine have made significant changes in people's lives, especially in developed nations; however, these changes also have created the potential for new disasters, ranging from global warfare to localized technological disasters at toxic waste sites.

D. Social Institutions and Change: During the past twentieth century, many changes occurred in the family, religion, education, economy, and political system; at the beginning of the new century, the U.S. government seemed less able to respond to the needs and problems of the country.

E. Changes in physical environment, population, technology, and social institutions operate together in a complex relationship, sometimes producing consequences we must examine by using our sociological imagination.

F. A few final thoughts: one purpose of this text is to facilitate understanding of different viewpoints in order to deal with the issues of the twenty-first century, thereby, producing a better way of life in this country and worldwide.

ANALYZING AND UNDERSTANDING THE BOXES

After reading the chapter and studying the outline, re-read the boxes and write down key points and possible questions for class discussion.

Sociology and Everyday Life: How Much Do You Know About Collective Behavior and Environmental Issues?

Key Points:

Discussion Questions:

1.

2.

3.

Sociology in Global Perspective: "Flash Mobs": Collective Behavior in the Information Age

Key Points:

Discussion Questions:

1.

2.

3.

Sociology and Social Policy: The Fight Over Water Rights

Key Points:

Discussion Questions:

1.

2.

3.

You Can Make a Difference: It's Now or Never: The Imperative of Taking Action Against Global Warning

Key Points:

Discussion Questions:

1.

2.

3.

PRACTICE TESTS

MULTIPLE CHOICE QUESTIONS

Select the response that best answers the question or completes the statement.

1. All of the following are factors that contribute to the likelihood that collective behavior will occur, **except**:
    a. the presence of deviant behavior
    b. structural factors that increase the chances of people responding in a particular way
    c. timing
    d. a breakdown in social control mechanisms and a corresponding feeling of normlessness

2. A _____ is a number of people who share an interest in a specific issue but are not in close proximity to each other.
    a. crowd
    b. mass
    c. riot
    d. mob

3. Casual crowds are
    a. comprised of people who specifically come together for a scheduled event and thus share a common focus.
    b. situations that provide an opportunity for the expression of some strong emotion.
    c. comprised of people who happen to be in the same place at the same time.
    d. comprised of people who are so intensely focused on a specific purpose or object that they may erupt into violent or destructive behavior.

4. Revelers assembled at Mardi Gras or on New Year's Eve at Times Square in New York are an example of:
   a. protest crowds
   b. expressive and acting crowds
   c. conventional crowds
   d. panics

5. When the residents of Love Canal burned both the governor and the health commission in effigy, they were engaging in a(n):
   a. acting crowd.
   b. casual crowd.
   c. conventional crowd.
   d. panic.

6. Convergence theory focuses on:
   a. the social-psychological aspects of collective behavior, including how moods, attitudes, and behavior are communicated.
   b. how social unrest is transmitted by a process of circular reaction.
   c. the importance of social norms in shaping crowd behavior.
   d. the shared emotions, goals, and beliefs many people bring to crowd behavior.

7. All of the following statements regarding the emergent norm theory are true, **except**:
   a. Emergent norm theory is based on the symbolic interactionist perspective.
   b. Sociologists using the emergent norm approach seek to determine how individuals in a given collectivity develop an understanding of what is going on, how they construe these activities, and what types of norms are involved.
   c. Emergent norm theory points out that crowds sometimes are irrational.
   d. Emergent norm theory originated with sociologists Ralph Turner and Lewis Killian.

8. Rumors, gossip, fashions, and fads are examples of _____ behavior.
   a. mob
   b. mass
   c. irrational
   d. casual

9. A form of dispersed collective behavior that occurs when a large number of people react with strong emotions and self-destructive behavior to a real or perceived threat is known as:
   a. mob behavior
   b. mass behavior
   c. mass hysteria
   d. contagious behavior

10. "Streaking" – students taking off their clothes and running naked in public – in the 1970s is an example of a
   a. panic.
   b. trend.
   c. fashion.
   d. fad.

11. According to the text, _____ is information provided by individuals or groups that have a vested interest in furthering their own cause or damaging an opposing one.
   a. propaganda
   b. public opinion
   c. political rhetoric
   d. a press release

12. Which of the following statements regarding social movements is true?
   a. Social movements are more likely to develop in preindustrial societies where there is an acceptance of traditional beliefs and practices.
   b. Social movements have become institutionalized and are a part of the political mainstream.
   c. Social movements make democracy less accessible to excluded groups.
   d. Social movements offer "outsiders" an opportunity to have their voices heard.

13. All of the following are types of social movements, except _____ movements.
   a. alternative
   b. dissident
   c. religious
   d. revolutionary

14. According to relative deprivation theory,
   a. people who are satisfied with their present condition are more likely to seek social change.
   b. certain conditions are necessary for the development of a social movement.
   c. people who feel that they have been deprived of their "fair share" are more likely to feel that change is necessary and to join a social movement.
   d. some people bring more resources to a social movement than others.

15. _____ theory is based on the assumption that six conditions, including structural conduciveness and structural strain, must be present for the development of a social movement.
   a. Value-added
   b. Relative deprivation
   c. Resource mobilization
   d. Emergent norm

16. _____ theory focuses on the ability of members of a social movement to acquire resources and mobilize people in order to advance their cause.
   a. Value-added
   b. Relative deprivation
   c. Resource mobilization
   d. Emergent norm

17. _____ is a social movement based on the belief that patriarchy is a root cause of environmental problems.
   a. Ecology Today
   b. Conflict Ecologists
   c. Environmental Justice
   d. Ecofeminism

18. In the _____ stage of a social movement, people begin to organize and to publicize the problem.
    a. preliminary
    b. coalescence
    c. institutionalization
    d. deinstitutionalization

19. The belief that a disproportionate number of hazardous facilities are placed in low-income areas populated by people of color is known as:
    a. environmental racism
    b. environmental justice
    c. reverse environmentalism
    d. racial pollution

20. All of the following statements regarding natural disasters are true, **except**:
    a. Major natural disasters can dramatically change the lives of people.
    b. Trauma that people experience from disasters may outweigh the actual loss of physical property.
    c. Natural disasters are not affected by human decisions.
    d. Disasters may become divisive elements that tear communities apart.

21. Unlike _____ behavior (for example, in education, religion, or politics), collective behavior lacks established norms to govern behavior.
    a. organizational
    b. group
    c. government
    d. institutional

22. _____ refer(s) to rumors about the personal lives of individuals.
    a. Mass Hysteria
    b. Blog
    c. Slander
    d. Gossip

23. In discussing rumors, the text points out that:
    a. while rumors may spread through an assembled collectivity, they also may be transmitted among people who are dispersed geographically.
    b. although they initially may contain a kernel of truth, as they spread, rumors may be modified to serve the interests of those repeating them.
    c. rumors thrive when tensions are high and little authentic information is available on an issue of great concern.
    d. all of the above are characteristics of rumors.

24. Sociologist _____ asserted that fashion serves mainly to institutionalize conspicuous consumption among the wealthy.
    a. Thorstein Veblen
    b. Pierre Bourdieu
    c. William A. Gamson
    d. Robert E. Park

25. _____ consist(s) of the attitudes and beliefs communicated by ordinary citizens to decision makers.
   a. Poll
   b. Public opinion
   c. Masses
   d. Gossip

26. According to sociologist John Lofland, the _____ refers to the "publicly expressed feeling perceived by participants and observers as the most prominent in an episode of collective behavior."
   a. casual crowds
   b. dominant emotion
   c. aggregate opinion
   d. expressive crowds

27. _____ crowds are collectivities so intensely focused on a specific purpose or object that they may erupt into violent or destructive behavior.
   a. Acting
   b. Casual
   c. Expressive
   d. Conventional

28. A(n) _____ is violent crowd behavior that is fueled by deep-seated emotions but not directed at one specific target.
   a. riot
   b. mob
   c. fight
   d. aggregate

29. Sociologist Riley Bunlap found that public awareness of the seriousness of environmental problems and support for environmental protection
   a. decreased precipitously during the 1980s.
   b. increased dramatically between the late 1960s and the early 1990s.
   c. has remained unchanged for the past three decades.
   d. is extremely difficult to measure accurately.

30. When sociologist Kai Erikson pointed out that people face a "new species of trouble," he was referring to
   a. new diseases for which there is no known cure.
   b. inhabitants of other planets that may be discovered by space exploration in the twenty-first century.
   c. natural disasters, such as hurricanes and earthquakes that have more deadly force than previous ones.
   d. technological disasters, such as toxic chemical pollution or radiation leakage.

31. Examples of _____ movements are groups organized since the 1950s to oppose school integration, civil rights and affirmation action legislation, and domestic partnership initiatives.
   a. resistance (regressive)
   b. revolutionary
   c. reform
   d. religious

32. The text identifies _____ as the most widely known resistance (regressive) movement.
   a. the Christian Temperance Union
   b. the National Rifle Association (NRA)
   c. MADD
   d. pro-life advocates (operation rescue)

33. In the _____ stage of a social movement, widespread unrest is present as people become aware of a problem. At this state, leaders emerge to agitate others into taking action.
   a. coalescence
   b. preliminary
   c. secondary
   d. institutionalization

34. In the _____ state of a social movement, people begin to organize and to publicize the problem. At this state, some movements become formally organized at local and regional levels.
   a. secondary
   b. institutionalization (bureaucratization)
   c. preliminary
   d. coalescence

35. Some religious movements are _____ – that is, they forecast that "the end is near" and assert that an immediate change in behavior is imperative
   a. foreseers
   b. Scientologists
   c. millenarian
   d. prophetic

36. People must become aware of a significant problem and have the opportunity to engage in collective action. This illustrates sociologist Neal Smelser's value-added theory condition of
   a. spread of generalized beliefs.
   b. mobilization for action.
   c. structural strain.
   d. structural conduciveness.

37. The _____ reflects the influence of sociologist Erving Goffman's "Frame Analysis", in which he suggests that our interpretation of the particulars of events and activities is dependent on the framework from which we perceive them.
    a. social constructionist theory
    b. value-added theory
    c. relative deprivation theory
    d. political opportunity theory

38. Sociologists have identified ways in which grievances are framed. _____ framing provides a vocabulary of motives that compel people to take action.
    a. Motivational
    b. Frame Alignment
    c. Bridge
    d. Transformation

39. According to _____ theory, people will choose options for collective action that are most readily available to them, producing the most favorable outcome.
    a. social constructionist
    b. relative deprivation
    c. political opportunity
    d. value-added

40. In a study of a contaminated landfill in the Carver Terrace neighborhood of Texarkana, Texas, sociologist Stella Capek found that residents
    a. saw no relationship between their plight and that of residents in areas such as the Love Canal.
    b. were able to mobilize for change and win a federal buyout and relocation.
    c. were powerless when confronted with the federal bureaucracy.
    d. believed that they had been included in previous environmental cleanups.

TRUE/FALSE QUESTIONS

1. Collective behavior lacks an official division of labor, hierarchy of authority, and established rules and procedures.
   T   F

2. People are more likely to act as a collectivity when they believe it is the only way to fight those with greater power and resources.
   T   F

3. People gathered for religious services and graduation ceremonies are examples of casual crowds.
   T   F

4. Panics often arise when people believe that they are in control of a situation.
   T   F

5. Protest crowds engage in activities intended to achieve specific political goals.
   T   F

6. Sociologist Robert E. Park was the first U.S. sociologist to investigate crowd behavior.
   T   F

7. According to Clayman, people like to "boo" a speech independently.
   T   F

8. For mass behavior to occur, people must be in close proximity geographically.
   T   F

9. Public opinion does not always translate into action by decision makers in government and industry or by individuals.
   T   F

10. Grassroots environmental movements are an example of reform movements.
    T   F

11. The 2001 terrorist attacks in New York City and Washington, D.C., proved that revolutionary terrorists can originate from outside the country.
    T   F

12. Revolutionary movements are also referred to as expressive movements.
    T   F

13. Movements based on relative deprivation are most likely to occur when people have unfulfilled rising expectations.
    T   F

14. Social construction theory is based on conflict theory.
    T   F

15. William Gamson believe social movements borrow, modify, or create frames as they seek to advance their goals.
    T   F

16. According to Box 20.1, scientists are forecasting a global warming of between 2 and 11 degrees Fahrenheit over the next century.
    T   F

17. According to the chapter introduction, freelance journalist Jill Carroll was never released by her kidnappers.
    T   F

18. In Box 20.2, sociologists suggest that flash mobs have become a fashion and will continue to remain as long-term activities.
    T   F

19. 19.   According to Box 20.3, riparian rights refers to the rights of a person or group to use water by virtue of owning or occupying the bank of a river or a lake.
    T   F

20. According to 20.4, there is nothing we can do individually to reduce the amount of greenhouse gases.
    T   F|

FILL-IN-THE-BLANK QUESTIONS

1. The publication of *Silent Spring* in 1962 by biologist _____ directed at demanding a _____.

2. The person most connected to the grassroots campaign of exposing sites of the chemical dump at Love Canal is _____.

3. Nonviolent action that seeks to change a policy or law by refusing to comply with it is known as _____.

4. _____ are unsubstantiated reports on an issue or subject.

5. The radio dramatization, The War of the Worlds, produced a type of _____.

6. A(n) _____ is a temporary but widely copied activity enthusiastically followed by large numbers of people.

7. The attitudes and beliefs communicated by ordinary citizens to decision makers are defined as _____.

8. Information provided by individuals or groups that have a vested interest in furthering their own course or damaging an opposing one is defined as _____.

9. The most common type of crowd behavior wherein people seek to escape from a perceived danger is known as _____.

10. Rumors about the personal lives of individuals are defined as _____.

11. Behavior that is voluntary, often spontaneous, and engaged in by a large number of people is known as _____.

12. The alteration, modification, or transformation of a public policy, culture, or social institutions over time is identified as _____.

13. A(n) _____, which is short for web log, is an online journal maintained by an individual who frequently records entries that are maintained in a chronological order.

14. Sociologist _____ divided crowds into four categories: casual, conventional, expressive, and acting.

15. A(n) _____ is a currently valued style of behavior, thinking, or appearance.

SHORT ANSWER/ESSAY QUESTIONS

1. What is collective behavior and what causes people to engage in collective behavior?
2. What are some types of crowd behavior? What causes crowd behavior? How could behavior influence individual behavior?
3. What are the five major theories that explain social movements? Which do you think is most applicable to solving future social problems? Why?
4. What is a social movement? What are the major types of social movements?
5. Why do people engage in social movements? Are social movements affective avenues for social change? Explain.

STUDENT CLASS PROJECTS AND ACTIVITIES

1. The uprising against socialism in the Soviet Union in 1991 is an example of a revolutionary movement; it sought, and apparently produced an entire restructuring of the Soviet political and economic system by overthrowing the existing social order and creating a new one. This could be considered the second of the revolutions that have occurred in this country since the Russian Revolution of 1917. Choose one of these two revolutions; trace the development of the movement, including the conditions necessary for the development of a social movement according to Smelser as mentioned in the text. In your paper, provide examples of the six conditions of the revolution and critique the success of the movement. Submit your paper to your instructor at the appropriate time, following any additional instructions you have been given.
2. Collect at least 10 articles dealing with some form of collective behavior discussed in the text. The articles could focus on fads, fashions, mobs, the public or specific types of social movements. Provide a copy of each article in your paper. You are to: (1) summarize each article; (2) explain the significance (if any) of the article; (3) write a personal reaction to each article; (4) provide a conclusion and personal evaluation of the project; and (5) provide a bibliographic reference for each article selected. You should submit your paper to your instructor at the appropriate time, following any additional instructions given to you.
3. Write a paper on the topic of the impact of artificial intelligence in our society today and upon the societies for tomorrow. Provide at least three bibliographical references in your paper. Submit your paper to your instructor at the appropriate time; following any additional instructions you may have been given.
4. Research the impact of the U.S. war in Iraq upon the political, economic, religious, educational, and medical institutions in both the U.S. and in the country of Iraq. What are the present national and international reactions to the war? What U.S. and international social movements have been created in response to the war? What are the levels of support for the war both nationwide and in other international communities? What has been the economic cost in both the U.S. and Iraq? What is the political cost? How have the major social institutions in both countries been affected? What is the level of support on your college campus? In addition to providing the information above in your paper or project, conduct a survey of thirty or more people on your campus to gain some insight of the level of support. Write up your research and findings according to your professor's requirements. Submit your paper on the required date.

INTERNET ACTIVITIES

1. To learn about international terrorist groups and guerilla groups, go to the following sites and write a brief summary of the major terrorist organizations functioning today. Be sure to focus on the revolutionary aspects of each. Follow links on each site: **http://www.foreignpolicy-infocus.org/briefs/vol3/v3n38terr_body.html**, **http://www.state.gov/s/ct/**, and **http://www.infoplease.com/spot/terrorism1.html**.

2. **Civil disobedience** is a form of **collective behavior**. This website contains a history of **civil disobedience** in America as well as a training manual for those wishing to organize their own social movement: **http://www.actupny.org/documents/CDdocuments/CDindex.html**

3. **PETA** (People for the Ethical Treatment of Animals), **http://www.peta-online.org/**, represents an animal rights activist movement. Members of PETA often engage in tactics that border on violent behavior as they seek to eliminate cruelty to animals. Discern the "cruelty-free" companies that do not use animals in testing their products.

4. Environmental terrorism is one of the topics covered in this chapter. Examine the following websites that give conflicting evidence to support their claims: **http://users.aol.com/DaveMcCall/terror.htm**, and **http://www.greenpeace.org/**, **http://www.sierraclub.org/**.

5. These are just a few of the sites on the Internet dealing with riots. Conduct a more extensive search for historical accounts and news stories of riots here in the U.S. and around the world: **http://www.chipublib.org/004chicago/disasters/riots_1966to1977.html**, **http://www.cnn.com/WORLD/asiapcf/9903/21/indonesia.borneo/**, and **http://www.cato.org/pubs/journal/cj14n1-13.html**.

INFOTRAC COLLEGE EDITION EXERCISES

Visit the **InfoTrac College Edition** website at: **http://www.wadsworthmedia.com/webtutor/infotrac.htm**. You will arrive at a screen that enables you to search topics.

1. Search for references on InfoTrac about **environmental movements**. Bring to class an article or brief report on a specific movement. Focus attention on the social aspects of these movements.

2. There are periodical references on the "social aspects" of **collective behavior**. Read one and then answer this question: *How can this article add to my knowledge about **collective behavior**?*

3. Search for the word **propaganda**. There are a number of categories such as "propaganda in television," "communist propaganda" and "war propaganda." Explore the different categories of propaganda.

4. Look up the concept **urban legend**. InfoTrac has a number of periodical references to this phenomenon. Bring an example to class. How do urban legends fit into the study of **collective behavior**?

SOLUTIONS

MULTIPLE CHOICE QUESTIONS

1. A, p. 662
2. B, p. 666
3. C, p. 666
4. B, p. 666
5. A, p. 667
6. D, p. 668
7. C, p. 669
8. B, p. 669
9. C, p. 671
10. D, p. 671
11. A, p. 673
12. D, p. 674
13. B, p. 675
14. C, p. 676

15. A, p. 677
16. C, p. 677
17. D, p. 680
18. B, p. 676
19. A, p. 681
20. C, p. 682
21. D, p. 662
22. D, p. 671
23. D, p. 670
24. A, p. 672
25. B, p. 673
26. B, p. 666
27. A, p. 666
28. A, p. 667

29. B, p. 676
30. D, p. 674
31. A, p. 676
32. D, p. 676
33. B, p. 676
34. D, p. 676
35. C, p. 675
36. D, p. 677
37. A, p. 678
38. A, p. 679
39. C, p. 680
40. B, p. 681

TRUE/FALSE QUESTIONS

1. T, p. 662
2. T, p. 665
3. F, p. 666
4. F, p. 667
5. T, p. 667
6. T, p. 668
7. F, p. 669

8. F, p. 669
9. T, p. 673
10. T, p. 675
11. T, p. 675
12. F, p. 675
13. T, p. 677
14. F, p. 678

15. T, p. 678
16. T, p. 664
17. F, p. 660
18. F, p. 672
19. T, p. 684
20. F, p. 687

FILL-IN-THE-BLANK QUESTIONS

1. Rachel Carson, p. 662; clean environment, p. 662
2. Lois Gibbs, p. 674
3. civil disobedience, p. 667
4. Rumors, p. 670
5. mass hysteria, p. 671
6. fad, p. 671
7. public opinion, p. 673
8. propaganda, p. 673
9. panic, p. 667
10. gossip, p. 671
11. collective behavior, p. 662
12. social change, p. 661
13. blog, p. 666
14. Herbert Blumer, p. 666
15. fashion, p. 671